Other Quality Books of Interest

Total Quality Essentials

**Using Quality Tools
and Systems
to Improve and Manage
Your Business**

Sarv Singh Soin

Second Edition

McGraw-Hill

New York   San Francisco   Washington, D.C.   Auckland   Bogotá
Caracas   Lisbon   London   Madrid   Mexico City   Milan
Montreal   New Delhi   San Juan   Singapore
Sydney   Tokyo   Toronto

Library of Congress Cataloging-in-Publication Data

Soin, Sarv Singh.
    Total quality essentials : using quality tools and systems to improve and manage your business / Sarv Singh Soin.—2nd ed.
        p.   cm.
    Updated ed. of: Total quality control essentials. 1992.
    Includes bibliographical references and index.
    ISBN 0-07-059551-8
    1. Total quality management. 2. Quality control. 3. Consumer satisfaction. I. Soin, Sarv Singh. Total quality control essentials. II. Title.
    HD62.15.S633   1998
    658.4′013—dc21
                                       98-37083
                                         CIP

McGraw-Hill

A Division of The McGraw·Hill Companies

  2 3 4 5 6 7 8 9 0   IBT/IBT   9 0 3 2 1 0 9

ISBN-13 978-0-07-059551-4
ISBN 0-07-059551-8

The sponsoring editors for this book were Robert Esposito and Linda Ludewig, the editing supervisor was Patricia V. Amoroso, and the production supervisor was Sherri Souffrance. It was set in Century Schoolbook by Dina John of McGraw-Hill's Professional Book Group composition unit.

Printed and bound by R. R. Donnelley & Sons Company.

 This book is printed on recycled, acid-free paper containing a minimum of 50% recycled, de-inked fiber.

McGraw-Hill books are available at special quantity discounts to use as premiums and sales promotions, or for use in corporate training programs. For more information, please write to the Director of Special Sales, McGraw-Hill, 11 West 19th Street, New York, NY 10011. Or contact your local bookstore.

Contents

Preface to the Second Edition

I have been deeply satisfied with the success of my original book entitled *Total Quality Control Essentials*. I have received numerous suggestions for improving the original text. The crux of the suggestions has been to make the book more business oriented—a book that shows how to use quality tools to manage a business. That is the purpose of this book.

I am using the experience gained in managing a business during the past six years to change the tone and direction of the original book to that of *Total Quality Essentials: Using Quality Tools and Systems to Improve and Manage Your Business*. Specifically, more details have been provided on increasing customer satisfaction, long-range planning, and process management, and the section on total quality reviews has been completely revamped.

The great companies of the world continue to generate a wave of high-quality products and services that are attractive, satisfying, and affordable: companies such as Sony, Toyota, Matsushita (Panasonic), Boeing, Hewlett-Packard, Motorola, Daimler Benz, and Singapore Airlines. These companies have strong total quality programs. While total quality may be less important for a young and nimble start-up company, which relies on one innovative product or technology, it is vital for a larger company or organization that needs to better focus its efforts and increase its effectiveness and efficiency.

This book will help you implement or improve the total quality effort in your organization. The total quality review procedure provided will help you to continually improve your total quality effort and to lay a strong foundation for generating a stream of attractive products and services.

How to Use This Book

This book is dedicated to chief executives, senior managers, and professionals. Chapter 1 gives an introduction to the latest thinking on total quality and defines its critical elements. Chapters 2 to 6 give details on each of these elements. Chapter 7 gives advice on starting a total quality effort. In Chapter 8 we discuss a total quality review procedure that provides a means for discovering and encouraging the use of quality tools and systems to manage a business in order to make it more efficient, productive, successful, and better prepared to face the future. *This is the focus of the book—to provide tools and systems to make a business more successful.* Chapter 9 discusses the essence of total quality, and we finish the book with an afterword on the chief executive's role in the total quality effort. We also provide appendixes, where we give a review of several quality tools and processes for the eager professional.

With a proper implementation of a robust total quality effort, every company or organization will be able to improve and manage its business better and be ready to face the future. This book will help to meet that goal.

Onward.

Acknowledgments

I am extremely indebted to the numerous people who have contributed to, or guided me, in my effort to write this book. Katsu Yoshimoto, for his inputs over the years; Tan Bian Ee and Paul Ow for permission to use material from Hewlett-Packard; and special thanks to Khushroo Banu and Mike Ward for editing. Also, to Craig Walter for his encouragement; Neoh Kah Thong for his valuable inputs; Walt Sousa for his continual inspiration; Dr. Noriaki Kano for his valuable insights and permission to use some of his material; Dieter Legat for his observations on management's role in total quality; Karamjit Singh for inputs on Singapore Airlines; Keith Watson for providing linkage between quality and the business world; and Mona for her infinite patience and encouragement.

Total Quality Essentials

1

The Quality Revolution

The right quality and uniformity are
foundations of commerce, prosperity and
peace.   W. EDWARD DEMING

Overview—The Genesis of Total Quality Efforts

Quality! Quality is on everyone's lips these days, because it can make the difference between success and failure in a very competitive and tumultuous world. Today quality means more than product reliability; today it means a total quality effort—an effort in which everybody and every function in an organization participate.

The term *total quality* originated in the United States, but the current concept of total quality was further developed in Japan and then improved in the United States. After World War II, Japanese managers realized that they had to export or perish. And export they did, but the quality of their products was shoddy. Two renowned U.S. statisticians and quality experts, Edward Deming and Joseph Juran, spent several years educating the Japanese on improving quality. Japanese product quality improved; the Japanese then went on to develop the current concept of total quality, which goes beyond the ideas of these great teachers to embrace quality throughout the entire company—in manufacturing, administration, marketing, sales, after-sales support, business planning, and services.

The introduction of the prestigious Deming Prize in Japan in 1951, awarded to Japanese companies that show a high commitment to quality, helped to motivate many Japanese manufacturing and service companies to improve their quality. Since then many other countries have introduced similar quality awards to encourage quality awareness and improvement in their industries.

Keen competition inside and outside Japan has allowed many companies throughout the world to improve. Today Japan is renowned for its manufacturing skills and quality. Failure to learn from Japanese companies, and from other successful companies, will greatly reduce the competitive ability of any company.

Let us start our discussion by reviewing the benefits of quality.

Benefits of Quality

What are the benefits of a total quality effort? Many people will speak enthusiastically of the benefits of quality. But what is the value of high quality to the business enterprise? Let us look at a generic model of total quality and then review some industry data.

A total quality effort properly implemented will focus on providing the best possible products and services via robust processes; this will positively impact productivity, customer satisfaction and loyalty, market share, and profits. The impact will be seen both internally and externally. We illustrate the impact in Fig. 1.1 and in the following:

- *Internally.* We aim for zero defects in our products or services. This high quality will give us better productivity, which allows us to lower prices (in which case we are competing with price), market share increases, and we get higher profits. Alternatively, the lower cost gives a direct increase in profits.

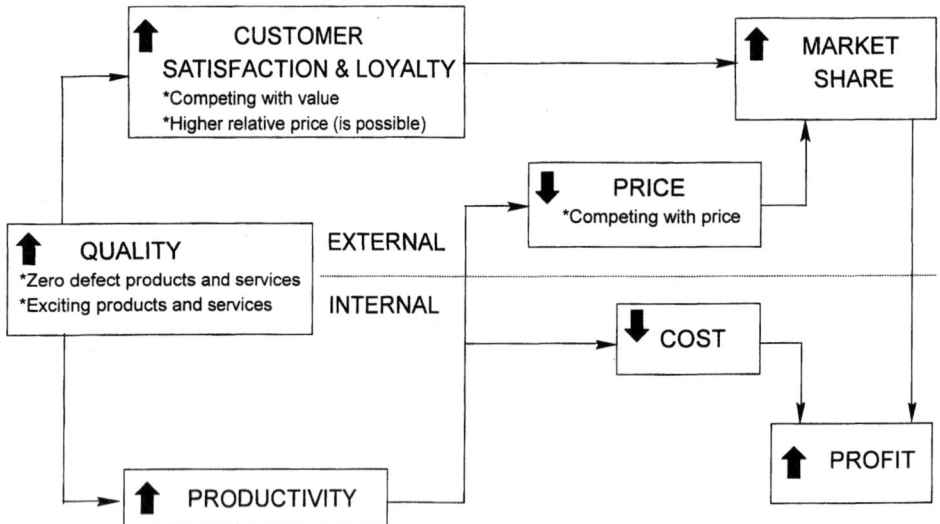

Figure 1.1 How a total quality effort contributes to profits.

- *Externally.* Higher quality, via a zero-defects mentality, allows us to increase customer satisfaction, increase customer loyalty, and get more repeat purchases. This results in increased market share and higher profits. Alternatively, we could compete on a value basis, charging a higher relative price for our higher quality. Many companies do this, for example, Sony in Japan, Boeing Aircraft in the United States, Mercedes-Benz in Germany, and Singapore Airlines in Singapore. The result, again, is higher profits, although market share may remain about constant.

PIMS study

Let us look at some data on the impact of quality. Figure 1.2 shows the Profit Impact on Marketing Strategy (PIMS) study done by Buzzel and Gale.[1] These data from the Strategic Planning Institute show the relationship between market share, relative quality, and return on investment (ROI). You can see that increased market share gives a higher ROI; you probably expected that. But an increase in relative quality also gives a higher ROI. The best situation is when both high market share and high quality exist—resulting in the highest ROI. This study supports our generic model.

QUALITY AND MARKET SHARE BOTH DRIVE PROFITABILITY

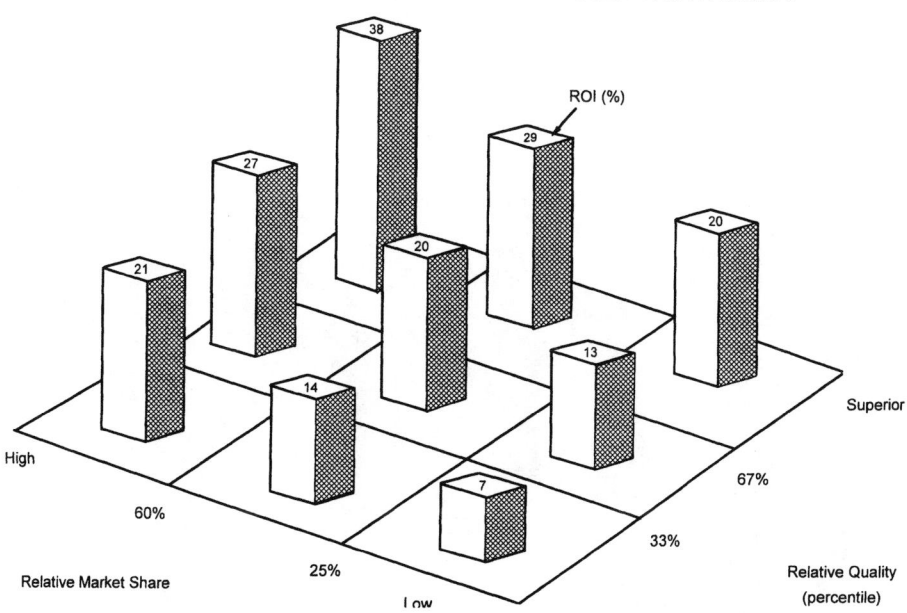

Figure 1.2 The impact of quality and market share on profits. (*Reprinted with permission of the Free Press, a division of Macmillan, Inc. from* The PIMs Principles: Linking Strategy to Performance *by Robert D. Buzzel and Bradley T. Gale. Copyright 1987 by The Free Press.*)

Specific data from one company

Let us now look at one company's quest for quality. In 1982, Yokogawa Hewlett-Packard (YHP), the Japanese subsidiary of Hewlett-Packard Company, won the coveted Deming Prize. John Young, the CEO of Hewlett-Packard, quotes the following gains for YHP during a 5-year period (as reported in the *Journal for Business Strategy* and Hewlett-Packard Company presentations): "This is an excellent example of how a strategic focus on total quality can maximize the attainment of a company's key objectives—profit and growth." These improvements are shown in Fig. 1.3.

Benefits of a total quality effort

By now the data should be convincing. The question you may have is, How do I make this work for my company? Answering that is the purpose of this text, and we will go into great detail, but first let us list some benefits of a good total quality effort:

- Higher employee morale
- More efficient processes
- Higher productivity
- Less fire-fighting, resulting in more time for innovation and creativity
- Lower costs
- Zero-defect products and services

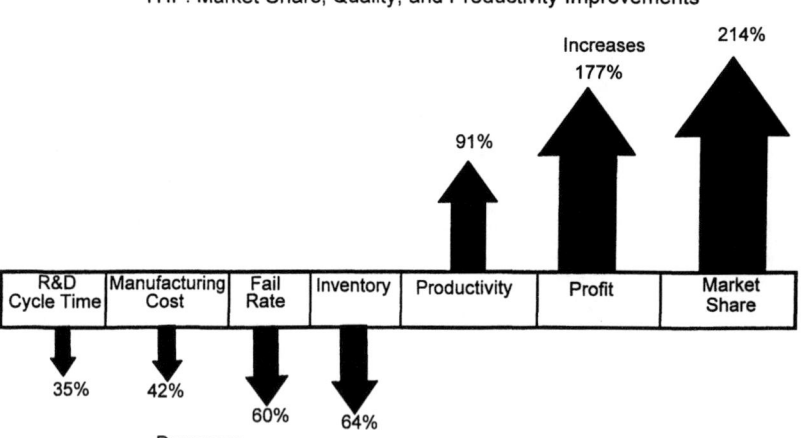

Figure 1.3 Improvements at YHP.

- Increased customer satisfaction and loyalty
- Increased market share
- Higher profits

Given some of the data we just reviewed, quality has to be part and parcel of a company's business strategy. This is the key to survival and growth—there is no alternative.

Objective of a Total Quality Effort

There are many definitions of total quality efforts, but here's one definition that includes the objective of total quality:

An effort of continuous improvement that uses quality tools and systems to manage a business, in order to make it more efficient, productive, and successful and better prepared to face the future

In our context, total quality goes beyond traditional product quality and even recent total quality control (TQC) or total quality management (TQM) programs. It includes efficiency, productivity, customer satisfaction and loyalty, zero-defect products and services, and good management of key areas, such as planning and human resources.

Definition of quality

At this juncture it is appropriate to review some definitions of *quality*.

- A definition that gained much popularity in the 1980s was Phil Crosby's "conforming to specifications." The difficulty with this definition is that the specifications may not be what the customer wants or is willing to accept.
- A much better definition is one that Joseph Juran has proposed for many years: "fitness for use." Fitness is to be defined by the customer.
- Noriaki Kano et al.[2] have proposed a two-dimensional model of quality. Refer to the discussion in the accompanying box. They suggest that quality has two dimensions: "must be quality," or a set of expected features, such as reliability, and "attractive quality," or the unexpected that goes beyond the customer's needs; these are extra features that the customer would love to have but has not desired because he or she has not yet thought about them.
- By now we can come up with a comprehensive, yet simple, definition of quality such as "products and services that meet or exceed customers' expectations."

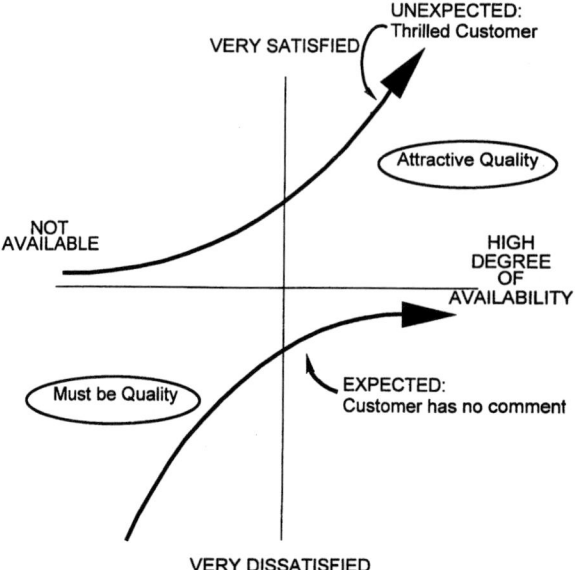

Figure 1.4 The two dimensions of quality.

The Two Dimensions of Quality

Noriaki Kano and others have proposed the concept of two dimensions of quality: "must be quality" and "attractive quality." See Fig. 1.4.

■ *Must be quality* is that aspect of a product or service which the customer expects. If the customer does not get it, there will be extreme dissatisfaction. An example is a reliable, safe, and easy-to-use product. This is the minimum acceptable standard—very like Joseph Juran's fitness-for-use concept.

■ *Attractive quality* is that aspect of a product or service that goes beyond current needs. If a special feature is available, the customer will be thrilled and excited; but if this feature is not available, the customer has no comment. An example could be a sunroof, antilock brakes, or a safety air bag in a car. With time, such an attractive quality becomes a must be quality. An example is the safety air bag which is fast becoming a must be quality item. Other items that were once attractive quality are remote control and multiprogramming. They are now standard features in most video cassette recorders (VCRs). So, higher quality is a never-ending quest.

A well-designed product or service should have both dimensions of quality—these can strongly influence the customer's buying decision.

Need for a System

Recently there has been a blizzard of new quality initiatives and methodologies. Introducing them ad hoc without a basic total quality system in place would result in a less than optimum benefit. Even without new initiatives, all ongoing quality efforts—product quality, improvements, processes, employee morale, customer issues and needs, planning, and a host of other issues—must run like clockwork. This is even more crucial in today's turbulent times and highly competitive environment. Small start-up companies with unique or innovative products do not need a system, but big organizations do. The alternative is inefficiency, complacency, and declining performance—just look at the current demise and incredible waste at many large corporations.

In the succeeding pages we will elaborate on what a good total quality system is and how to manage it. But let us first determine the elements of total quality.

Key elements of a total quality effort

There are many ways to cut the total quality pie. We need an approach that is simple to comprehend. Here is our approach to separating the various elements of total quality:

1. *Customer obsession.*   This will include all activities required to keep your customers happy, satisfied, and—whenever possible—thrilled!

2. *Business planning.*   This is the best way to show and implement management commitment to customers, employees, improving quality, and planning for the future. This is surely one of the most important processes in any organization.

3. *Managing improvements and breakthroughs.*   This element is to ensure a rigorous, effective, and systematic method of improving processes and managing new or breakthrough products and services. This element, when properly executed, will help you move toward a mentality of creating zero-defect products or services.

4. *Process management.*   This will ensure good day-to-day management of your key processes, resulting in efficient and predictable processes internally as well as with your partners. The result will be a lower-cost and more efficiently managed organization.

5. *Employee development, participation, and leadership.*   All employees must be educated to participate in the total quality effort. In addition, management must show strong leadership and prepare for the future by moving the organization toward a common goal.

The separation between the elements is not clean, nor can it be, but it is sufficient for our discussion. In particular, business planning, managing improvements and breakthroughs, and process management are linked closely, while all other items will overlap. In the remainder of this text, we discuss these five elements and go a little beyond them.

Reengineering versus a Total Quality Effort

We have often been asked to explain the difference between reengineering and total quality. Reengineering came after total quality and is one of many techniques proposed by consultants. Its key ideas are nothing new—an employee focus (team activity, empowerment, rewarding employees) and improving processes. These ideas have been conventional wisdom for many years, and they originated from total quality and other management techniques. So why introduce them again? Partly we feel consultants do so because managers like new ideas. While reengineering has many good ideas, it has had both successes and failures. Its weakest point is that it is divorced from an organization's strategy or future direction. This can result in an organization's doing the wrong things. In total quality, however, we do much more than reengineering, while ensuring that all the activity is linked to or driven by the organization's strategic direction.

Summary: The Quality Revolution

Let us summarize some of the concepts discussed—objectives, benefits, and elements of a total quality effort. The objective of a total quality effort is an effort of continuous improvement that uses quality tools and systems to manage a business, in order to make it more efficient, productive, and successful and better prepared to face the future.

We have discussed the benefits of a total quality effort. A short list includes the following:

- Higher employee morale
- More efficient processes
- Higher productivity
- Less fire-fighting, resulting in more time for innovation and creativity
- Improved quality of products and services
- Lower costs
- Zero-defect products and services

- Increased customer satisfaction and loyalty
- Increased market share
- Higher profits

We defined quality as "products and services that meet or exceed customers' expectations." We went on to develop a model of total quality that consists of the following five elements:

1. Customer obsession
2. Business planning
3. Managing improvements and breakthroughs
4. Process management
5. Employee development and participation, and leadership

In the remainder of this text, we take an approach to total quality that encompasses these five elements; we also go beyond these elements and discuss how to manage and measure a successful total quality effort.

References

1. Robert D. Buzzel and Bradley T. Gale, *The PIMS Principles: Linking Strategy to Performance* (New York: The Free Press, 1987). In this book, Buzzell and Gale provide some very convincing data that relate quality with profitability. The relationship is summarized in our Fig. 1.2. To develop this relationship they used a database of about 200 strategic business units (SBUs) from a total database of about 3000 SBUs. The relative perceived quality axis in Fig. 1.2 was determined by getting groups of managers to identify nonprice product attributes that affected customer buying decisions. From these a quality score was constructed.
2. Noriaki Kano et al., "Attractive Q vs. Must Be Q," *Hinshitsu (Quality)*, vol. 14, no. 2, 1984, pp. 39–48.

2

Customer Obsession

*Satisfying customers is the only reason we're
in business.*   JOHN YOUNG, CEO,
                         HEWLETT-PACKARD COMPANY

The customer is king.
                         ANONYMOUS, UNITED STATES

The honorable customer is God.
                         ANONYMOUS, JAPAN

Overview

In most companies and cultures, the customer is considered very
important. But how far would anyone go to help the customer? Read
the following news article:

Japan's Got Us Beat in the Service Department, Too

My husband and I bought one souvenir the last time we were in Tokyo—
a Sony compact disk player. The transaction took seven minutes at the
Odakyu Department Store, including time to find the right department
and to wait while the salesman filled out a second charge slip after mis-
spelling my husband's name on the first.

My in-laws, who were our hosts in the outlying city of Sagamihara,
were eager to see their son's purchase, so he opened the box for them the
next morning. But when he tried to demonstrate the player, it wouldn't
work. We peered inside. It had no innards! My husband used the time
until the Odakyu would open at 10 to practice for the rare opportunity in
that country to wax indignant. But at a minute to 10 he was pre-empted
by the store ringing us.

My mother-in-law took the call, and had to hold the receiver away from
her ear against the barrage of Japanese honorifics. Odakyu's vice president was on his way over with a new disk player. A taxi pulled up 50 minutes later and spilled out the vice president and a junior employee who
was laden with packages and a clipboard. In the entrance hall the two

11

men bowed vigorously. The younger man was still bobbing as he read from a log that recorded the progress of their efforts to rectify their mistake, beginning at 4:32 p.m. the day before when the salesclerk alerted the store's security guards to stop my husband at the door. When that didn't work, the clerk turned to his supervisor, who turned to his supervisor, until a SWAT team leading all the way to the vice president was in place to work on the only clues, a name and an American Express card number. Remembering that the customer had asked him about using the disk player in the U.S., the clerk called 32 hotels in and around Tokyo to ask if a Mr. Kitasei was registered. When that turned up nothing, the Odakyu commandeered a staff member to stay until 9 p.m. to call American Express headquarters in New York. American Express gave him our New York telephone number. It was after 11 when he reached my parents, who were staying at our apartment. My mother gave him my in-law's telephone number. The younger man looked up from his clipboard and gave us, in addition to the new $280 disk player, a set of towels, a box of cakes, and a Chopin disk. Three minutes after this exhausted pair had arrived they were climbing back into the waiting cab. The vice president suddenly dashed back. He had forgotten to apologize for my husband having to wait while the salesman had rewritten the charge slip, but he hoped we understood that it had been the young man's first day.[1]

An incredible story! But it is true. Surely this is what any customer wants—to have excellent service and products, to be treated well, and to purchase more from a customer-obsessed company. And yet, how few are the managers, companies, or organizations that show this type of customer obsession.

In today's environment, as competition increases and more sophisticated goods become commodities, the main distinguishing factor becomes quality—the quality of the product or service. In most cases this means the product must be of high quality and provided via excellent service; and if the customer does not get this, you will lose the customer. In small mom-and-pop managed or family-run stores, you will still find excellent service. Many small organizations or young companies are quick to give the customer what he or she wants—a new product, a new service, or both. But what can larger organizations do to avoid the bureaucracy, complacency, and impersonal attitudes that may eventually cause them to lose touch with the customer?

In this chapter we offer a few methodologies that will help ensure a strong customer focus—and preferably a customer obsession. These, together with a well-educated and trained workforce, will help you get closer to your customers—to hear their voices and to meet their needs.

Customer Obsession via a Systematic Approach

Any organization that is serious about quality and customers must take a systematic approach to ensure a customer-obsessed workforce.

An educated workforce is a good start, but it must also have the right tools. Given below is a short list of several tools or activities that will foster customer obsession. The list is separated into reactive and proactive activities.

Reactive activities

1. A system to manage and resolve customer complaints
2. Customer satisfaction surveys and follow-up corrective action
3. Development of a customer satisfaction model
4. Competitive benchmarking: learning from world-class companies to compete better

Proactive activities

1. Capturing the customer's voice or needs for new product and services via a systematic process
2. Developing and implementing the "total product or service" concept

The reactive approach is necessary to understand and resolve challenges and problems arising from current products and services. The proactive approach is essential to help influence and create new products and services.

Customer Complaint and Feedback Management System

> One of the strongest signs of a bad or declining relationship is the absence of complaints from the customer. Nobody is ever that satisfied, especially not over an extended period of time. The customer is either not being candid or not being contacted.—*Theodore Levitt*

Have you ever lost your luggage on an airplane? Or had a problem with your personal computer or its software? Or had terrible service at your bank? Did you complain, and how was your complaint handled? Was your complaint handled quickly and effectively? If, for example, you lost your luggage and the corrective action by the airline was sloppy, tardy, and unsympathetic, you will probably never fly this airline again.

In the lean and mean United States, this seems to be the trend: Profits decline, costs are cut, and people are fired, resulting in deteriorating service, and customers go elsewhere—to a company that understands that service and customer satisfaction are essential for success. This pattern is very visible in the airline, banking, and retail sectors, but it is also common in all other sectors.

When a customer is upset, you may get a complaint coupled with a demand for compensation, as illustrated in the above news article. On other occasions customers may give you feedback, because they genuinely want you to improve your product or service. *Each customer complaint should be treated as an unpolished gem; a gem that needs to be captured, examined, and polished.* Your company or organization can be only richer and wiser as it collects and polishes each of these gems of insight and wisdom. Having a system that captures these gems of wisdom and polishes them is extremely important. If you do not, the wronged or unhappy customer may not return to you again.

Consider the following information, released by the U.S. Office of Consumer Affairs, on March 31, 1986:

The High Cost of Losing a Customer

- In the average business, for every customer who bothers to complain, there are 26 others who remain silent.
- The average "wronged" customer will tell 8 to 16 people. (Over 10 percent tell more than 20 people.)
- Of unhappy customers, 91 percent will never purchase goods or services from you again.
- If you make an effort to remedy customers' complaints, 82 to 95 percent of them will stay with you.
- It costs 5 times as much to attract a new customer as it costs to keep an old one.

The most shocking statistic is that 91 percent of unhappy customers will *never* purchase goods or services from you again. The following box contains an interesting memo from Kent Stockwell of Hewlett-Packard (HP) Company, entitled "A Customer Lost Is a Customer Lost," that supports this statistic. Notice that customers have very long memories—they remember both the good and the bad that you do for a long time!

A Customer Lost Is a Customer Lost

I called an HP calculator customer in Los Angeles to ask him some survey questions, following up on our recent customer satisfaction survey. This person was a very busy civil engineer who runs his own consulting company. Each time I called, I was asked to call back at a specified time. The third time I called, I asked to wait, and I finally got through.

After going through the questions for 10 minutes, I asked the final question: "Is there anything else you would like to say about HP calculator products and service?" This customer said, "No, your calculators are great but your computers have a

---

problem." Then he proceeded to talk for another 10 minutes (this very busy person) about his negative experiences with an HP desktop computer that he had purchased 12 years ago. The computer he had purchased was delivered late, with no explanation. And he was unable to get help from the sales office regarding missing cables and questions he had about the manuals and the system. Finally he called the factory in Loveland, where he got help and found out about training classes he could arrange through the sales office.

Today, 12 years later, he still remembers the story in detail; most important, this technical computer user owns IBX equipment, not HP. Because of some minor product problems and some major service problems, this former HP customer has not looked to HP for his technical computer needs for over 12 years.

The need for a system to manage customer complaints

Every company or organization is always receiving customer complaints. The difference between organizations is the frequency and intensity of the complaints. Some organizations ignore the complaints, while other organizations with a well-informed and educated workforce attend to complaints. But this alone is insufficient. The better organization must keep a record of these complaints—the frequency, intensity, location, and so on.

To track lost luggage, an airline's management needs to know the following: where, when, and how often luggage was being lost. Was it in one airport or in every airport? Only with such data could the airline move away from solving lost luggage on an ad hoc basis toward repairing or correcting the entire system.

Hence, it is vital that your workforce record most, if not all, complaints. This must be followed with resolution of the complaints and elimination of the root cause of each complaint.

Figure 2.1 shows a flowchart of a customer complaint and feedback system. The flowchart is simple and self-explanatory. The key points include the following:

1. *Complaints are collected from all sources.* Complaints from letters, phone calls, meetings, and verbal inputs are collected routinely.

2. *Data are collected via a customer complaint and feedback form.* Data collection and transmission at the local entity are done via a customer complaint and feedback (CC&F) form. Refer to Fig. 2.2. This same form can go to the central coordinator, who will use a formal corrective action request (CAR) (see Fig. 2.3). More about this later.

3. *There is quick resolution.* The complaint is resolved as quickly as possible, and the customer is contacted and informed.

4. *All customers get a response.* This can be a thank-you note or resolution of a complaint.

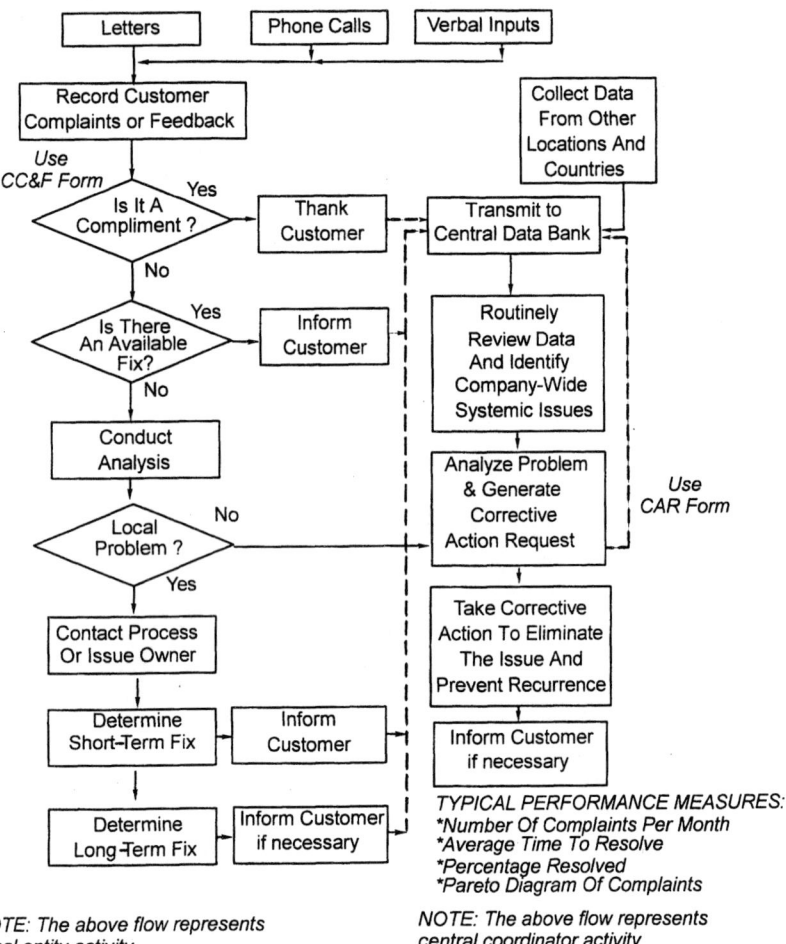

Figure 2.1 Flowchart of a customer complaint and feedback system.

5. *Issues are resolved locally.* For a locally generated issue (that is, generated at the site which collected the complaint), the issue is resolved locally.

6. *Issues are resolved by the organization.* For an issue beyond the control of the local entity (that is, generated elsewhere in the company), the information is sent to a central coordinator or analyst. This coordinator will further analyze the issue and propose a solution to the process owner in the organization.

7. *There is resolution of systemic issues.* On a regular basis, data are analyzed and systemic issues are identified, resolved, and eliminated. Referring to our earlier news article, the causes of lost baggage, wrong phone numbers, and inattentive personnel must be under-

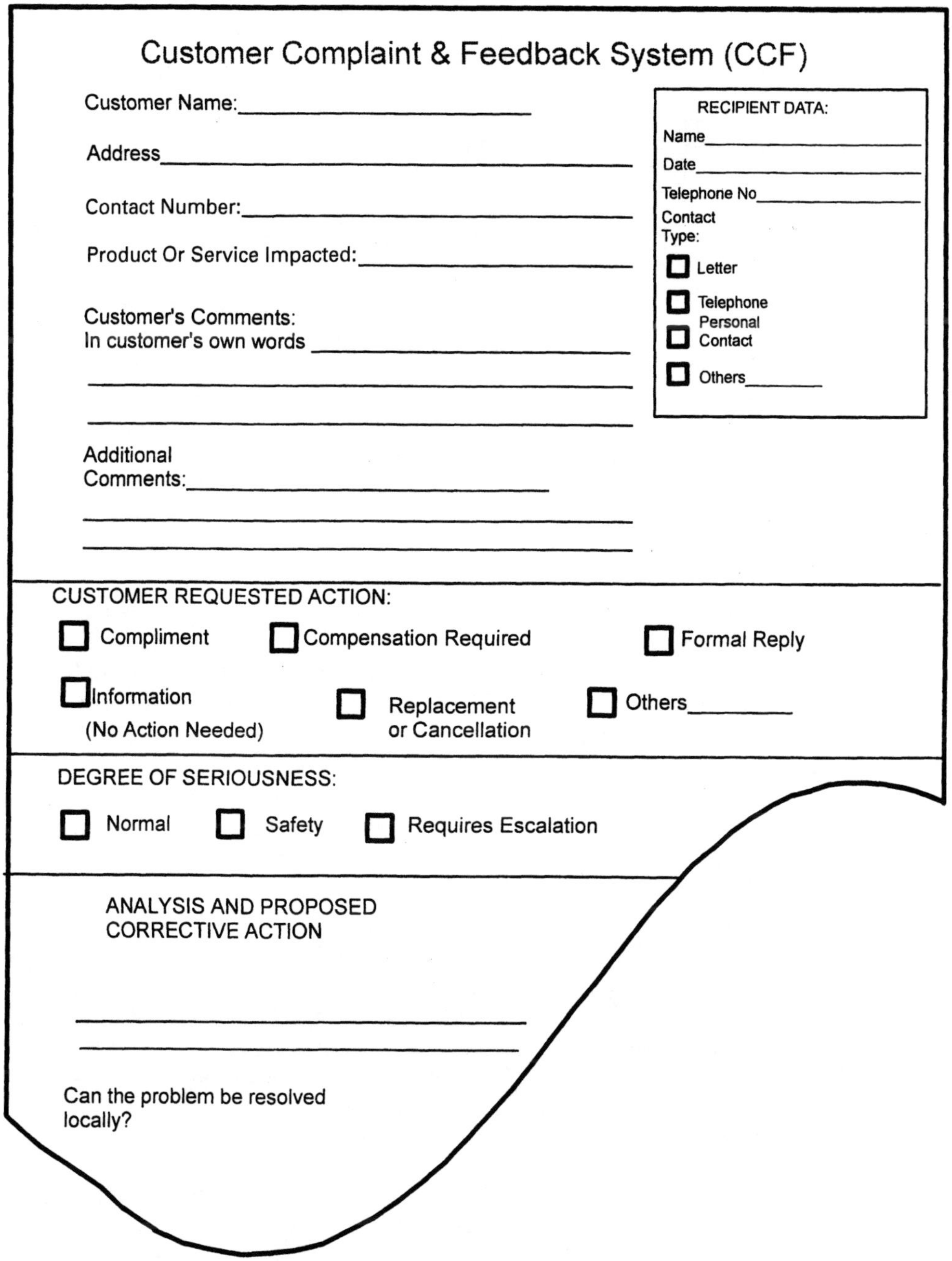

Customer Complaint & Feedback System (CCF)

Customer Name:_____

Address_____

Contact Number:_____

Product Or Service Impacted:_____

Customer's Comments:
In customer's own words _____

_____

_____

Additional
Comments:_____

_____

_____

RECIPIENT DATA:
Name_____
Date_____
Telephone No_____
Contact Type:
☐ Letter
☐ Telephone
☐ Personal Contact
☐ Others_____

CUSTOMER REQUESTED ACTION:

☐ Compliment   ☐ Compensation Required   ☐ Formal Reply

☐ Information   ☐ Replacement   ☐ Others_____
(No Action Needed)   or Cancellation

DEGREE OF SERIOUSNESS:

☐ Normal   ☐ Safety   ☐ Requires Escalation

ANALYSIS AND PROPOSED
CORRECTIVE ACTION

_____

_____

Can the problem be resolved
locally?

Figure 2.2 Typical form for recording customer complaints.

stood and eliminated. Not doing so can impact future purchases from current and potential customers. It can result in lost business and reduced market share. Here, again, you can use the CAR form.

8. *Identify and monitor performance measures.* Regularly review performance measures, such as number of complaints each month, time to resolve complaints, Pareto diagrams of complaints, and so on. Keeping track of these and setting appropriate targets will help provide proper management of the system and give an indication of its health. Targets should be set only after sufficient data on the current status are available. For example, time to resolve complaints could be an average of 20 days. A target of 10 days can be set, and this target can be gradually reduced over months and years. Another example of appropriate targets would be the elimination or gradual reduction of certain categories in the Pareto diagram of types of complaints. We discuss this point in greater detail later in this section.

Recording customer complaint data. Customer data can be recorded on a paper form such as a CC&F form, shown in Fig. 2.2; or the data can be keyed into a computer-based form. It will not be easy to record the issue on paper or in a computer database. A busy sales representative or manager at best may be able to phone in the customer's complaint. A busy counter clerk may solve the problem immediately, but may only have time to scribble a few words. Other very analytical employees may insist on writing down a detailed history of what happened.

Hence a flexible system is needed, one which provides easy ways of collecting inputs. For example, inputs can be taken via telephone to a central voice mailbox, directly to a complaint specialist, via an electronic mail system, or on the recommended form shown in Fig. 2.2. As the system matures and gains credibility, recording the data will become easier, but employees will always appreciate several options.

Data collection via a telephone voice mailbox. If you decide to experiment with a telephone voice mailbox, a cue card can be provided to all employees. This card can be used during discussions with customers and to phone in the problem to the customer complaint voice mailbox. An example of a cue card is given in Table 2.1.

Corrective-action requests. An issue may be resolved locally depending on the situation or type of complaint. But when a class- or companywide problem is being resolved, it will be sent to a central coordinator for further analysis. In such cases it will be necessary to ensure that the problem is resolved and eliminated throughout the organization. If this process is done properly, the problem should never recur

TABLE 2.1   Cue Card for Customer Complaint and Feedback System

Cue card for voice mail

Phone number of telephone voice mailbox: XXX-YYYY
Please have the following information available:
  Your name, department, and phone number
  Customer's name, title, address, and phone number
  Customer's actual words describing the problem
  What the customer wants
  Any additional information about the problem
  Model and product number, serial number, sales order number, or service impacted
  Action already taken
  Specific help needed, from whom, when, or how
  Any other comments

in the organization. One way of proposing a companywide fix is to send the proposal, via a CAR form, to the process or problem owner. This, like the CC&F form, can be transmitted electronically. But initially you should try this on a manual system.

A hypothetical example of a completed CAR form is shown in Fig. 2.3. This, when completed, is sent to the process owner within the organization. This example is based on an actual problem that a major automobile manufacturer had. Unfortunately, this manufacturer took about 4 years to correct the problem. That is approximately the product life cycle of some cars and possibly more than the average development time for a new car at Toyota Motor Company. Obviously, the manufacturer we quote did not have a good system to manage customer complaints.

Promoting and facilitating the system

Promotion activity. A customer complaint and feedback system will require constant nurturing and promotion, especially in the first few years after it is launched. Here is a list of procedures to explain the system and its benefits:

Promote during new employee training. New employees are typically most receptive to such a system. Their participation should be requested.

Promote at monthly department meetings. This needs to be done on a very regular basis and reduced, but not eliminated, as the system becomes accepted. At these sessions, encourage participation, review issues, and convey success stories of resolved customer complaints.

Encourage managers to use the system. Managers should use the system and become role models for their employees. Their role is

CORRECTIVE ACTION REQUEST (CAR)	Reference No:

PRODUCT/PROCESS OWNER: ABX | **MAIL TO:** ABX Operations Mgr

Item   Windshield Wiper
Product: _____Xetta Automobile_____   **Category** _All 1984 Models With R.H.Drive_
Requested by: ____CC & F Analyst____   **Date** ____April, 1984____

SERIOUSNESS: Is This a Class Problem: ☒ Yes   ☐ No   Dealers grumbling and calling in
What is the current impact?   Unhappy customers, sales are slowing down. Estimated loss in Sales:$XXXXX.
What is the future Impact?   Loss of reputation, loss of future sales, $ impact unknown.

SUMMARY OF REQUEST: (ADD MORE DETAILS, WHERE APPROPRIATE)

We have received over 350 complaints via the CC&F system from several countries.
(See attached data). For all 1984 models shipped to right-hand drive countries,
the windshield wiper does NOT provide sufficient clear vision during rain.

ANALYSIS:

Our Analysis, and that of dealers and customers is as follows: The windshield wiper is in the
position required for left-hand drive cars (Continental Europe). But the cars are shipped to
right-hand drive countries with right-hand steering. However position of wipers was not
changed. As a result, during rain, the vision is clearer on the front passenger's side, but
poor on the driver's side.

PROPOSAL: (ADD DETAILED PROPOSAL, WHERE APPROPRIATE)

The solution is obvious. We need to retro-fit wipers on current cars to the
correct position. And, ensure future production is done correctly.

COMMENTS BY PRODUCT/PROCESS OWNER (Including $ impact):

This product was designed correctly. However the manufacturing
documentation missed this point. We have done the update.
We estimate retrofit will cost $YYY. And tooling costs of $NNN

ACTION:
Short-Term Fix   Immediate retrofit kit to dealers
Long-Term Fix: Stop R.H.Drive production,until
                          documentation and tooling
                          is redone. Estimate is 3
                          weeks.

Figure 2.3 A completed corrective action request form.

crucial as they can help to break down resistance to change. For example, a general manager at a hotel can talk to hotel guests and understand their gripes or compliments; or a general manager at a large corporation can call several customers and find out how the new system is performing at the customer's site. The data obtained can be fed into the formal customer complaint and feedback system.

Facilitation of the process. The organization's quality managers must help to manage the entire process. The specific activities for the quality managers and their staff include the following:

Promote use of the system. This is discussed above.

Oversight at the organization level. The organization's president or director of quality can oversee the entire system and also provide a central coordinator (see the flowchart in Fig. 2.1).

Function at each entity within the organization. The entity's quality manager can manage the customer complaint and feedback system and help resolve locally generated problems and issues (see Fig. 2.1). The requisite activities include these:

- Record all complaints and monitor their ongoing status.
- Analyze the root cause of the complaint—why it occurred and how to avoid recurrence in the future.
- Ensure proper routing of customer complaints to the responsible process or issue owner.
- Request follow-up if an employee, manager, or entity is lax.
- Acknowledge receipt and inform on corrective action taken to the individual who reported the complaint.
- List all complaints and corrective action as part of a monthly quality report. Document the Pareto diagram of complaints.
- Review complaints, especially outstanding complaints. Refer to the flowchart and discussion provided in the quality assurance system in Chap. 5, "Process Management."
- Forward a complaint to senior management if corrective action or progress is deemed unacceptable. Recommend appropriate action.
- When necessary, ensure that customers are informed of the progress of their complaint.

Extending the system to customers. The customer complaint and feedback system can be extended to external customers—but only after it is working well inside the company.

Customers can dial a hot line telephone number. The number should not be a telephone voice mailbox, but a number attended by a warm body. The quality department can manage this service. Initially, the service can be extended to preferred customers; later it can be provided to all customers.

The benefits of extending the system to customers include the following:

- It will serve as a safety net to catch customer complaints that are managed poorly by the normal system.
- It will convince customers that you care.
- It will give you a competitive advantage.

Even the best company will benefit by doing this; and we know of several companies that have done this successfully. In the United States, for example, Marriot Corporation and American Express have installed hot lines to collect customer complaints. In Britain, British Airways has installed video booths at London's Heathrow Airport to collect customer complaints—this is a very innovative way to collect "hot" customer inputs, immediately after their trip. And yet, as far as we know, nobody else has imitated British Airways' approach.

When the system is extended to customers, employees should be encouraged to go out of their way to help customers. A way of doing this is to empower employees to make decisions that cost, say, $10 to $100. The impact of such a decision can be electrifying to customers and employees. The result can be a quantum jump in customer satisfaction. We suggest you try this, after the customer complaint and management system is running smoothly.

What Can You Expect If You Start a Complaint System?

The various categories of complaints that you can expect, in a design, manufacturing, and sales organization, are discussed in the next section. Here we give a random list of complaints from the automobile, computer, and instruments industries. They range from thoughtful to funny. All represent problems that need resolution; and all have been sanitized.

- Your component failure rate is 150 parts per million (or 0.015 percent). You need to improve it! (*Comment:* Imagine the response if the failure rate were as high as in the complaint that follows. This is a good example of a customer and supplier working to get to zero defects.)
- About 15 percent of your products are defective on arrival!
- Your parts and service are too expensive; and the service is slow! Therefore I go elsewhere for support.

- You sold me a car with the windshield wipers fixed incorrectly, and the dealer says it is a manufacturing defect. Now, whenever it rains, I cannot see clearly. (*Comment:* This is discussed elsewhere in the section on corrective-action requests.)
- The instrument arrived with the glass tube broken. This has happened before. Please replace or repair.
- Why bother to survey again? You never did anything to help me since the last survey! (*Comment:* This was scrawled across a blank survey form.)
- You promised a delivery date of xxxx. Now you say you cannot make it. Let it be known that if you do not, there will never be any more business for you!
- The invoice is illegible! I am not going to pay you anything until you send me a proper invoice.
- Using my car's turn signal is like breaking a chicken's leg!
- Take back your product! (*Comment:* This was scrawled across a blank customer survey form.)
- You shipped a product and I never ordered anything. I am not going to open the box!

Trends in a customer complaint and feedback system

Will complaints occur—always? The answer is yes. Humans are not perfect, and our knowledge is incomplete. Hence errors and complaints will occur. But we should manage complaints by doing the following: Document, resolve, control, minimize recurrence, and reduce the intensity of complaints. This will result in a change in the trends of complaints.

Trends of complaints in a design, manufacturing, and sales company. Consider a design, manufacturing, and sales company, large or small. When initially it starts to record customer complaints, it will find that the majority of the issues relate to the interface between the company and the customer. Interface issues are those involving employees and the customer—issues such as administrative problems, rude or careless employees, cosmetic problems (dirty or damaged products), miserable after-sales service, and delivery problems. As these interface issues are eliminated, product (or service) design issues become paramount. In addition, new categories of complaints may appear as the company's business strategy changes.

Look at Fig. 2.4. This figure is constructed from data gathered after implementing a customer complaint and feedback system in a country operation of a large manufacturing and sales company. In the initial stages, interface issues, such as product delivery (late delivery, damaged items, etc.), and product support issues (after-sales support, repair quality, costs, etc.) abounded. But as awareness of these issues

Typical Complaints Recorded When System Is Launched

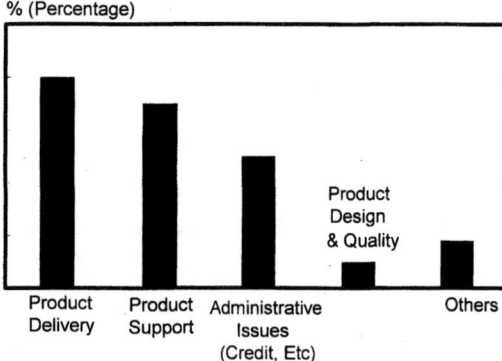

Typical Complaints Recorded Several Years Later

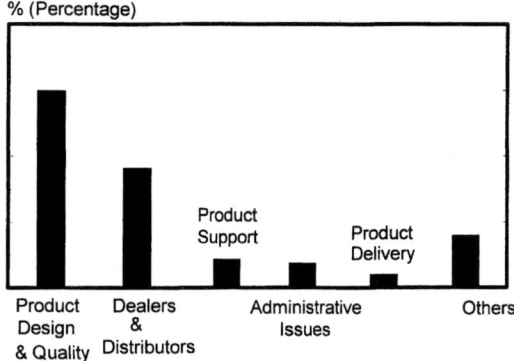

Figure 2.4 Pareto diagrams of trends in a customer complaint and feedback system at a design, manufacturing, and sales environment.

increased, the systemic root causes of these issues were gradually eliminated. With time the paramount customer issue became product design issues.

As shown in Fig. 2.4, after several years, the percentage of product design issues increased. Typically the number of product design issues will also increase. This is not because there are more design problems, but because the company is now more sophisticated—proper analysis, training, and new methods have eliminated the interface issues, while allowing the detection and understanding of product design problems facing the customer. The company can now focus on reducing design problems via several tools that are discussed in Chap. 5, "Process Management."

In Fig. 2.4, the second issue recorded several years later turned out to be problems caused by the company's dealers and distributors, who sell this company's products. This category did not exist earlier, because of the nature of the business and product strategy.

Trends of complaints in the service industry. The same situation would occur at a service organization such as an airline or a bank. For example, at an airline, initially complaints on interface issues will abound. From our experience, these interface issues will include telephone courtesy, counter service, baggage handling, and passenger management on busy flights. Later, as these issues are resolved (and this is the case with a well-managed and mature airline), complaints on the design and sophistication of services will increase—complaints concerning the quality and variety of meals, internal decor, design of seats, and special handling of baggage on first and business classes.

Summary: Customer complaint and feedback system

Errors will occur, hence every company or organization will get customer complaints. These problems must be monitored, managed, and eliminated—with the help of a formal customer complaint and feedback system. Elimination or reduction of many categories of complaints can occur if systemic root causes are removed.

As the systemic root causes of the complaints are removed, the trend of the complaints will change. Complaints that pertain to design of products and services will become evident and increase, especially in a design and manufacturing company.

Customer Satisfaction Surveys

"Talking to customers tends to counteract the most self-destructive habit of great corporations...that of talking to themselves."[3] There are several kinds of customer satisfaction surveys. We will discuss just a few here.

A *postpurchase survey.* This is conducted after a product is purchased by the customer. This can be done via a response card that comes with the product, which the customer completes and returns to the manufacturer or distributor of the product. This can provide a wealth of information about the product—its quality, ease of use, the product handbook or manual, ease of doing business with the company, and so on.

A *postinstallation survey.* This is conducted after a product is delivered and installed at the customer's site. This survey can be done via mail or preferably via telephone. This survey is for a sophisticated product that requires a manufacturer or distributor to install the product. Some topics covered in the survey are delivery commitments, courtesy of the staff, speed of installation, and the quality of the solution.

A customer satisfaction survey. This survey measures the customer's satisfaction level with a company's products and services. There are numerous specialist companies that do third-party surveys. For example, in the United States, Datapro conducts surveys on customer satisfaction for computer companies, and John D. Powers does the same for the automobile industry. In addition to these third-party surveys, which are very helpful in competitive analysis, a survey which measures your customer's satisfaction with your company is required. Many large corporations such as Hewlett-Packard, Florida Power & Light, and Xerox conduct such surveys in the United States. We briefly discuss this type of survey.

Conducting customer satisfaction surveys

Such a survey should measure your customer's satisfaction with the various attributes of your products and services. For example, if yours is a design, manufacturing, and sales organization, you would measure satisfaction in areas such as

- Marketing programs
- The sales force's interaction with customers
- Administrative services (credit, invoicing, etc.)
- Product delivery and installation
- Product education and training
- Product documentation and information
- After-sales service and support
- Product quality
- Effectiveness and value of the solution provided
- Overall cost of ownership (from purchase to use)
- Ease of doing business with the company

Each of the attributes would break down into a group of questions. For example, the sales force's interactions with customers would include the following:

- Availability of sales representative
- Technical knowledge of sales representative
- Understanding of the customer's real needs
- Consulting skills of the sales representative
- Prediction of the customer's future needs

All questions can be measured on a scale of 1 to 5, or 1 to 10. Consistency of measurement is important, to allow trends to be observed over several years. In addition to measuring satisfaction, each question or group of questions should be measured for importance.

Customers who complete the survey should be encouraged to provide comments for each item, as well as random comments. We recommend doing a survey on a random sample of customers. But for the sample you should try to get close to a 100 percent response. If the response rate is low, research on nonrespondents must be done.

Setting priorities for issues and customer comments

The issues that customers face which must be resolved can be prioritized in several ways:

In order of survey score. You will resolve the items with the lowest scores, that is, lowest satisfaction.

Selected from a scattergram. You can construct a scattergram showing both score and importance of each attribute. This can be done if you also measure each question or attribute (group of questions) by its importance to the customer. Hence, the customer may give you a low satisfaction score for administrative services, but may also rank them low in importance. Your scattergram may look like the one in Fig. 2.5. For clarity, we show only three attributes.

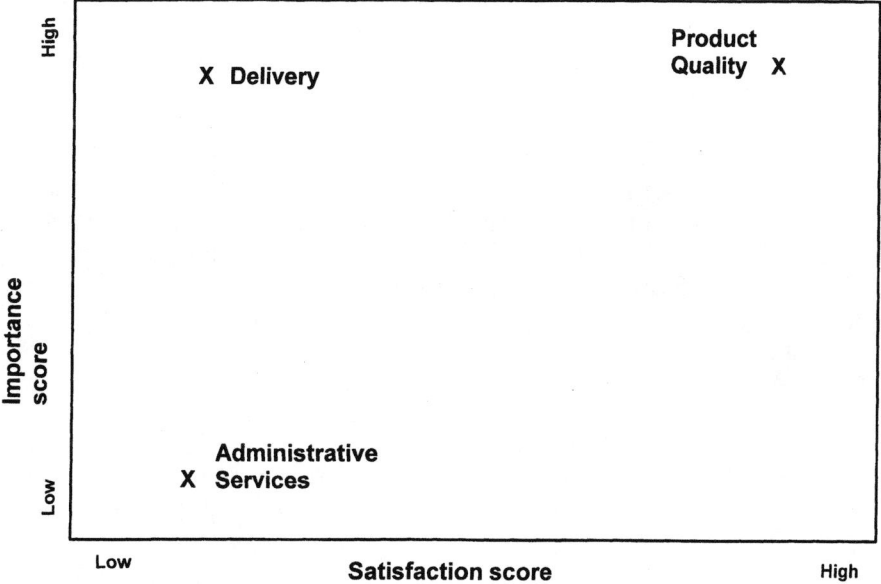

Figure 2.5 Scattergram of satisfaction versus importance.

Look at the scattergram: Both administrative services and delivery score low in satisfaction. But delivery has a higher importance level. If resources were limited, you would work on delivery only. If resources were plentiful, you would work on both. At the same time, product quality, although important, is providing high satisfaction—no additional resources need to be diverted to this attribute. But the current quality must be maintained. This is where process management becomes important. There is more on this in Chap. 5.

Selected from a Pareto diagram.   A Pareto diagram of customer comments can be prepared. Many customer comments will require further contact with the customer. But numerous customer comments can be converted to quantitative data. How? We recommend the use of the KJ method. If KJ is unfamiliar to you, refer to the section "Seven New Quality Control Tools" in App. B. This methodology will allow you to group verbal data with affinity. For example, complaints on delivery and installation or about poor dealers and distributors can be grouped. The resultant groups can be displayed on a Pareto diagram. This will provide another list of priorities.

Customer satisfaction versus customer loyalty

How much customer satisfaction is enough? It must be enough to ensure a repurchase—because the purpose of achieving customer satisfaction is to create customer loyalty, which increases the probability of a repurchase.

But even when there is a high degree of satisfaction, the customer may still defect and not repurchase. Why? Recent research shows that a customer must be completely satisfied to ensure repurchase. A study by Jones and Sasser[4] dispels some common myths about satisfaction, such as that in the real world, service levels below complete satisfaction are enough, or that the investment required to move a customer from satisfied to completely satisfied may be unwise. Instead they found that complete satisfaction is the key to securing customer loyalty and will generate long-term superior financial performance.

The authors reviewed studies done by Xerox Corporation, which discovered that totally satisfied customers were 6 times more likely to repurchase Xerox products over the next 18 months than its satisfied customers. This was followed by the author's research which showed a satisfaction-loyalty link. They found that in highly competitive industries, such as automobiles, the link was the strongest, and a slight drop from completely satisfied (5 on a survey scale of 1 to 5) to satisfied (4 on the survey scale) caused a steep drop in loyalty and repurchases. For business personal computers there was a similar trend. In the airlines industry, however, loyalty also falls off quickly, but at a customer satisfaction level below satisfied—3 or below on the survey scale. Presumably the loyalty is influenced by artificial constraints such as frequent-flier programs. But nevertheless dissatisfied customers, they conclude, will someday defect, given the right conditions, conditions such as technology changes, deregulation, or new competitors.

In summary, the basic elements of a product or service must be managed proficiently in order to obtain completely satisfied customers, who will repurchase. We believe that our customer satisfaction models will help you get there—by providing and measuring the basic elements of a product or service, a support element when there is a product or service failure, and a proactive element that looks at customer needs. We discuss the creation and generation of customer needs later in this chapter.

Other types of customer satisfaction data

We mention here one other kind of customer satisfaction data—a list that your customer keeps. For example, many companies keep a regular score card of their suppliers' performance. You may want to ask your major customers to tell you what they think of you. If they do not have such data, then in your next visit to a major account you should ask for them. Since this could be a broad-brush assessment of your performance, you could start the discussion by asking for information in the following areas of your products and services:

- Technology
- Reliability and quality
- Effectiveness and value of the product or service
- Cost of ownership (from purchase to use)
- Ease of doing business and responsiveness
- Delivery

This is a shortened and modified list of what we show in Fig. 2.6, in our discussion of a customer satisfaction model. Typically this kind of information can be rated on a scale of 1 to 5 points and displayed on a radar chart, similar to the one we discuss in the chapter on conducting quality reviews. Comparisons to competitors can also be made on the chart.

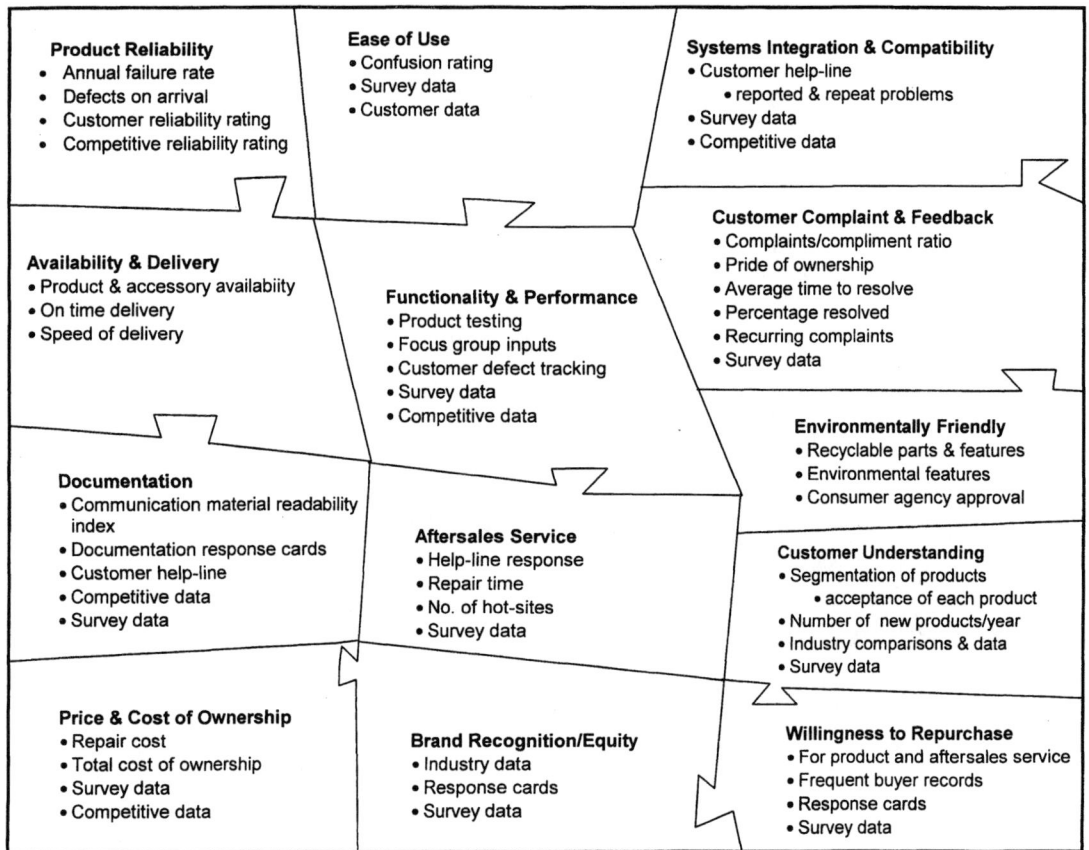

Figure 2.6 Customer satisfaction model for a product.

Ensuring follow-up to customer satisfaction surveys

The most crucial part of a customer satisfaction survey is to take corrective action after the data are collected and analyzed. In App. A we document an example of corrective action taken after a customer satisfaction survey is conducted.

Follow-up is the responsibility of management. Xerox Company in the United States has an extremely customer-obsessed management that tracks customer satisfaction diligently. They tie part of management compensation to customer satisfaction as measured in a survey. Many managers would disagree with the Xerox method—we certainly do. The better way is to plan for it, by setting the correct expectations. This will be discussed in the next chapter, "Business Planning." Nevertheless, Xerox has had success in increasing customer satisfaction. It recently won the coveted Malcolm Baldrige Award in the

United States. It is also gradually increasing its market share after years of decline.

Customer Satisfaction Model

Product satisfaction model

The surveys we mentioned earlier can be used as a starting point to prepare a customer satisfaction model for each product. Figure 2.6 shows a model listing several elements of a customer's satisfaction for a sophisticated product, such as a printer, computer, railway engine, or airplane. Notice that the elements go beyond the traditional reliability of a product. Instead, we measure a full spectrum of product attributes or elements, such as those in Table 2.2. Each element can be divided

TABLE 2.2 Customer Satisfaction Model for a Product, with Performance Measures

Product: Essex 400X, 420X, 480XE			
Item (to be measured every 6 to 12 months)	Target	Actual	Owner
1. Reliability			
Annual failure rate (%)	1.0	1.1	Manufacturing
Defective on arrival (%)	0.2	0.1	Manufacturing
Customer's reliability rating	1	1	Quality Assurance
Competitive reliability rating	1	2	Quality Assurance
2. Ease of use			
Customer rating of confusion factor*	>8.5	8.9	Research & Development
Competitive rating*	>8.5	8.6	Research & Development
3. Availability to dealers and retailers			
Product and accessory availability (%)	>95	93	Distribution
On-time delivery (%)	>97	88	Distribution
Speed of delivery (after order) (h)	29	27	Distribution
4. Documentation			
5. Price and cost of ownership			
6. After-sales service			
7. Systems integration or compatibility with other products			
8. Functionality and performance			
9. Customer complaints and feedback			
10. Environmentally friendly product			
11. Customer understanding			
12. Brand recognition			
13. Willingness to repurchase			

*Measured on a scale of 1 to 10.

Note: This list is derived from ideas generated by Dr. Juran's discussion on product fitness for use.[5]

into subelements, with appropriate targets. Examples are given in the table for reliability, ease of use, and product availability and delivery. Other elements are listed without elaboration. Such data will have to be collected from several sources, both internal and external.

Service satisfaction model

In a similar vein to the product satisfaction model, we can develop a service satisfaction model. Figure 2.7 shows a model listing several elements of a customer's satisfaction for a sophisticated or complex service, such as an airline or hotel chain. Notice that the elements go beyond the traditional attributes of a service and include elements such as credibility and understanding of customer needs. We must measure a full spectrum of service attributes, and each element can be divided into subelements, with appropriate targets, such as shown in Table 2.3.

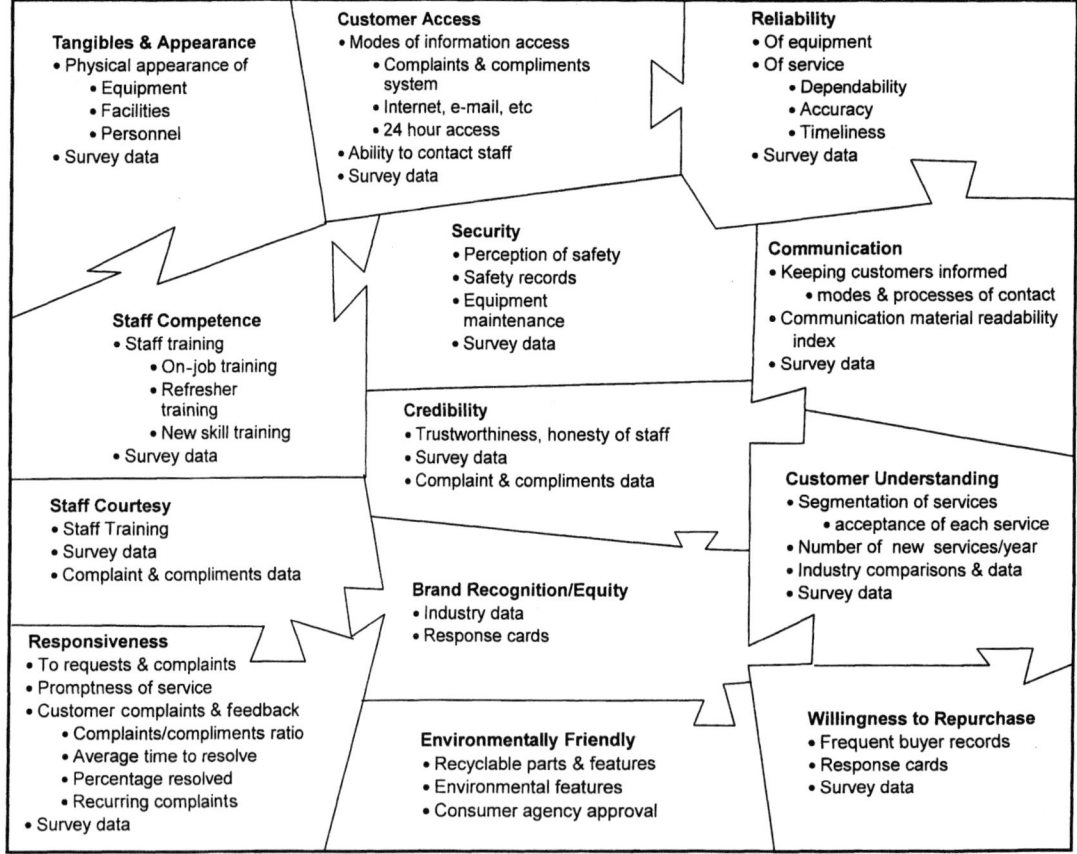

Figure 2.7 Customer satisfaction model for a service.

TABLE 2.3 Customer Satisfaction Model for a Service, with Performance Measures

Item	Target	Actual	Owner
Service: Trans Pacific Airline			
1. Tangibles			
Physical appearance (internal audit)			Maintenance
▪ Aircraft equipment	4.5	4.1	
▪ Facilities/lounges	4.5	4.2	
▪ Front-line personnel	4.5	3.9	
Food taste/appearance (internal audit)			Kitchen quality assurance
▪ First class	4.5	4.7	
▪ Business class	4.25	4.4	
▪ Economy class	4.0	2.5	
Survey rating	4.1	3.6	
2. Reliability			Maintenance
Aircraft equipment (% downtime)	99	95	
Service dependability			
▪ On-time arrivals (%)	95	92	
▪ On-time departures (%)	95	89	
Survey rating	90	87	
3. Responsiveness (all-station average)			Customer service
Request and complaints by phone			
▪ Number of phone calls per day	0	570	
▪ Average phone wait time (s)	30	33	
▪ Average phone call duration (s)	45	58	
▪ Phone calls resolved at first call (%)	70	65	
▪ Abandoned calls (%)	1	2.5	
Promptness of service (key stations sample)			
▪ Average queue time for first class (min)	1.0	1.0	
▪ Average queue time for business class (min)	3.0	3.5	
▪ Average queue time for economy class (min)	7.0	8.1	
Customer complaints and feedback (all-station average)			
▪ Complaints:compliments ratio	0.3	0.34	
▪ Average time to resolve (days)	7	9.2	
▪ Percentage resolved (%)	90	88	
▪ Recurrence of complaints	25	26	
Survey rating	4.5	4.6	
4. Competence			
5. Courtesy			
6. Credibility			
7. Security			
8. Access			
9. Communication			
10. Environmentally friendly			
11. Customer understanding			
12. Brand recognition/equity			
13. Willingness to repurchase			

Note: Items are measured on a scale of 1 to 5. These ideas are derived and elaborated from a discussion on dimensions of service quality from Zeithaml.[6]

Our example is for an airline service. Three elements are shown in detail, while others are listed without elaboration. Such data will have to be collected from several sources. The list may seem long, but for a complex service business it may need to be even longer. British Airways (BA), an award-winning airline that has made a tremendous comeback and posts healthy profits, has over 300 performance measures that it tracks and manages. Singapore Airlines (SIA), one of the safest airlines in the world and an award winner rivaling British Airways, has a service and performance index (SPI). It tracks the SPI quarterly and aims to achieve or exceed its targets. Both BA and SIA take corrective action when ratings fall. These data and trends influence the addition of new processes, systems, and services. In the case of SIA, gradual declines in its ratings for in-flight entertainment drove it to introduce exciting interactive video and audio entertainment for all classes of travel; customer feedback for this new service has been outstanding.

Managing the customer satisfaction model.   Targets can be set for the various elements or performance measures, after appropriate data analysis. These targets then become business fundamentals or daily management items for the products and services. These targets must be reviewed regularly and achieved. Refer to Chap. 3, "Business Planning," for a discussion on business fundamentals or daily management.

Who owns and manages the customer satisfaction model? The entire company does, but the company's quality or customer services department can facilitate this activity and ensure that targets are set, results measured, and corrective action taken. This customer satisfaction model, if managed well, can become a powerful tool to collate, measure, and achieve customer satisfaction at any entity that designs, manufactures, and sells products or services.

Competitive Benchmarking

Competitive benchmarking has been around for thousands of years. It is a process that allows us to measure our products, services, and processes against the toughest—preferably world-class—or leading-edge companies. This methodology forces us to have an external view that can catalyze internal change in our company or organization, and helps us to be more efficient and competitive.

Purpose of benchmarking

Before we continue, we wish to stress that the purpose of competitive benchmarking is not to conduct industrial espionage or to do anything unethical. Rather, the purpose is to look at processes used by other leading companies and to use that learning to improve our own processes. This will provide us with competitive advantages for a short duration until another company arrives with a better way or new benchmark—because breakthroughs and improvements keep happening. At the same time, we can expect other competitors and leading companies to emulate successful new arrivals.

What can be benchmarked?

Almost everything can be benchmarked—both products and services. We give examples of both. A partial list for a design, manufacturing, and sales environment is shown below, but the same concept can be applied to the service industry, as shown in our examples.

1. Product development process
   - Understanding and collecting data on customer needs
   - Design tools and technologies
   - Process of quick design and introduction to market
2. Manufacturing methodologies
   - Production technologies
   - Automation techniques
   - Miniaturization techniques
3. Product design and performance
4. Quality systems
5. Supply chain management
6. Order management
7. Sales and marketing techniques

Benchmarking process

Shown below is a suggested benchmarking process. This is fairly generic and applies to benchmarking processes or product performance. We then elaborate on some of the steps, using a product and process example. The process is segmented into two sections: the benchmarking study and a proposal to implement lessons learned from the study. This second step is crucial; otherwise, you should question why the benchmarking study was done.
   I. Benchmarking study
      A. Identify subject (product or process) to be benchmarked and objective of the benchmarking study.

1. The objective of the benchmarking study must be clearly stated to ensure that the study is productive and successful. That is, state what practice you wish to understand or what you want to measure. The objective also helps state the minimum expectations of the study.
2. Examples of subject and objective:
   a. Benchmark printer model X from company Y, and determine the manufacturing cost, assembly process, and design technology used.
   b. Benchmark the supply chain in the packaged food (suppliers to supermarkets) industry, and determine the distribution cost at each stage, inventory turns, delivery speed, and information technology used to manage the supply chain.
   c. Learn how leading-edge companies use the Internet, intranet, extranet, and other technology to streamline the order management process. Understand cost structure, technology used, productivity gained, and customer reaction and satisfaction to these processes.
B. Set up a team with a leader.
   1. The team will conduct the benchmarking study.
   2. A very viable alternative is to subcontract the study to a consultant or to work with organizations that do benchmarking and best-practice studies across industries. This can save time and effort. Note, however, that sometimes these studies give average data in the industry, but best results are usually achieved by a leading-edge company in that industry.
C. Identify world-class or leading-edge companies that have a product or process you want to study.
   1. It is important that you look within and outside your industry for the best companies.
   2. Approach the world-class or leading-edge company, or obtain competitive product.
D. Prepare and implement the data collection strategy.
   1. List the data you wish to get. This will include a series of items or processes that will help you get the information you stated in your objective.
   2. Obtain the data and understand the best practices that generate these data to give superior results.
E. Collect and analyze data after you visit the company or examine competitive product.
F. Determine performance gaps between the benchmarked subject and your product or process.

II. Proposal for improvement
A. Discuss and determine areas for improvement with performance targets to be achieved.

    1. Prepare list of recommendations and estimated costs, resources, and time line needed to achieve them.

    2. Prepare realistic proposals with estimated budget and time lines.

    3. Get management approval to implement them.

B. Prepare an implementation plan for the recommendations. Ideally, this will be incorporated into the annual plan, unless the issue is critical and requires immediate implementation.

C. Implement the plan.

D. Monitor progress and ensure successful implementation.

E. Review results and recalibrate the benchmarked subject to determine if results were achieved.

Most of the steps are quite straightforward, but let us review some of them. Look at step IC. We must look at world-class or leading-edge companies. This is an important step, because the world-class process we are trying to emulate may not be that of a competitor. If we are distributing computers, the best distribution process may be in the supermarket sector. If we manufacture laptop computers, the best miniaturization technology may be in another industry, such as cameras. Therefore, it is important to look at the best external practice and not just within an industry or competition.

Next, look at step ID—prepare and implement the data collection strategy. The actual data required in this step will vary; it will depend on the type of product or service being analyzed. The review team meets and discusses these requirements in great detail before setting forth on this step. If we were analyzing an electrical or electronic product, here is what we could look for, if the objective were to benchmark printer model X from company Y and determine the cost, assembly process, and design technology used.

1. Delivery and cost information
   - Availability of product
   - Price of product
   - After-sales maintenance costs
   - Overall cost of ownership
2. User-friendliness
   - Simplicity and usefulness of manuals
   - Product packaging quality
   - Ease of setup and use of product
3. Performance review
   - Speed of operation
   - Feature set

4. Design review
   - Style of product
   - New concepts
   - New technology
5. Manufacturing methods review
   - Design for manufacturing (ease of assembly)
   - Automation and process technology
   - Cost of materials and manufacturing
6. Stress testing
   - Electrical characteristics
   - Vulnerability to shock and vibrations
   - Vulnerability to temperature and humidity
7. Environmental testing (optional)
   - Radiation and interference generated
   - Environmentally friendly attributes, such as packaging, noise generated, and energy costs
8. Executive summary (list of all items required in the initial objective)

If we were benchmarking a process, we would arrange to visit a friendly company or organization with a world-class process. So for a review of the supply chain process, here is what we would look for, if the objective were to benchmark the supply chain in the packaged food (suppliers to supermarket) industry and to determine the distribution cost at each stage, inventory turns, delivery speed, and information technology used to manage the supply chain. *Note:* Supply chain management in the packaged food industry in the United States is reputed to be highly efficient through a program known as efficient consumer response (ECR).

1. Products distributed
   - Markets and customers served
   - Types of products
   - Volumes of products
   - Shipment dollars
2. Review of the entire supply chain
   - Process steps from suppliers to supermarket
   - Inventory turns at each step
   - Transit times at each step
3. Warehouse facilities

- Locations and size
- Inventory levels and management techniques
- Automation methods
- Information systems
- Costs

4. Shipments
   - Packaging methods
   - Shipping methods
   - Shipping costs
   - Shipment transit times

5. Use of third-party services
   - Service providers used
   - Services provided
   - Costs and savings incurred

6. Order management
   - Methods of ordering by customers (supermarkets)
   - Information systems and processes used
   - Ease of ordering by customers
   - Order management costs

7. Other useful data on efficiency and quality of services in the supply chain
   - Overall inventory turns per year at each step
   - Quality of shipments: damage, delays, failure to meet commitments, etc.
   - Ability of coordinating several orders into one shipment
   - Supplier response time: average time taken to ship after the order is received
   - Line order fill rate: percentage of time items are in stock when an order is received
   - Return on assets

8. Executive summary (list of all items required in the initial objective)

Once this information is gathered, performance gaps of the benchmarked subject can be better understood. The resulting improvement plan can yield a more competitive product or a world-class process in your company. The outcome can be higher productivity, lower expenses, more satisfied customers, and increased profits.

Other Thoughts on Competitive Benchmarking

Competitive benchmarking analysis, at best, helps you be the equal of the toughest competitor or industry leader. But it will not enable you to leapfrog a competitor or predict what the future holds.

Was the U.S. automobile industry able to predict the success of the Japanese car industry in the United States? And was it able to predict that Japanese manufacturers would move upscale into the luxury car market, via Toyota's Lexus and Nissan's Infiniti? Probably not on both accounts.

A recent article in *The Economist* magazine[7] discusses the need to go beyond competitive benchmarking. The article quotes Pankaj Ghemawat, an associate professor at Harvard Business School, as saying that most companies only benchmark price and performance of their product with a competitor or the entire industry. This methodology will fail to spot potential rivals.

One option is to take the approach of NEC. When it plans for the future, NEC (as quoted in the *Harvard Business Review*[8]) concentrates on honing its "core competencies"—its key skills and technologies. In planning for the future, NEC decided that its communications, computer, and component business would increasingly overlap. It predicted that success in these businesses would require increasing skills in one specific core competence—semiconductor manufacturing. It invested heavily in this technology. In fact, NEC's semiconductor division recently won the coveted Deming Prize in Japan, thus establishing itself as a high-quality, customer-oriented supplier. This strategy has put it in a commanding position to succeed in computers and telecommunication.

The article goes on to give another option by quoting the example of Chaparral Steel, a profitable U.S. steelmaker. This company does not waste time trying to improve on its existing products. Instead, it visits competitors and university research departments with the purpose of detecting trends—this allows it to predict the next successful product.

Capturing the Customer's Voice and Needs

Before products are designed and built, it is important that the customer's voice and needs be captured. This information can lead us to provide the right product for our customers. And yet, numerous manufacturers design and build products that are never released or do not sell, resulting in canceled projects, wasted resources, higher costs, and often lower morale.

Why do products succeed in the marketplace?

Why do some products, which were expected to be blockbusters, fail in the marketplace and yet other products, which were developed at the whims of a chief executive, succeed beyond all expectations? Two examples that come to mind are the Hewlett-Packard HP 35, the first electronic scientific calculator; and the Sony Walkman, the first personal stereo tape cassette player.

The difference is customer acceptance. But how do we get customer acceptance? The chief executives of Hewlett-Packard and Sony were willing to bet, by gut feel, that what they proposed would be block-busters; and they were right. Unfortunately, this process is not easily repeatable; and we need to use a more systematic process that minimizes failure. Is there a better way?

Project Sappho study. A study called *Sappho*[9] (Scientific Activity Predictor from Patterns with Heuristic Origins) was conducted to understand successful industrial innovation. The study was published by the University of Sussex in 1972. The title of the report was "Success and Failure in Industrial Innovation." The report describes the key factors for success in designing new products and new processes (in the chemical and like industries). In descending statistical order of importance, the factors are

1. Successful innovators were seen to have much better understanding of user needs.
2. Successful innovators pay much more attention to marketing.
3. Successful innovators perform their development work more efficiently than unsuccessful ones, but not necessarily more quickly.
4. Successful innovators make more effective use of outside technology and scientific advice, even though they perform more of the work in-house. They have better contacts with the scientific community in the specific area concerned.
5. The responsible individuals in the successful attempts are usually senior and have greater authority than their counterparts who fail.

Unfortunately this vital research was not well utilized to increase the success ratio of new products or services by the western industrialized nations such as the United States or Great Britain. In addition, the research itself has remained a well-kept secret. But we can clearly see that the most vital ingredient is an understanding of user needs. Let us now look at another study that probes why products fail.

Why do products fail in the marketplace?

A study at Hewlett-Packard came up with many reasons why products fail following release to market. Among them were the following, quoted by John Doyle, executive vice-president.[10]

1. Lack of project endorsement by upper management.
2. The better mousetrap that nobody wanted.
3. The "me, too, product" that meets a competitive brick wall.
4. The technical dog—a product so technically sophisticated that no one understood it.

5. The price crunch—the market wanted a Chevrolet, but we gave them a Cadillac.

These data are not prioritized. We do not know the biggest contributors. But items 2, 4, and 5 relate to the customer's needs and perceptions.

Why are products terminated during the design stage?   Very often product introductions are canceled because the expectation is that customers will not buy them. In addition, companies have their own reasons for terminating products during the design stage. The result is wasted product development effort. In this area there have been some studies. Let us review one of them.

The Union Carbide Industrial Gases (UCIG) Division did an analysis of wasted product development.[11] They estimated that 12 percent of UCIG's annual development effort was wasted on terminated projects. There were three main reasons for termination:

1. *Customer/market.*   The project did not meet customer needs, the potential customers were too few, or alternate products do the same thing more cheaply.

2. *Business.*   Project is outside UCIG's domain. That is, the project was not endorsed by upper management.

3. *Technical.*   Project failed to achieve technical objective.

The study has similarities to the Hewlett-Packard study. According to UCIG, the first item listed above, customer/market reasons, contributed to the bulk—86 percent—of the wasted development effort. In fact, its analysis defines the root cause of the customer/market problems as follows: "Many projects are selected and continued without management and the developers having sufficient market intelligence and/or understanding of customer needs and benefits. This results in a needless expenditure of effort on projects for which termination is inevitable."

Wasted product development and product failures will always occur, because risks have to be taken; but these must be minimized. This is possible if the reasons are understood; the data from UCIG give some of the reasons and possible solutions. The solution that UCIG proposed includes the following:

- A written information-gathering plan
- Agreed upon responsibilities and tradeoffs
- Milestones and decision points
- Customer feedback for verification and periodic reviews

Summary: Why products succeed in the marketplace. A common thread runs through the three studies that we have quoted: To have a successful product, we must understand the user's needs. There will always be a place for the brilliant chief executive or designer who dreams up a blockbuster product. But for most product designs, we will need to rely on a better way—by using a process that attempts to capture user needs. We discuss this next.

Quality function deployment

A good product definition process is crucial for ensuring that a product or service meets customer expectations and needs. The product definition process is part of the product development process, which starts with the customer's needs and ends with delivery to the customer. The same concept applies to a service, where we will need a good service definition process.

The quality function deployment (QFD) methodology is a process that helps to provide better product definition and product development. QFD is a systematic process that helps a design team to define, design, manufacture, and deliver a product or service that meets or exceeds customer needs. The key features of QFD are as follows:

- *It captures the customer's voice.* The customer's voice is captured in order to define product or service specifications.

- *It ensures strong cross-functional teamwork.* Teamwork is ensured between the various functions involved with the design, such as marketing, R&D, and manufacturing. Numerous studies have shown that such teamwork is essential for a company's success.

- *It links the main phases of product development.* A more thorough use of QFD ensures the generation of four matrices that link the four main phases of product development: (1) product planning, after customer needs are understood; (2) part deployment; (3) process planning; and (4) production planning.

The QFD methodology has been around for several years. It seems to have been first used by Bridgestone Tire Company. It has been successfully used by Toyota Motor Company in product development and has helped reduce product costs, rework, and design changes after product release.[12] Toyota builds cars that better meet customer's needs. Toyota has also reduced the overall time to design and market its cars. Currently, Toyota has about the shortest product development time in the automobile industry—about 3 years.

We recommend use of QFD methodology; details are provided in App. C.

The Total Product or Service Concept

The various methodologies discussed so far will help you to have a more customer-sensitive organization. There is, of course, another ingredient that is necessary—creativity. We believe that the customer-oriented activities that we have recommended will form a strong foundation, which provides a good basic system and allows creativity to be channeled to the right areas.

One such area lies in providing "attractive" quality or exciting features in your products and services. This is what the customer wants: exciting products and services. This concept of attractive quality can be integrated into the QFD matrix after customer needs and competitive offerings have been assessed. Refer to the discussion in Chap. 1 on definitions of quality, specifically, must be and attractive quality.

Attractive quality in products

For products, attractive quality can be easily seen. In automobiles, for example, it includes features such as antilock brakes, safety air bags, sun and moon roofs, electronically adjusted seats, quiet, and luxurious interiors. And, most important, all this is expected at a reasonable price. The Toyota Lexus is a stunning example. Its attractive quality features threatened luxury car leaders such as Mercedes-Benz and BMW, who woke up from complacency and then raced to catch up. The same concept can be applied to all products and services.

Attractive quality in services

How can we apply this concept to services? Think about a favorite restaurant or airline—its excellent and special services come to mind. Today, as we write, Singapore Airlines is planning an "office in the sky" for all its Boeing 747 aircraft. This will be a first and will include video conference facilities, desktop computers, photocopying equipment, and easy payment. This is *attractive* quality for the busy executive. In addition, SIA has gone ahead to provide interactive video and audio entertainment for all classes of airline travel. Why? It needs to stay ahead of the competition; it needs to retain and expand its customer base; and it needs to keep growing its revenue base and profits. So it must continue to stretch the limits of its services and expand its current offerings. We mentioned in Chap. 1 that your attractive quality features will eventually become must-be quality features; so you need to keep growing. Another way to look at this is via the total product concept. We discuss this further and review how one department store does this.[13]

Harvard Business School marketing professor Ted Levitt provides a handy idea, which he labels the *total product concept*. Picture four

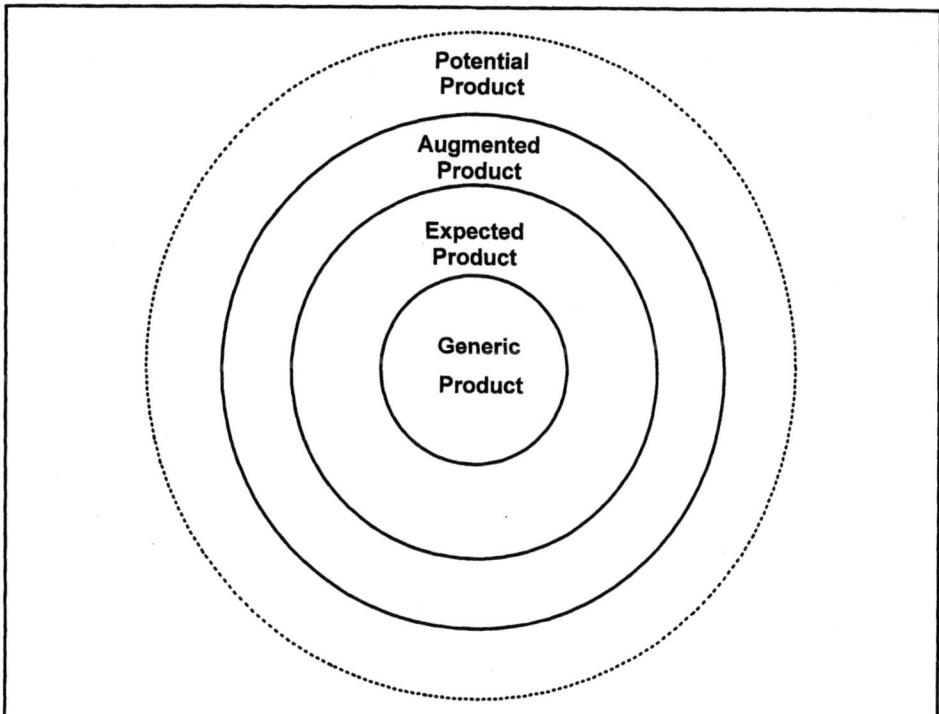

Figure 2.8 Total product concept.

concentric circles. The inner one is labeled *generic product,* next comes *expected product,* then *augmented product.* The last, which has no boundary, is labeled *potential product.* Refer to Fig. 2.8.

Take Nordstrom, whose fairly high-price specialty retail goods comprise the firm's generic trait. The expected trait includes stores that stay open normal hours and carry styles that are contemporary. Nordstrom invests heavily in its remarkable service—its augmented attribute—such as very high availability of odd sizes and colors and high pay, by industry standards, for an exceptionally large number of salespersons on the floor.

The unlimited potential trait is a vast number of touches—from regular personal notes to customers from salespersons, to the especially clean and colorful dressing rooms, to a "no-questions asked" return policy, all of which help Nordstrom live up to its "no problem at Nordstrom" motto. By its unconventional emphasis on the outer circles—the augmented and potential—Nordstrom has virtually redefined retailing. To quote a friend at a computer company, it is not a specialty retail store, but "a user-friendly, entertaining, total experience" that has something to do with the purchase of garments.

How does one develop the total product concept?

First, ensure that the generic and expected products are in place. You can do this via the customer satisfaction model for product or service, which we have discussed in detail. This ensures that all the basics are in place, tracked, and managed—it must include a good support or reactive process to respond to failures in the system and a measurement to discover whether the current processes are meeting customer needs. Most important, a process must be in place to measure satisfaction, ensure a high degree of satisfaction, and implement corrective action when there is a failure.

Next, ensure that you have augmented products. This will ensure that you are beginning to differentiate yourselves from the competitors. This can be via superior service, facilities, and features.

Finally, start to work on creating the potential product or service. This will distance and differentiate you from the competition. For a product, provide a superior product that BMW, Mercedes, or Toyota Lexus strive to do. For a service, provide the total experience that Nordstrom, British Airways, and Singapore Airlines do so well. This requires you to be constantly on the move, stretching the limits of what you have today and generating a constant stream of new products or services. Here is a short checklist to get you thinking and moving:

1. What new features can we provide?
   - Can we harness new technology?
   - Can we ensure that products and services are hassle-free?
2. What new customer segments can we reach out to?
   - Refer to this discussion in Chap. 3, "Business Planning."
   - What products or services can we offer?
3. Create product or brand managers to manage products and services in each segment.
   - Give them the responsibility to develop and grow a stream of products and services.
4. Remember that you are no longer competing on cost and price. Most customers will pay a premium for a better product or service. So give it to them—they deserve it—and charge a premium.
5. What must we do to create a total experience?
   - How do we ensure that the customer will repurchase?
6. A convenient tool for managing a total product or service project is the *proactive PDCA cycle,* which we discuss in Chap. 4.

In Fig. 2.9, we illustrate the use of the total product concept for a business-class service at Singapore Airlines, an award-winning air-

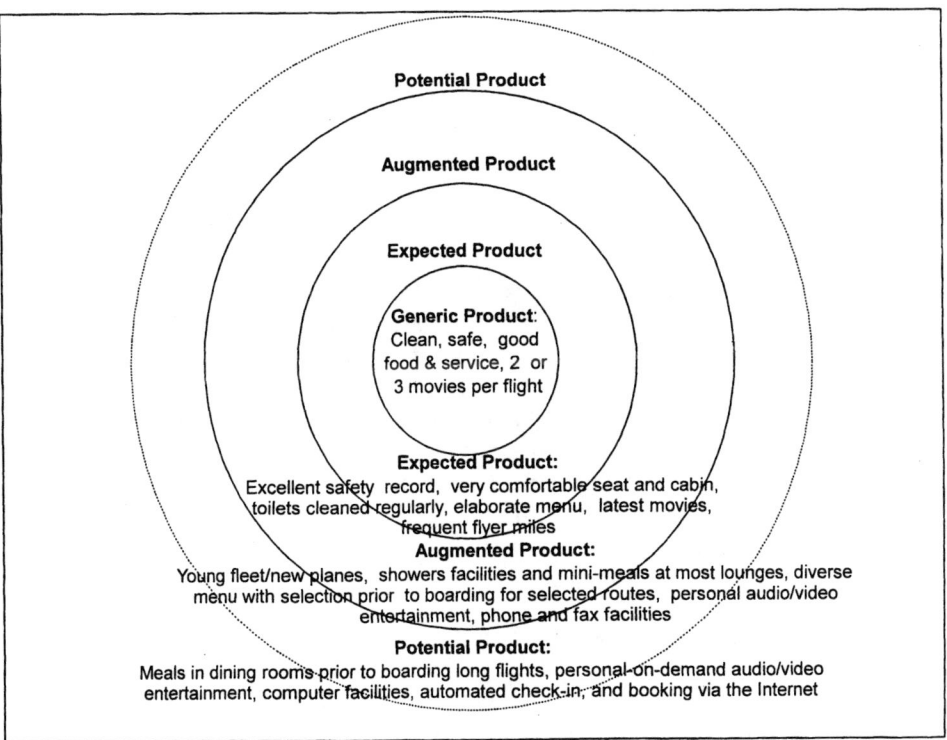

Potential Product

Augmented Product

Expected Product

Generic Product: Clean, safe, good food & service, 2 or 3 movies per flight

Expected Product: Excellent safety record, very comfortable seat and cabin, toilets cleaned regularly, elaborate menu, latest movies, frequent flyer miles

Augmented Product: Young fleet/new planes, showers facilities and mini-meals at most lounges, diverse menu with selection prior to boarding for selected routes, personal audio/video entertainment, phone and fax facilities

Potential Product: Meals in dining rooms prior to boarding long flights, personal-on-demand audio/video entertainment, computer facilities, automated check-in, and booking via the Internet

Figure 2.9 Total product concept, for business class service at Singapore Airlines.

line. The generic and expected traits list the basic needs and expectations from a business-class service. Today, not all airlines provide what is available in the expected trait. Most of the leading airlines however, provide this but not much more. But passengers would love more. The augmented and potential traits are what SIA is focusing on. The services in the augmented section are already available, services such as a young fleet with new planes, computer and showers facilities and minimeals at most business lounges, a diverse menu with meal selection prior to boarding for selected routes, personal audio and video entertainment, and phone and facsimile facilities. This is already very good, but SIA is going further and is planning potential offerings such as meals in dining rooms prior to boarding long flights, personal on-demand audio and video entertainment, computer facilities, automated check-in, and bookings via the Internet.

By focusing on the augmented and potential circles, SIA is redefining business-class travel and providing an exciting total flying experience. It is no surprise that SIA gets voted as the best business-class provider every year, not to mention its other awards.

As you plan to provide the total product concept, note the following important points.

1. Other airlines will try to compete with SIA by moving into the outer zones or circles. But this takes time. For example, we have yet to see toilets cleaned regularly in business class, although SIA has been doing it for all classes of travel for more than 10 years! Nevertheless, competitive pressure will require that SIA keep moving forward with new products and services. More important, services need to be refreshed to keep, impress, and attract new customers. This requires a constant stream of new products and services by an alert management team who stays on their toes constantly.

2. Once a new product or service is introduced, it must be managed and maintained consistently. That is, every item in the total product concept from generic to potential product must be maintained at the planned or promised level. This is a common area of failure with even the best organizations. We recommend managing all products and services via the customer satisfaction model, discussed earlier in this chapter. We have also discussed how British Airways and Singapore Airlines manage their products and services by identifying, tracking, and managing performance measures. In Chaps. 3 and 5, we discuss the concept of the business scorecard to manage processes by using a daily management, or business fundamentals, plan.

Questions and Answers on Customer Obsession

Let us review a few questions on customer obsession.

Question. The customer complaints and management system you propose is very complex. My employees are well trained and very customer-conscious. Why should I start a formal system and hire more resources?

Answer. An informal system has some weaknesses. You cannot be sure that every complaint is attended to, because sometimes an employee is busy or pressured by other priorities. Often, the recipient of the complaint has to go to someone else for help, and that process does not always work. Also, if you have a large organization, the same problem may be occurring in other parts of the organization—in this case, you have a class- or companywide issue that needs to be addressed, so that the problem does not repeat across the organization. All these are reasons to invest in a formal system.

Question. We recently conducted a customer satisfaction survey and got a very high satisfaction score, yet our market share is low. What is happening?

Answer. The reasons can be complex. You may be asking the wrong questions. That is, the things you ask about may not be important to the customers; your competitors may be providing those additional needs.

In measuring satisfaction, you may be talking to a captive audience. That is, customers who own your products may get the best satisfaction for your product and for the given set of features that your product provides; yet other competitive products may have better features and a higher value.

You must measure the entire spectrum of features and activities that a customer may want. Consider capturing and collating satisfaction data in a customer satisfaction model similar to the one shown in Figs. 2.6 and 2.7. This will require collating data from several sources and can reveal flaws in your product or service.

If you compete at a higher price or have poor marketing efforts, your market share will be lower, too. To understand the root cause of low market share, you need to perform a detailed analysis. For example, you could do a cause-and-effect analysis. You can get some guidance on cause-and-effect analysis from Chap. 4 and Appendix B.

Finally, you must aim for complete satisfaction (5 on a scale of 1 to 5). Otherwise, you may be getting some repeat purchases, while other customers may be defecting. Refer to the discussion in this chapter on customer satisfaction versus customer loyalty.

Summary: Creating a Customer-Obsessed Organization

We have discussed the importance of customer obsession. Having well-trained employees with good intentions will not be sufficient to provide this obsession. We have proposed several methodologies: reactive and proactive. They include the following:

- A customer complaint and feedback system
- Customer satisfaction surveys
- A customer satisfaction model
- Competitive benchmarking
- Capturing the customer's voice and needs via quality function deployment (QFD)
- The total product or service concept

Some will make you more customer-sensitive—specifically, the customer complaint and feedback system and customer surveys. The customer satisfaction model will allow you to keep track of all facets of customer satisfaction and will alert you when performance is declining or customers are becoming dissatisfied. Competitive benchmarking will help ensure that you are at least equal to the world's best companies. The use of QFD methodology will help you design the right products and reduce the frequency of failed products and wasted resources. QFD can be integrated with the concept of designing in an attractive quality. Finally, the total product concept will start to stretch your organization to move toward a potential product that does not exist today.

All these activities will propel you ahead in providing exciting products and services, give you a competitive advantage, and help create a successful, customer-obsessed organization. You need to plan for these activities, and that is our next topic.

References

1. Hilary Hinds Kitasei. Extracted with permission of the *Wall Street Journal,* July 30, 1985.
2. Printed by permission of Kent Stockwell, quality manager, Hewlett-Packard. *Note:* The name of the competing company has been disguised.
3. John Brooks, "Telephone—The First 100 Years."
4. Thomas. O Jones and W. Earl Sasser, "Why Satisfied Customers Defect," *Harvard Business Review,* Nov./Dec. 1995, pp. 88–99.
5. J. M. Juran, Frank M. Gyrna, and R. S. Bingham, *Quality Control Handbook* (New York: McGraw-Hill, 1951).
6. V. A. Zeithaml et al., *Delivering Quality Service: Balancing Customer Perceptions and Expectations* (Toronto: The Free Press, 1990).
7. "Competing with Tomorrow," *The Economist,* May 12, 1990.
8. C. K. Prahalad and Gary Hamel, "Core Competencies," *Harvard Business Review,* May/June 1990, pp. 79–91.
9. "Success and Failures in Industrial Innovation," by the Science Policy Research Unit, University of Sussex (Edinburgh: Bishop and Sons Ltd., 1972).
10. John Doyle, quoted from a speech for Hewlett-Packard quality managers, Santa Clara, Calif., May 1990.
11. E. Scott Timmins and Jack Solomon, "Quality in R&D: How to Involve the Customer and Like It," 1990 ASQC Quality Congress, San Francisco.
12. L. P. Sullivan, "Quality Function Deployment," *Quality Progress,* June 1986.
13. Excerpted from Tom Peters's weekly syndicated column. Copyright 1987, TPG Communications. All rights reserved.

3

Business Planning

For I dipped into the future, as far as human
eye could see, saw the vision of the world, and
all the wonders that would be.
ALFRED LORD TENNYSON
Locksley Hall

Overview

Preparing for the future is crucial. In today's highly competitive and changing marketplace, the margin for error is decreasing; hence planning for the future is necessary for survival and success. Planning and executing the plan are key activities in a company, and these are done by its managers. If done well, this process will give the desired results.

Formal planning provides many benefits, including systematic thinking, better coordination, sharper objectives, improved performance standards, and management involvement. All these result in a planned approach to tackling the marketplace that can end in higher sales and profits.

The following incident indicates what can happen when planning is not formalized in a company.

I visited a large sales operation to conduct a quality review. On the second day I met with a district sales manager and his staff. He was extremely intelligent, young and enthusiastic. During the current fiscal year he had lost his major account—they had been acquired by another company and their purchases from his company declined very rapidly. What to do? He decided to prepare a plan and purchased a copy of a book, entitled *One Page Management*. He read it and started training his staff in it. In addition, he started a process to manage his sales funnel— the performance measures and goals to manage this funnel were entered into his one page management plan.

He had been experimenting with this plan for several months and was still fine tuning the performance measures. I asked him why he did this. He replied that he knew he was in trouble when he lost his major account and would not meet his quota for that year; he decided he had to plan—do something different—in order to get his sales up and his team motivated. The book helped him get organized and he felt he was now well on the way to meeting quota.

I remarked that a planning process, better than what he had, existed in the company; also I personally had written a book that addressed how the sales funnel was managed in one of the countries in his company. To my astonishment he had never seen or heard any of this, which is why he had been forced to develop his own planning process.

The moral is simple: He managed by experience; the lucrative major account made him complacent. With the loss of the account he had to change his ways—he went into a panic mode and then developed a plan. He had wasted several months of valuable and irreplaceable time. Most of this could have been avoided if the company had standardized the planning process, trained him in it, and taught him how to manage his sales funnel. If you do not have a good planning process, your employees will design their own—and it may not be the best!

In this chapter we discuss the essential elements of a good planning process. The discussion includes details and guidance on successful long-range and annual planning.

Making the Future Happen via Planning, Execution, and Control

The future can be shaped through careful planning and execution. Charles Knight, CEO of Emerson Electric, says, "We believe that companies fail primarily for non-analytical reasons: management knows what to do but, for some reason, doesn't do it."[1] Planning carefully is one way to do what needs to be done. But the trick is to keep it simple. This requires incredible discipline. Because it plans well, Emerson claims an enviable record of being the low-cost producer with improved earnings for over 30 years. We discuss how to keep plans focused, manageable, and successful.

Essentials of a Planning Process

A good planning process needs to exist throughout an organization. It should be a standard process, and everyone should be trained in it. Once in place, it provides a common planning language with common formats. As managers and employees transfer or move within the company, they need worry about only the planning content, the process having been standardized.

A good planning process will consist of a long-range, strategic plan and an annual plan. The long-range plan focuses on product and service strategies that will lead to market success over the next few years. It includes an analysis of the current situation (reviews of strengths and weaknesses, competitors, customers, opportunities) and then defines broad objectives and strategies that must be pursued. These objectives, which are fact- and data-based, are aggressive and crucial for success, and achievable given the right resources.

The annual plan is the last step of the long-range plan. *It is a statement of intent.* It includes the breakthrough objectives and implementation plans for these objectives, more specifically, the how, who, and when statements of each objective. For this portion, we recommend the use of policy management better known as *Hoshin Kanri.*[2]

The combination of these plans—long-range and short-term—forms a powerful and planned approach to succeeding in the marketplace. Our discussion focuses in detail on both long-range and annual plans.

The Long-Range, Strategic Plan

An organization must understand its purpose for existence, the marketplace, and the competition; and then it must determine its long-term direction. One company in particular, Matsushita of Japan (owner of the Panasonic, Technics, and National brands), is reported to have a 250-year plan. This was developed in 1932, and by all accounts it is progressing according to plan; the plan is reputed to include company growth, products, markets, and countries covered. In our context, we are proposing a 3- to 5-year strategic plan.

The long-range plan should cover the following steps and areas:

1. Company's or organization's vision
2. Customer's needs, issues, and channels of distribution
3. Competitive situation
4. Necessary products and services
5. Description of product and service strategies, development of partners, and purchase plan
6. Financial analysis
7. Potential problem analysis
8. The 3- to 5-year objectives
9. The annual plan

Company's or organization's vision. Before any planning starts, the organization's vision should be prepared and clearly understood. The vision describes the fundamental set of reasons for the organization's existence; it should be inspiring, give a clear sense of direction, and provide a basis for decision making. It should also indicate where it will be in the future and how it expects to provide a competitive advantage. It should be a concise and simple set of beliefs. Let us discuss this further.

New research indicates that a well thought out vision consists of two items: core ideology and an envisioned future. In an article "Building Your Company's Vision" in *Harvard Business Review,* James Collins and Jerry Porras[3] suggest that core ideology and envisioned future are like yin and yang; see Fig. 3.1. Core ideology is the yin, which defines an organization's values and why it exists. The envisioned future is the yang, which is what the organization aspires to become, to create—something that will require significant change and effort to attain.

Core ideology. This defines an organization's values or core beliefs—values that stay with an organization through good and turbulent times, through high growth and stagnation, and through management fads and leadership change. An excellent example of this is Hewlett-Packard Company. Bill Hewlett commented after Dave Packard's death, "As far as the company is concerned, the greatest thing he left behind was a code of ethics known as the HP Way." This has guided the company for over 50 years—this author can attest to that. Collins and Porras maintain that great company builders such as David Packard of Hewlett-Packard, Paul Gavlin of Motorola, George Merck of Merck, William McKnight of 3M, and Masura Ibuka of Sony understand that it is more important to

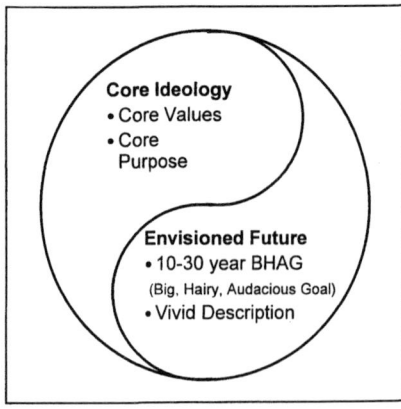

Figure 3.1 Shown here are the two components of a vision: core ideology and an envisioned future. They are linked tightly like yin and yang. (*Copyright 1996 by the President and Fellows of Harvard College; all rights reserved.*)

TABLE 3.1 Vision Components of Hewlett-Packard and Toyota

Vision component	Hewlett-Packard	Toyota
Core ideology: *Core values*	The HP way: Trust in people, open-door policy, commitment to innovation, product and service quality, growth from profits.	Total quality management, manufacturing and design excellence, comfortable work environment.
Core ideology: *Purpose*	To create information products that accelerate the advancement of knowledge and fundamentally improve the effectiveness of people and organizations.	Be number 1 in quality and offer products to respond to requests of society and trust of customers.
Envisioned future: *Goal*	Goals are set by product groups. One product group's goal: Be number 1 in channel revenue and customer satisfaction.	Global 10 (10% market share worldwide by 1999)
Envisioned future: *Specific description*	One product group's description: When information touches paper, HP is there.	We are recognized as the world's premier car maker.

know who you are than where you are going, because where you are going will change with time.

Core ideology values can be segmented into two items: core values and a purpose. See Table 3.1. Core values are, for example, the HP way, creative and innovative products at Sony, wholesome entertainment at Walt Disney, customer service and satisfaction at Nordstrom, and exceptional quality at Toyota. The other component of core ideology is the purpose statement, which states the reason why the organization exists. For example, at Hewlett-Packard the purpose is to provide information products that improve productivity, and at Walt Disney it is to make people happy. A purpose statement should last a long time. Hence the core ideology (value and purpose) is timeless.

Envisioned future. The other half of the vision framework is the envisioned future. Like core ideology, it can be segmented into two parts—a 10- to 30-year aggressive goal and a specific description of what it will be like when this goal is achieved. The goal should be audacious or BHAG (big, hairy, audacious goal), as Collins and Porras call it. It should be very aggressive and be a motivator of the entire organization. For example, NASA's goal was to reach the moon. Or Winston Churchill's goal (yes, politicians and nations need a vision, too; see box entitled A Nation's Vision) during World War II was to " brace ourselves to our duties and so bear ourselves that if the British Empire and its Commonwealth last a thousand years, men will say, 'This was our finest hour.'"

Does this imply that often such aggressive goals may not be achievable? Yes. Nevertheless, it is such goals that can inspire, galvanize, and drive an organization. Consider the case of Toyota Motor Company.

Toyota's purpose statement is, "Be number 1 in quality and offer good products to respond to requests of society and trust of customers." And its goal, set in the early 1980s, was "Global 10"—a goal to gain 10 percent market share worldwide by 1999. Toyota is close to this goal (at 9.5 percent in 1997), has been too successful, and has had to relent. Read the following excerpt.[4]

Toyota Seems to Relent Amid Industry Woes

...The company is under heavy fire for its reluctance to back off and let competitors at home and abroad catch their breath. Toyota is under increasing pressure from many quarters in Japan to take more initiative and leadership...."Leadership" is a fuzzy term, but labor unions, government officials and rival manufacturers say that what they want Toyota to do is *lengthen its product cycles, reduce work hours, raise prices, and stop its drive for market share overseas.* They say that until Toyota does these things, Japan's other manufacturers can't. In effect, Toyota is being asked to overhaul the strategies that made it the powerhouse it is today. The surprise: Toyota seems to be coming around. Repudiated, at least for now, is Toyota's notorious "Global 10" strategy...

Incredible! This simply shows the power of a bold vision and the resolve to achieve it. Refer to Table 3.1 for a discussion of Toyota's and Hewlett-Packard's vision. *Note:* This was the situation several years ago, but today the pressure on Toyota has eased, and it is once again aggressively going for market share and speedier design cycles (less than 15 months) and is even diversifying into other businesses.

Consolidating, stating, and conveying the vision. To summarize, a vision statement should have the components listed above. Most successful organizations have these components in their culture, but do not attempt to consolidate them to understand the secret of their success or as a tool for organizational and employee development. Even Hewlett-Packard and Toyota Motors do not explicitly put all the components together, but their top leaders understand that these are the reasons for their success. We had to seek the various components from both companies through research and interviews.

Successful companies have visions that embrace both customers and employees—to lose sight of the vision, to move in the wrong direction, can often lead to business blunders. Our suggestion, then, is for you to understand, consolidate, and state the vision of your company and perpetuate it.

A Nation's Vision

Can a nation have a vision? Can such a vision drive it to success? Absolutely. Here are two examples from newly industrialized countries (NICs) with strong leadership. Note, however, that a vision must be supported by a detailed plan of action—the methodology is discussed in our long-range plan. Nations need similar, detailed plans, and the following nations have them.

Singapore

Goal: To achieve the Swiss 1984 per capita income by 1999

This was set in the early 1980s. This goal is supported with a targeted economic growth, which in turn is supported by detailed plans, including

- An industrialization plan—mainly light industry moving quickly into high technology
- Transport infrastructure—local, regional, and international
- Human resource development, including education in schools, business, and technology
- Health services at a reasonable cost
- Savings, pension, and investment plans
- Tourist development
- Information technology to convert Singapore to an "intelligent" island by year 2000 (a new item added in 1995)

Singapore has seen GNP growth of over 7 percent for the last 25 years, has become an industrialized country (removed from developing-country status in 1995), and continues to be one of the most successful, envied, and emulated NICs. Because of the Asian economic crisis in 1998, Singapore has announced that it will miss its goal by several years, but it will press on.

Malaysia

Goal: Vision 2020

This catchy phrase, first used in 1991, is to have a growth of 7 percent every year, thus achieving an 800 percent growth (over 30 years), which will allow Malaysia to become a developed nation by the year 2020. This is supported by plans which include

- An industrialization plan—in both light and heavy industry
- Transport infrastructure within and without Malaysia, with linkages to south and east Asia
- Role of private and public sectors to create "Malaysia Incorporated"
- Savings and investments plans
- Human resource development, including education in schools, business, and technology
- An environmental plan
- Information technology growth via the multimedia supercorridor

(Continued)

> This project calls for massive investments in new technologies and is attracting worldwide attention and involvement. It is Malaysia's way of leapfrogging into the 21st century. This is a new item added in 1996, reflecting the need to update strategies but not the goal.
>
> Malaysia has been seeing an average growth of 8 percent per annum from 1970 to 1979, 5.7 percent from 1980 to 1989, and now has leaped forward with a growth of 8.8 percent during 1990 to 1996. Malaysia went through some major hiccups in 1998 because of the Asian economies crisis. It is struggling to recover and will miss its goal by at least 5 years.
>
> *Was this growth inevitable?* Some critics may say that vision or no vision, these countries would have grown. Perhaps. In the lesser developed nations, there is a tremendous opportunity for high growth, because they are starting from a low base, there is a large pool of human resources, and costs are low. Nevertheless, they need to educate their population, build the infrastructure, increase productivity, and minimize corruption—and in this, only some nations have been successful, but many more are struggling or show inconsistent growth. The difference is the nation's bold vision and the political will to achieve it.

Customer's needs, issues, and channels of distribution.   Here we conduct a rigorous analysis, based on data, of customer needs and trends. For example, what problems are customers trying to solve? What are their dreams? This ensures the building of a data-based model of the market and allows discovery of new market opportunities. When marketplace needs are understood, they can be grouped by market or customer segments with areas of opportunities identified. In addition, the market segments must be mapped against current and future channels of distribution. The following is a list of key activities for this section:

- Collect information on customers and distribution channels.
- Analyze the market, and determine areas of focus or for further investigation.
- Conduct research into customer needs.
- Confirm new market opportunities—product and service positioning.
- Analyze market attractiveness.

Here is an explanation of the above items:

Collect information on customers, distribution channels, and competitors.   Get data from customers; understand the market situation and the competition. This is an ongoing process throughout the year, and the most recent data go into the plan. For an existing business, get data for current product and service offerings. Get data about current customers, the existing channels of distribution, and the competition. Note that although we have indicated above that the competitive situation should be analyzed in the next step, you may find

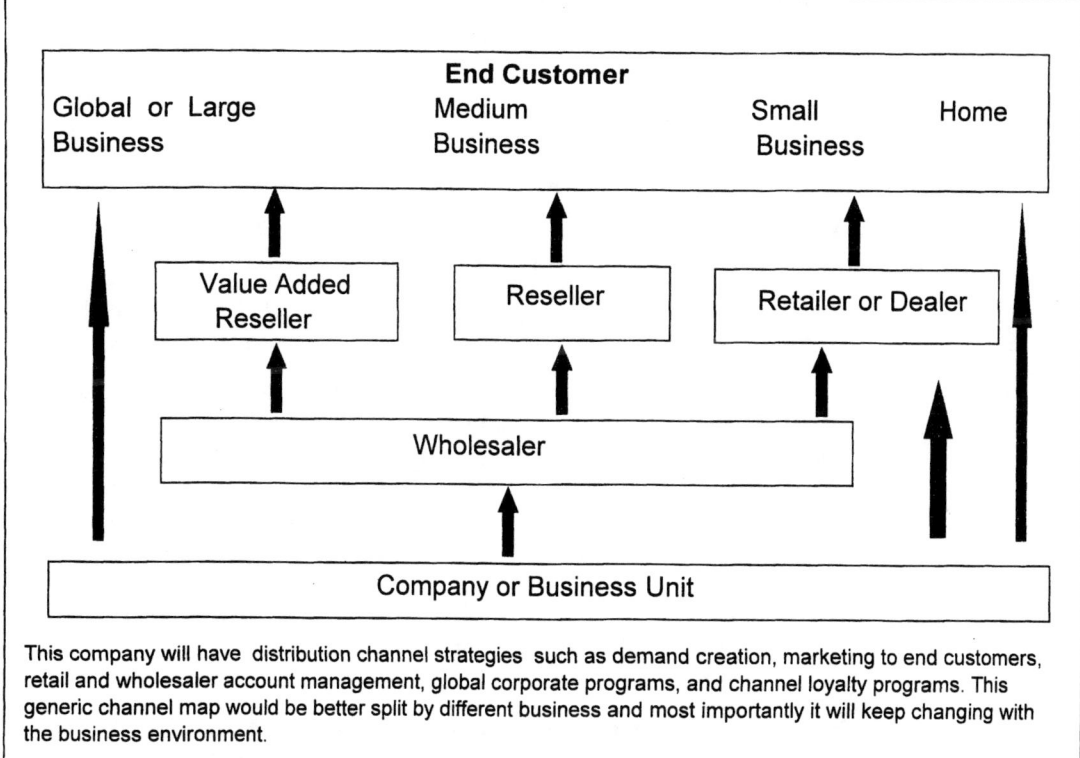

Figure 3.2 Channel distribution model.

that these data are available as you go through this step. Attempt to understand the current channels of distribution, their strengths and weaknesses.

Figure 3.2 shows a distribution channel map for a company that manufactures and distributes commercial, industrial, and consumer products. In this case the company does both direct selling and indirect selling (via wholesalers and retailers). Typically, the direct and indirect businesses would be managed separately, with more detailed distribution channel maps. As the business grows or changes, these distribution channels must be improved and alternatives and partnerships sought.

Analyze the market and determine areas of focus or for further investigation. The purpose here is to look for new market or service categories that offer prospects for opportunity, growth, and competitive advantage. Alternatively, the desire may be only to improve current products and services. It is important, however, to always look for breakthroughs and new opportunities.

Conduct research into customer needs. Once a segment has been chosen for further research, the purpose is to determine needs. First determine customer needs. What is the customer trying to solve or do? What is the customer's dream? How can we make life easier for the customer? Customers' needs can be stratified into three areas:

- Basic needs—what we called *must-be quality* in Chap. 1
- Spoken needs—what we called *attractive quality* in Chap. 1, also issues from satisfaction surveys and dreams
- Unspoken needs—items in the outer zone of the total product or service concept, discussed in Chap. 2

Next, determine what factors will influence the customer's desire to buy, who decides to buy and when, how much will the customer pay, how many a customer will buy, and the potential market size. With regard to distribution, will it be direct, via an existing channel or a new channel? The same questioning applies for new services that will be provided. In determining customer needs and responding with appropriate strategies, a useful tool is quality function deployment (QFD), discussed in App. C. This tool can be used for both products and services. A QFD application, to determine or improve strategies to respond to customer needs for a service is shown in Fig. 3.3.

Confirm new market opportunities—product and service positioning. The previous sections should give plenty of information about new market opportunities. Specifically, look for and identify new product and service positioning that gives advantages over competitors. According to Michael Porter,[5] there are three types of positioning that one can strive for:

- *Needs-based.* Needs-based positioning looks at needs of a certain customer segment and fulfills those needs, such as a special color printer for poster designers or graphic artists. This is the classic customer segmentation model.
- *Variety-based.* Variety-based positioning provides choices in products and services, such as a store that focuses on only car wheels (sells tires and fixes brakes) or oil and fluid changes.
- *Access-based.* Access-based positioning focuses on customers confined to certain geography or who require specific activities to reach, such as minimarkets along highways or a special no-frills airline service between two cities.

When the required positioning is determined, then confirm what offering to provide.

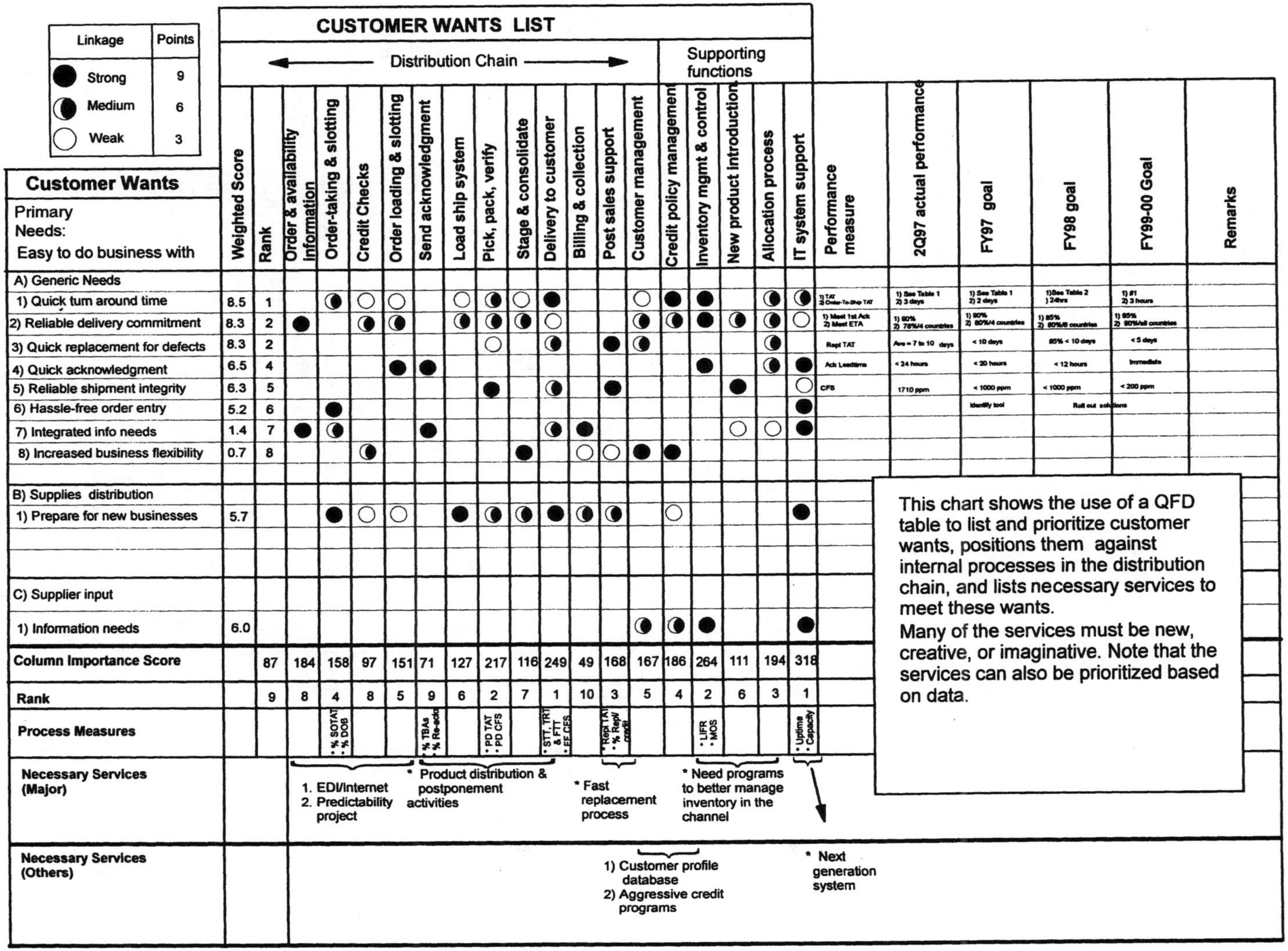

CUSTOMER WANTS LIST

This chart shows the use of a QFD table to list and prioritize customer wants, positions them against internal processes in the distribution chain, and lists necessary services to meet these wants. Many of the services must be new, creative, or imaginative. Note that the services can also be prioritized based on data.

Linkage	Points
● Strong	9
◐ Medium	6
○ Weak	3

Customer Wants

Primary Needs:

A) Generic Needs — Easy to do business with
- 1) Quick turn around time — Weighted Score 8.5, Rank 1
- 2) Reliable delivery commitment — 8.3, Rank 2
- 3) Quick replacement for defects — 8.3, Rank 2
- 4) Quick acknowledgment — 6.5, Rank 4
- 5) Reliable shipment integrity — 6.3, Rank 5
- 6) Hassle-free order entry — 5.2, Rank 6
- 7) Integrated info needs — 1.4, Rank 7
- 8) Increased business flexibility — 0.7, Rank 8

B) Supplies distribution
- 1) Prepare for new businesses — 5.7

C) Supplier input
- 1) Information needs — 6.0

Distribution Chain — Supporting functions

Column Importance Score / Rank:
- Order & availability information — 184 (8)
- Order-taking & slotting — 158 (4)
- Credit Checks — 97 (8)
- Order loading & slotting — 151 (5)
- Send acknowledgment — 71 (9)
- Load ship system — 127 (6)
- Pick, pack, verify — 217 (2)
- Stage & consolidate — 116 (7)
- Delivery to customer — 249 (1)
- Billing & collection — 49 (10)
- Post sales support — 168 (3)
- Customer management — 167 (5)
- Credit policy management — 186 (4)
- Inventory mgmt & control — 264 (2)
- New product introduction — 111 (6)
- Allocation process — 194 (3)
- IT system support — 318 (1)

Performance measure / 2Q97 actual performance / FY97 goal / FY98 goal / FY99-00 Goal / Remarks

Necessary Services (Major)
1. EDI/Internet
2. Predictability project
- Product distribution & postponement activities
- Fast replacement process
- Need programs to better manage inventory in the channel
- 1) Customer profile database
- 2) Aggressive credit programs
- Next generation system

Figure 3.3 A QFD chart used for developing needs and services for product distribution

61

Analyze market attractiveness. Based on the information collected, determine and prioritize the attractiveness of new market segments; then figure out the competitor's position, if any. If we position a new product or service creatively, there may be no competition. On the other hand, we may want to enter new segments that are attractive but dominated by competitors, or to continue in current segments where we are strong. We need to determine costs of manufacturing, implementation, or purchase; cost of distribution; projected sales and profits; and competitive threats. In addition, we need to know the competition's strengths, weaknesses, and strategies.

The output of this section is a list of prioritized customer needs for current market segments, new market segments, and attractiveness of the new segments. This will drive the section on necessary products and services.

Competitive situation. Much of this section will have been covered in the previous step, since competitive information will unfold as we look at customer needs, issues, and channels of distribution. Nevertheless, it is important to have some specific competitive information, such as

- Competitive position and advantage
- Competition's strengths and weaknesses
- Competition's directions

In analyzing competitive position and advantage, we need to understand market growth versus our growth rate; market share—by product and service—and the trends; and overall product contributions, for example, via a cash cow, dogs, stars, etc., bubble chart. In analyzing competitive strengths and weaknesses a good method is to use a radar chart to plot product and service attributes against those of the competition. Equally important, we need to understand the competition's current strategies, how it differentiates its products and services, how its strategies have evolved over the years, where it is expected to go in new products and services, and its financial strength.

Necessary products and services. Our purpose here is to gather what we have learned and list products and services that address customer's needs or new market opportunities. This includes improving current offerings and, more importantly, planning for new offerings that are creative, imaginative breakthroughs. Key technologies that must be identified are also planned for here. To ensure linkage, a good process is to list the needs and opportunities generated in steps 2 and 3, and opposite each, list the necessary strategic response—products, services, technologies, acquisitions, etc. Figure 3.4 gives an example of what to cover in necessary products and services.

4.0 Necessary Products and Services Page 15

Listed here is a summary of the customer & market needs and the opportunities they present.

Customer & market inputs

1. The current offerings are aging and perceived to be slow. A new printer product range is needed quickly. However, many customers cannot wait for introduction of the new range, which rolls in 15 months. There is danger of losing customers and market share.

   * Accelerate the R&D program. This is crucial. The R&D plan will reflect the next 3 year's offerings.
   *Review current printer offerings and decide which can be improved or modified until the new range is introduced. This is a very high priority item.

2. We need a lowend printer as our printer prices have been increasing with new feature sets.

   * Modify printer model zz and make it the basic workhorse, low cost, minimum features printer. Market potential and sales are excellent.

3. Market studies show there is a new segment of customer, i.e. poster designers and graphic artists.

   * Design a wideformat printer family for this segment, width will range from 120 to 180 cm, with speed of 1 page per minute for text & color.

4. Customers have been complaining about our ink quality—poster designers have complained of fading in the sun. Restaurant owners and secretaries have complained about the ink running when touched by moisture, coffee, etc.

   * Work with laboratory to provide special inks that are waterproof and fadeproof.

5.

6.

7.

8.

4.0 Necessary Products and Services Page 16

Listed here is a summary of the customer & market needs and the opportunities they present.

9. There is a large untapped market of customers who prefer to order direct from us—shopping on the Internet is crucial to increase market share, sales, and recognition.

   * Set up a system solution to take orders via the Internet.

Distribution channel needs

10. Computer superstores & large retailers want an EDI link for ordering products. Many would prefer communications via the Internet for order, delivery status, and billings.

    * Improve current systems to take orders, provide order & shipment information, and send billings via the Internet.

11. The new wideformat printers will require new channel partners.

    * Look for new partners for wideformat printer.

12. The channel needs consolidation into fewer wholesalers. New programs are needed for improving the skill-set of current partners. See figure 3-2.

    * Weed out wholesalers with annual sales below $10 million. Provide them access to larger wholesalers with special discounts for them. Review and improve channel training programs.

13. Due to heavy sales growth our distributors are having difficulty financing their businesses. Solutions will have to focus on root cause, which is high inventory due to overstocking and more new product stocking.

    * Source and provide better financing and look at methods of reducing inventory at distributors & wholesalers.

Internal Needs & Issues

14. Our internal order management systems will reach the limits of their capacity in 2 years; they also lack capabilities that are crucial for new business, e.g. promotions (special pricing, etc.), channel inventory management, and real time capability. Further enhancements will not be cost effective.

    * Seek out new order management system that meets all our upcoming needs.

Sales & Marketing Needs

Figure 3.4 Necessary products and services.

Description of product and service strategies, development of partners, and purchase plan. From the list of necessary products and services we must define a plan. These will include the product and service strategies that must be generated. In addition, we must determine plans for developing partnerships, such as joint ventures, technology sharing or acquisitions, and assisting or growing distribution channel partners. More specifically, we must determine the following:

- *Products and services.* What products or services need to be designed internally or acquired? What technologies need to be developed or bought? What partnerships do we need to develop?

- *Marketing and sales.* What is the sales forecast? What are the sales and marketing strategies? How will after-sales support be performed?

- *Distribution channels.* How will the product be distributed? Do the distribution channels need to be changed or improved; or will there be alternate distribution channels? Plans for distribution channels must include strategies for demand creation, marketing to end users, retail and wholesale account management, global and corporate programs, value-added dealer programs, channel profitability programs, channel satisfaction programs, and loyalty programs. Refer to Fig. 3.2 for more details; however, for most businesses a more specific distribution channel map, with channel growth and management plans, will be required. The distribution map shown is for a manufacturer of commercial, industrial, and consumer products. For a service operation, an entirely different distribution channel map is required.

- *Manufacturing.* Where will the product be manufactured? What are the short-term and long-term production capacity needs? Will the manufacturing ramp meet the marketing and sales requirements? What are the costs? What are the quality and reliability goals?

- *Services.* What services need to be deployed, and who will implement and manage the services?

- *Human resources.* What are the staffing needs? What are the training and education needs?

In Fig. 3.5, we show an extract of this step. In the figure we give a situational analysis and the proposed offerings with time lines. A recommended format is to list proposed strategies for the short term and the long term. The short-term list includes strategies that are most crucial and can or must be implemented immediately. The long-term list includes strategies that are not achievable quickly or will take more time, because of resource limitations, lower priority, etc. This

5.0 Description of Products and Services Page 18

5.1 Improve current offerings and introduce products for current and new customer segments
We continue to face increasing competition and price erosion. This will dramatically impact our profitability. Our response will be in several areas and will include plans to refresh our current printer offerings, development of new printers for the lowend market, development of a new range of printers based on the channel segmentation model (see chart xy), and improvement in our ink technology to provide waterproof and fadeproof printing.

199X Plans:
* Introduce new products at an accelerated schedule—first products in the XO range must roll out in 9 months.
* Refresh current printer offering: Improve current printers (xx and yy) in speed of color printing through software & hardware enhancements and changes in the physical chassis to provide a new look.
* Modify printer zz and offer it as our lowend printer to keep out competition that is expected to attack our low end. This strategy will allow us to keep our market share in the low & medium end market.
* Start a fast track program—12 months—to design a wideformat printing (WFP) family of 150 cm. width, refer to customer segmentation chart xy. Market studies indicate that this category of printers have a market size of $3 billion by year 2000. The customers targeted for this category are graphic artists and poster designers. Currently, no competitive and lowprice WFP printer is available for these customers.
* We must design or acquire improved ink technology to ensure waterproof and fadeproof printing. This will be necessary to support the WFP printers. The lab will start developmental work and in addition will look into a cross-licensing arrangement with the Rainier Chemical Company to speed up development within the year.

199Y - 199Z Plans:
* **Complete rollout of XO range products, and start preparing the EXO printer range design concept.**
* Roll out the series of WFP printers per the proposed schedule. Market share expectations are 75% the first year and 90% the second year once our technology gets accepted.
* Introduce new ink technology together with the introduction of the WFP printers.

5.0 Description of Products and Services Page 19

5.2 Dramatically Improve Distribution Channels and Supply Chain
Our current retail penetration is weak and we must move into new channels to have better access to untapped home, family, and small business customers. The new WFP printer will need different distribution channels. In addition specific improvements are needed to provide ordering, information, and billing via the Internet or EDI for our distribution channel partners; our channel partners need help from us to better manage their inventories to improve their cash flow.

Plans for 199X:
* Start work on Internet and EDI linkage into our current systems. This should cover orders and billings to current retailers and wholesalers this year, and prepare for customers buying from home next year.
* Evaluate current order management system capability and upgrade where possible to manage projected business growth. Explore next generation system technology and prepare for system roll-out if system is approved. System must cater for growth and needs for the year 2000 and beyond.
* Provide ability to track and manage distributor & wholesaler inventory in order to improve their cash flow. This must be incorporated into current system and be a requirement for the future system.
* Ensure training of current channel partners in areas of increasing product demand and promotion management, after sales support, and better inventory control (to improve their cash flow and ensure lower inventory write downs).
* Identify, train, and plan to manage new distribution channels for WFP printers.

Plans for 199Z:
* Complete work for Internet order and billings for end (final) customers. Timing to be coordinated with marketing & sales programs. Refer to strategy 5.3.
* Implement roll-out of identified order management system to prepare us for the year 2000 and beyond.

Figure 3.5 Necessary products and services.

process of segregating short- and long-term strategies will force a necessary prioritization of strategies and allow for easier preparation of the annual plan in the last step of the long-range plan.

Financial analysis.   Here we perform an analysis of product and service volumes, revenues, cost of goods, expenses, investments (including R&D), overhead costs, profits, and so on. We must also determine capital expenditures and return on investments for all products and services. Projections for the planning period should be available.

Potential problem analysis.   It is important that a risk analysis be conducted and appropriate contingency plans proposed. Areas of risk and the possible competitive response should be reviewed. Review items such as external or competitive threats, business environment changes, technological changes, and problems with suppliers. The focus here should be on potential problems. Once these problems are identified, determine contingency plans that could be implemented.

The 3- to 5-year objectives.   We are now ready to *synthesize the information and summarize* the plans into the 3- to 5-year broad objectives. This is a management summary, and it could be placed at the beginning of the plan. The objectives should be stated by key functions (R&D, manufacturing, sales and marketing, and so on). In addition, it is important to state the financial goals (revenue stream for the various business products and segments, costs, and profits), market goals (market share and position), and customer satisfaction goals. Here is an example of items to put in a summary; the list will vary with different businesses.

- Revenue goals: revenue, return on investment, cost, and profit projections
- Market goals: market share and unit sales projections by product or service
- Customer satisfaction goals: customer goals, quality and product reliability goals
- Breakthroughs: entry into a new market segmentation or new businesses, acquisitions of new technologies, or major product or service introductions and improvements

The annual plans.   Finally, we prepare the annual plan. The Hoshin planning format is recommended. The Hoshin plan will include items from the long-range plan as well as other areas, which are discussed in the next section on annual planning.

Completing the long-range plan

The output of this planning process is a long-range plan (3 to 5 years), given in steps 5 to 8, and the annual Hoshin plan. This entire process must be repeated every year—because markets and competitors change, or the economic environment may vary. Typically, there will be fine-tuning of the plans throughout the year with a major revision each year. In some years, if there is little change in the market or competition, the 3- to 5-year plan may require only minor adjustments. The more focused or well positioned the products and services of a company, the less change occurs from year to year in the long-range plan—but only a few companies or organizations will be so lucky! The annual Hoshin plan, however, is done anew each year.

Successful long-range planning with revolutionary strategies

Before we discuss annual planning, let us take a break to discuss some of the issues of long-range or strategic planning. Mintzberg[6] and numerous others have debated endlessly the pros and cons of planning (see the box for a more detailed discussion of the pitfalls of long-range planning). Some say, "Why bother to have a long-range plan? Management will get distracted with other issues. Corporate planners cannot plan for a business that is dynamic; besides, they are too far removed from the business or marketplace; or most companies are short-term-oriented because Wall Street only rewards short-term performance." The list is endless, and most of the criticism is valid.

If your company has an unsatisfactory long-range, strategic planning process, it could be due to a central or corporate planning model. This can be solved only by moving strategic planning to business unit managers, who should focus on adding value, should plan within a reasonable time period, and should be measured by their performance. Equally important, there must be flexibility in the long-range planning process. What we have given is some rigor and a guide or format to direct ideas and creativity.

You need to find the ingredients to make the process work—ingredients such as a bold vision and revolutionary strategies. Remember, you must create strategies, not another list of routine plans. So the business team must look for revolutionary ideas to help leapfrog the competition.

Pitfalls of Long-Range Planning

There has been much criticism and analysis of long-range or strategic planning; specifically, it is a waste of time, and long-range plans are seldom implemented. Mintzberg has written an entire treatise discussing "The Rise and Fall of Strategic Planning." He quotes 10 pitfalls to be avoided during long-range strategic planning:

1. Top management assumes that it can delegate the planning function to a planner.

2. Top management becomes so engrossed in current problems that it spends insufficient time on long-range planning, and the process becomes discredited among other managers and staff.

3. There is a failure to develop company goals suitable as a basis for formulating long-range plans.

4. There is a failure to assume the necessary involvement in the planning process of major line personnel.

5. There is a failure to use plans as standards for measuring managerial performance.

6. There is a failure to create a climate in the company which is congenial and not resistant to planning.

7. Corporate comprehensive planning is assumed to be somewhat separate from the entire management process.

8. So much formality is injected into the system that it lacks flexibility, looseness, and simplicity, and restrains creativity.

9. Top management fails to review with departmental and divisional heads the long-range plans which have been developed.

10. Top management consistently rejects the formal planning mechanism by making intuitive decisions which conflict with the formal plans.

In our planning methodology there are several tactics presented that help overcome these obstacles. Some of them are the preparing and sharing of the plan with the entire management team, an annual plan that must be implemented during the first year of the long-range plan, detailed implementation plans that ensure strategies are converted to action, regular reviews of plans with the management team with suitable accountability, corrective action when strategies are not achieved, and a learning process during each planning cycle. In addition, it is important that plans be prepared by the managers in an organization's business unit—with, perhaps, a facilitator—and not by a corporate planning unit. It is also important that the company's vision, long-term goals, and annual plans be tightly linked. Finally, there must be flexibility and a willingness to change plans when necessary—but this must be based on data, facts, or changes in the marketplace.

Planning for success—a perspective from Singapore Airlines

How do great companies plan? Let us look at one company, Singapore Airlines (SIA). SIA, based in tiny Singapore, with a population of 3 million, has a reputation for being the most consistently profitable airline in the world, despite various world recessions. It has also been voted as Asia's most admired company, best airline, best for business travel, and so on. So, what is its secret?

First, SIA believes it is in the service business, not the transport business. With this as its primary purpose, SIA does several things to prepare for and provide excellent service, including planning, listening to the customer, understanding the competition, designing revolutionary products and services, process management, and people development.

Planning. SIA has several planning methodologies. There is a 10-year fleet planning cycle in which it sets goals for growth and route expansion; the current plan looks out to the year 2004 and targets an 8 to 10 percent annual capacity growth. Then there is a 5-year detailed plan, which specifies routes, cities, and aircraft to be purchased; for example, in 1996 it committed to acquire 77 Boeing 777 aircraft, the largest aircraft purchase ever. Next, it has its annual operating plan, covering plans and the budget for one year. This is very detailed and includes new products and services in its aircraft, for its numerous subsidiaries, and what is required on the ground. Finally, SIA has long-range scenario planning sessions, using the Shell Company's methodology.

Listening to the customer. SIA is very good at this. It has a customer feedback system, and *all* complaints are reviewed by the managing director and acted upon. To its credit, SIA also tracks all feedback via a compliments-complaints ratio. Typically, the number of compliments exceeds the number of complaints by a large margin—this is something SIA management is very proud of. Another important aspect of the customer feedback system is that it focuses on service recovery and not damage control. Whenever possible, customers are usually compensated for genuine complaints caused by SIA. This has the effect of winning loyalty from customers because "SIA cares." It was this feedback system that gave SIA an insight into its in-flight entertainment system and motivated SIA to provide a personal audio and video entertainment system for all passengers. Even its in-flight magazine, *Silver Kris,* was restructured after passenger feedback. In addition, there are quarterly passenger surveys, focus group discussions with frequent fliers, and checks on competitive offerings. All this activity provides a wealth of information and influences future service strategies.

(Continued)

Designing revolutionary products and services. SIA runs a product and service development department that designs and develops new products and services. Inputs for new products come from the customer feedback system, competitive benchmarking, SWOT (Strengths, Weaknesses, Opportunities, Threats) analysis, and the desire to stay ahead of the competition. To this end, SIA has pioneered not only in-flight services but also ground services, something that most airlines have ignored. Overall, SIA believes that products and services must be refreshed every few years. In addition, SIA wants to convey an image of safety and high technology, which it does by having the youngest and best-equipped fleet in the airline industry (average aircraft age is 5 years).

Process management. All services, or processes, must be well managed via process controls. SIA does several things. The key is the service performance index (SPI), which measures key parameters of SIA's service (for example, check-in, phone efficiency, ticket office operations, and on-time arrivals). These are also compared to other airlines' performance. Refer to the customer satisfaction model in Chap. 2 for a discussion. Other quality controls are placed on food, cutlery, linen, and in-flight service. Over the long term, SIA sets goals and measures its progress using the SPI. Typically, SIA believes in providing a high standard product on all classes of travel. There may, however, be a premium provided on certain flight segments based on cultural or competitive needs, such as the Japanese market.

People development and training. Being a service organization, SIA recognizes the need to be people-centered, to empower its staff, and to provide training. In fact, training is crucial for success, and to this end SIA has invested in a US $55 million training center. The center provides functional training (such as ticketing and in-flight training), skills improvement (such as supervisory or technical), and management development.

Can other airlines catch up? British Airways and several other airlines are attempting to catch up with SIA, but in the short term most U.S. and European airlines will have difficulty. While U.S. and European airlines try to compete on the lowest price, SIA believes that customers will pay a premium for good products and services. SIA's success confirms that point.

Planning for success—another perspective from Hewlett-Packard Company

Hewlett-Packard (HP) started as an instrument company, has evolved into the world leader in printers, and is now the world's number 2 computer company. Currently HP is organized into several business units (such as commercial printers, consumer products, personal computers, and medical products).

Overall, the chief executive has an annual Hoshin plan and a business fundamentals plan that addresses common business needs of the company. The business units, however, operate independently, preparing their own product plans and strategies, while specific platform strategies such as sales and distribution are shared across several business units. Within each business unit there are several product (or service) divisions. The product divisions focus on long-range and annual planning, listening to the customer, developing successful products, and people development.

Planning. HP puts a premium on planning. Each product division will typically have a long-range and an annual plan (HP uses annual Hoshin planning). The long-range plan used to cover 5 years, but now with the fast-changing business environment it has evolved into a 3-year plan, called a *business strategy review* (BSR). This long-range plan looks at the vision, customer needs, market needs and opportunities, the competition, channels of distribution, and the necessary products and services required to move to a better position in the market. The last step of the long-range plan is the annual Hoshin plan. This looks at what is needed for the upcoming year, including improvements and new products and services.

As mentioned earlier, product divisions do their own long-range product strategy planning. This can be both a strength and a weakness. While this autonomy has generated tremendous creativity in new products and services, it can also result in overlapping products from different divisions, which have resulted in products from one division cannibalizing products from another division. As product divisions grow, HP management tends to split them, and this often perpetuates the problem. Overall, however, this strategy and independent business planning work well—in fact, the entire printer business evolved when one visionary manager, Dick Hackborn, decided to experiment with laser printing and created a multibillion-dollar business!

Listening to the customer. This is done continuously via customer feedback systems (including a customer feedback center at the chief executive's office), customer response centers (which provide after-sales support), focus groups, customer response cards included in sold products, and competitive information. These activities help HP stay in touch with customer concerns and needs. Independent customer satisfaction surveys show very high satisfaction with its products and services. In a recent survey by *Fortune* magazine, HP was rated the most admired computer company.

Developing successful products. HP is a large company, but it manages to stay quick and nimble by spawning new product divisions from large, successful product divisions. The young divisions focus on new product categories that generate growth and move HP into new markets. Today, there are several initiatives aimed at creating new markets in the consumer business such as home PCs,

(Continued)

home printers, digital photography, and multifunction products. This focus on innovation results in HP's generating a continuous wave of new products, and in any given year, 80 percent of its revenues comes from products introduced in the last 3 years. In an effort to keep costs down, HP is tending to develop and source new products from original equipment manufacturers.

Process management.   HP uses a business fundamentals, or daily management, plan to keep track of its key processes. This is the business scorecard to monitor and measure key processes, and it is an excellent tool, providing daily feedback on key business metrics. The key processes vary by entity. In a sales entity (for example, a country sales organization), the processes tracked include orders, shipments, costs, customer satisfaction, customer feedback, and inventory. At a factory, processes tracked include orders, shipments, costs, customer satisfaction, manufacturing yields and quality, and customer feedback. The actual processes and performance measures used vary by operation; refer to the detailed discussion of process management in Chap. 5.

People development and training.   HP puts a premium on people development and training. The overall management philosophy is guided by the HP way, discussed in this chapter.

Can HP continue to grow profitably?   HP's management philosophy, business planning process, and ability to reinvent itself every few years have provided profitable growth since its inception. The future is going to be tougher because of its size and increased competition. To succeed it needs to focus better its diverse operations, put its creative ideas into products quickly, be less complacent, and be very paranoid.

Annual Plan

Every organization must have an annual plan. A good annual plan will move the organization to a better position in the marketplace each year. And yet, we have found a dearth of literature on a detailed annual planning process—details such as process flows, formats, planning guidelines, and training material. Even business schools do not put much emphasis on an annual planning process. Therefore, we discuss this topic in detail.

Note: This section on annual planning is a stand-alone section and can be used without reference to the previous section on long-range planning.

We recommend the use of the Hoshin Kanri and Nichijo Kanri methodology. This is a very successful planning process used by almost every Japanese company that has won the Deming prize. It originated in the 1960s at Bridgestone Tire Company in Japan. At that time, weaknesses in their planning process were becoming apparent.

The Hoshin Kanri philosophy originates with ancient military traditions and efficiency. Many large Japanese companies and several U.S. companies that adopted *management by objectives* (MBO) have given up that process and switched to Hoshin Kanri type of planning. Details of Hoshin Kanri and comparisons to MBO are given in the following discussions.

The terms *Hoshin Kanri* and *Nichijo Kanri* are loosely translated from the Japanese as follows:

Hoshin Kanri. *Hoshin* means objectives or directions, while *Kanri* means control or management. So in essence Hoshin Kanri means policy management or management of objectives (contrast this to MBO or management *by* objectives in the western corporate environment). Henceforth we just call it the Hoshin plan.

Nichijo Kanri. *Nichijo* means daily. Hence Nichijo Kanri means *daily management.* In the western corporate environment, a clearer definition could be *business fundamentals.* We use the terms interchangeably, but prefer the term *daily management*; the reason will become apparent.

A brief description of the Hoshin planning process is given below. In the succeeding pages the various formats to be used are described, and suggestions to develop a good plan are given with examples.

The Hoshin planning process

The process begins with a review of the following items (refer to the planning flow in Fig. 3.6):

Company or organization vision. The company's or organization's vision is reviewed.

Long-range plan. The long-range plans are reviewed. If the long-range planning process discussed earlier in this chapter is used, then it will already have a component that can immediately plug into the annual Hoshin plan.

Specific customer inputs. Specific customer inputs, issues, and needs that have been collected during the year are reviewed. Customer inputs are important—customer data gathered during preparation of the long-range plan should be reviewed, but other more recent inputs or ongoing pressing issues must also be reviewed. These can come from regular customer surveys, meet-

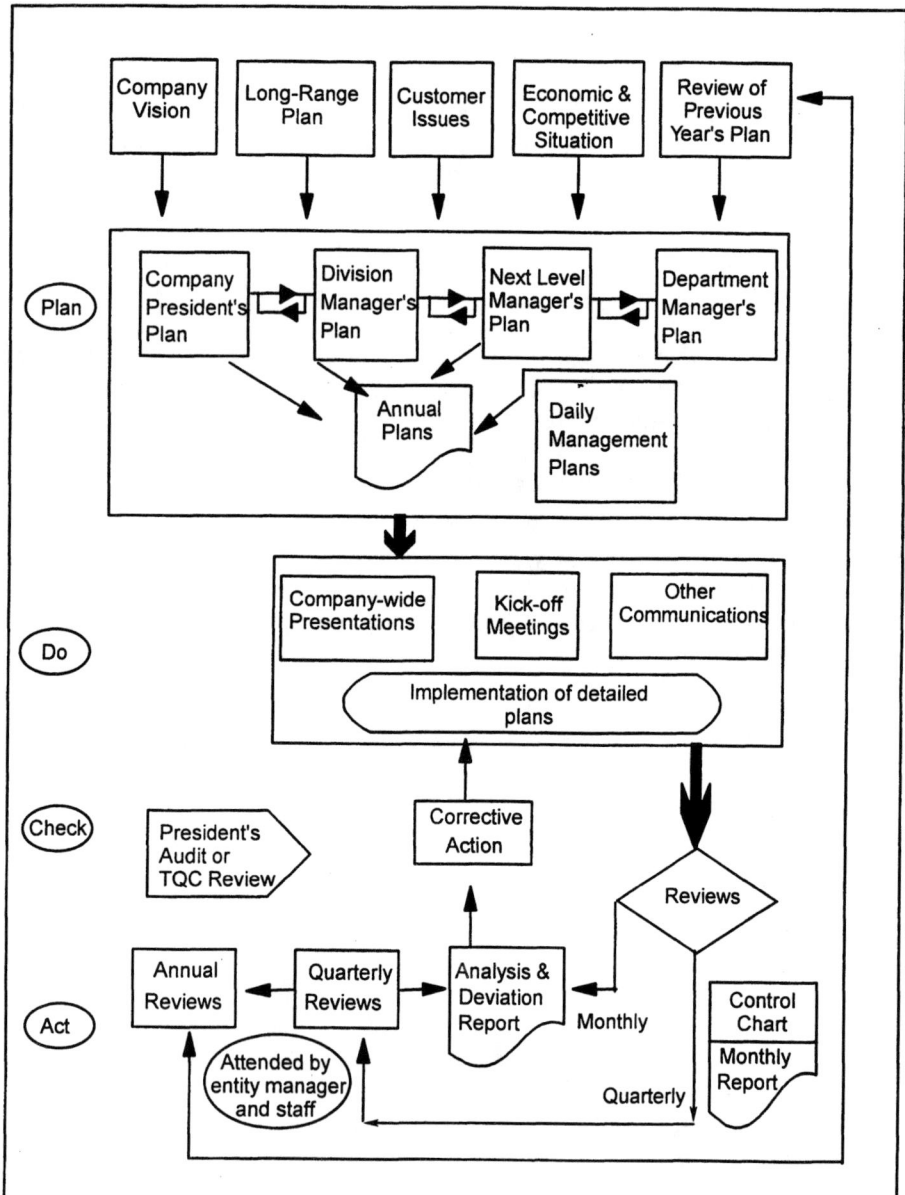

Figure 3.6 Hoshin planning flowchart.   This chart shows the sequence of activity for generating a Hoshin plan and a Daily Management or Business Fundamentals plan. Note the double arrow symbols ($\rightleftarrows$), which denote management discussion nodes. These are crucial in ensuring good deployment of plans.

ings, visits, letters, and the customer complaints and feedback system, discussed in Chap. 2. The need for closeness to customers can never be overemphasized.

Current economic situation.   The current economic situation and its impact are reviewed.

Review of the previous year's plan.   Successes and failures in the previous year's plan are reviewed at all levels in the organization. The lessons learned will influence the upcoming year's plan. This is an important step, and it is discussed later in the section entitled "Review for Hoshin and Daily Management, or Business Fundamentals, Plans."

All the above-mentioned items form a list of key issues from which the Hoshin plan will be prepared. Typically, the entity general manager (which can be the CEO and president or a product or service division manager) will prepare the annual Hoshin plan. This plan will provide broad objectives, strategies, and performance measures. The next-level managers—for example, marketing, finance, research and development, operations or manufacturing, and quality—will then prepare a more specific plan, after discussions with their department managers. Refer again to the Hoshin planning flowchart in Fig. 3.6. The department managers will then prepare more detailed and specific plans. In addition, the department managers will prepare an implementation plan.

Daily management, or business fundamentals, plan

So far the Hoshin plan covers the objectives that flow down from upper-level management. These indicate what the entity must achieve in the year. But what about day-to-day items—items that must be maintained and monitored? Examples include cost controls, employee morale, selling or manufacturing a product, maintaining or slightly improving a product or process, or some other annual repetitive task.

The daily management, or business fundamentals, plan addresses this issue and focuses on "keeping the house in order." There is very little higher-level management involvement in preparing this plan, but all levels of management should have it. The final plan should have both Hoshin and daily management plans.

Launching the plan

Next, the plan is launched. Refer again to Fig. 3.6. Often a facility-wide meeting is held to present the plans, but at a minimum the plans should be presented to all management and supervisory staff.

Hoshin plan reviews

Hoshin plan reviews are held regularly, preferably every quarter or 3 months. If everything is on track, it is business as usual. If not, plans may have to be changed or more resources acquired. Finally, the results and experiences of the current year are summarized in an annual Hoshin review. The annual review is done during the last quarter of the planning year; this sets the stage for starting the next year's annual planning cycle.

Illustrating the Difference between MBO and Hoshin Planning

We have discussed the difference between MBO and Hoshin planning. In Fig. 3.7 we illustrate the difference between alignment of objectives in MBO and Hoshin planning.

As the illustration shows, the use of Hoshin Kanri planning ensures better linkage and alignment of objectives at the various levels of management in the company. This is due to the Hoshin planning methodology, which ensures very tight cascading of objectives from one management level to the next. In fact the original Kanji (Japanese) characters of Hoshin Kanri mean "shiny needle or compass." Today, we know of several managers who refer to Hoshin Kanri planning as compass management; that is, everybody managing and moving in the same direction.

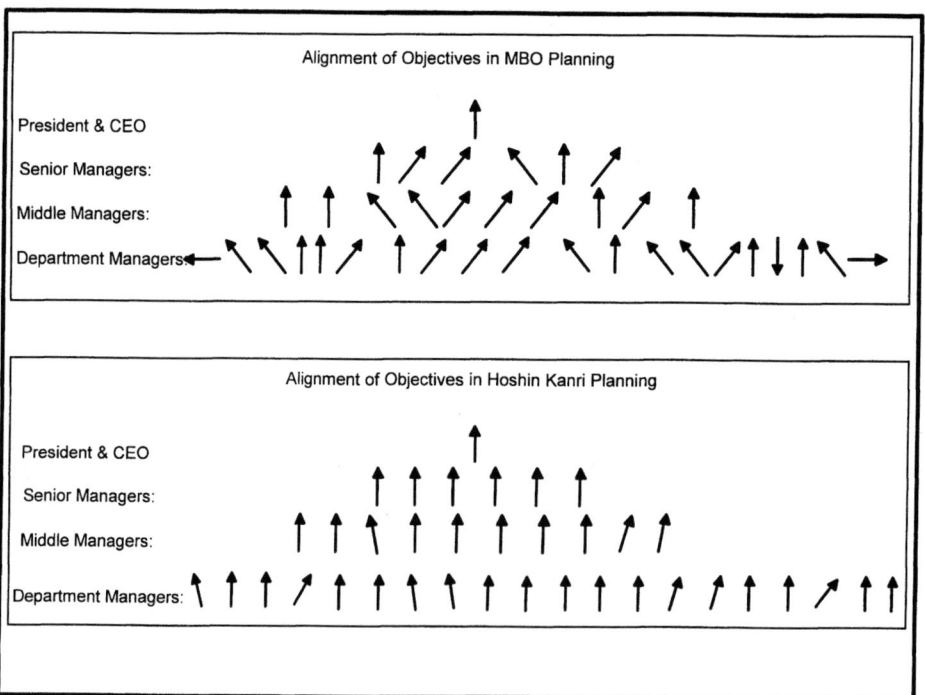

Figure 3.7 Illustrating the difference between MBO and Hoshin planning.

Hoshin plan: Ensuring success

The strength of the Hoshin and daily management planning process is that it is a systematic and tightly coupled process. This process ensures three things: First, the plan can be achieved because the next level has committed to it. Second, a hierarchy of objectives and strategies will ensue. Third, and most important, the plan is reviewed regularly, and corrections are made to get the entity back on course.

Relation between Hoshin plan and daily management plan

At this juncture, it is useful to discuss the difference and relationship between Hoshin planning and daily management planning.

Imagine a spaceship taking passengers on a journey from Earth to the planet Mars. Refer to Fig. 3.8. Its Hoshin plan addresses how it gets from Earth to Mars—planning and managing the journey through space and watching out for meteors and aliens! At the same time its daily management plan addresses the day-to-day task of feeding and entertaining the passengers, and keeping the spaceship well maintained and running smoothly. Obviously both types of plans are needed, and the separation into Hoshin and daily management is very useful for directing management focus, allocating appropriate resources, and setting priorities.

Figure 3.8 Shown here is the relationship between a Hoshin plan and a Daily Management or Business Fundamentals plan. Both plans are necessary, but each serves a different purpose, as illustrated: one manages the direction of the spaceship or entity and the other manages the routine or day-to-day items on the spaceship or entity.

MBO or Hoshin planning?

You may ask, "Why adopt Hoshin planning? After all it is very similar to MBO, and we have been successful with MBO." While MBO has many strengths, it has also many weaknesses. For example, there is a weak linkage between strategy and implementation; there is no detailed planning process; there is an insufficient consensus approach; a hierarchy of objectives, although apparent in theory, may not exist in reality; finally and most important, there is no framework for a formalized review procedure to monitor and ensure success.

Hoshin planning, on the other hand, has all the strengths of MBO and more, but none of its weaknesses. The strength of Hoshin planning is that it is a systematic and tightly coupled process. It does, however, require much more effort and consensus than MBO; but it helps provide a focus, a single-minded approach by the entire management team. The entire process is designed to ensure success. In the final analysis, Hoshin planning can be considered a more mature MBO process.

Formats and Guidelines

Next, let us review some recommended formats and guidelines for the Hoshin planning process and daily management plan.

Annual Hoshin plan

The annual Hoshin plan summarizes the breakthrough objectives for the entity or organization. These could be objectives that let the company leapfrog its competition. These objectives normally require more than the ordinary sustaining effort to accomplish and are likely to involve multidepartment collaboration. The plan generally comprises four elements:

1. Objective

2. Target or goal

3. Strategy

4. Performance measure

A recommended format to capture these elements is shown in Fig. 3.9, followed by an explanation of each element. Completed examples with comments are shown in Figs. 3.10 and 3.11.

Objective.   This is a purpose to be achieved, usually an aggressive or breakthrough statement.

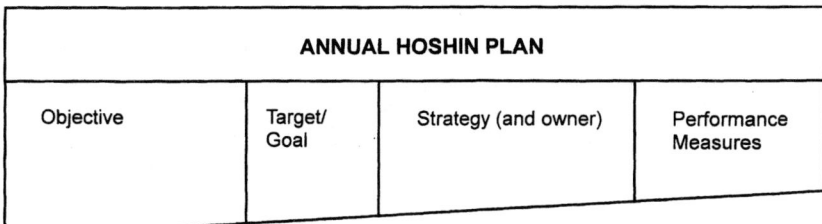

Figure 3.9 Annual Hoshin plan.

ANNUAL HOSHIN PLAN		Prepared by: General Manager	Division/Department: Apex	Fiscal Year 1998	Page 1 of 2
OBJECTIVE	NO.	STRATEGY (OWNER)		PERFORMANCE MEASURE	
1. Dramatically Increase business volume	1.1	Improve current printer offerings to get a sales increase. (R&D and Sales)		* Improve printer xx, yy and convert printer zz as our low cost printer. * Sales increase of 35%	
	1.2	Introduce products for new customer segments. (R&D)		* Design WFP printer within 12 months, plus develop new ink technology	
	1.3	Improve the supply chain in the areas of Reseller training and inventory reduction. (Sales & Marketing)		* Resellers rate us as the number one company to do business with.	
TARGET or GOAL	1.4	Develop a system strategy to handle current growth, electronic commerce (EDI & World Wide Web), and new needs beyond the year 2000. (IT)		* Solutions for growth and electronic commerce by 3Q * New system proposal by 3Q.	
* Orders at $ 1.5 Billion					

Here are some comments on the development of these Hoshin plans:
* The division manager develops his or her plans as shown on the left. The appropriate next level manager picks up whichever strategy is assigned to him/ her and makes it his/her objective.
* In this example the sales & marketing manager picks up the general manager's strategy no. 1.3, which becomes his/her strategy and its performance measure becomes the goal.
  In a similar fashion, all the general manager's strategies will be deployed to the next level.

ANNUAL HOSHIN PLAN		Prepared by: Sales & Mktg. Manager	Division/Department: Apex	Fiscal Year 1997	Page 1 of 2
OBJECTIVE	NO.	STRATEGY (OWNER)		PERFORMANCE MEASURE	
1.3 Improve the supply chain in the areas of Reseller training and inventory reduction.	1.31	Provide training to current channel partners in the areas of increasing product demand, promotion management, after sales support, and inventory management.		* Training for all Resellers by 2Q. * Channel increases sales by 35 %	
	1.32	Work with the IT operation to develop a system to automatically track both channel inventory and product sell through information.		* System pilot by 2Q, and implementation by 3Q.	
TARGET or GOAL	1.33	Set up a process to track and manage reseller inventory and keep it at optimum levels.		* Inventory reduction in the channel of 40% by 4Q.	
* Resellers rate us as the number one company to do business with.					

Figure 3.10 Preparation and deployment of a product division manager's plan.

ANNUAL HOSHIN PLAN	Prepared by: General Manager	Division/Department: Apex	Fiscal Year 1998	Page 1 of 2

Situation Analysis: The Apex Division profit has been declining. Mostly this has been due to an old product line, high cost structure, and delayed introduction of products. The new product line will increase sales and profits. However, analysis shows we must reduce current manufacturing & administrative costs by 30 & 15% respectively. Field failure rates must also be reduced below competitive levels. In all this will boost profits from 3 to 6%.

OBJECTIVE	NO.	STRATEGY (OWNER)	PERFORMANCE MEASURE
2.0 Increase profits	2.1	Introduce new products per schedule (R&D and Marketing)	* New products to provide 30% of total sales
	2.2	Reduce costs in manufacturing and administration. (Manufacturing Operations and Administrative Department)	* Reduction of manufacturing costs by 22% * Reduction in admin. costs of 15%
	2.3	Reduce product field failure rates and thereby minimize warranty costs. (Manufacturing Operations)	* Reduce failures from 4 to 1%
TARGET or GOAL			
2.0 Increase from 3 to 6 %			

Here are some comments on the preparation of these Hoshin plans: The division general manager has selected profits as his/her no. 2 objective. This is a corollary of C in the QCDE categories. In addition the strategies for the objective include Q, C, and D items.
* The operation manager has selected the strategies that are deployed to him/her.
* Note also that there is a numbering system that provides traceability of objectives and strategies. For example, the general manager's objective is no. 2.0; the strategies are no. 2.1, 2.2, etc.; the operation manager's objectives take on the number of the general manager's strategies - 2.2 and 2.3; and the strategies become 2.21, 2.22, etc.

ANNUAL HOSHIN PLAN	Prepared by: Operations Manager	Division/Department: Apex	Fiscal Year 1998	Page 1 of 2

OBJECTIVE	NO.	STRATEGY (OWNER)	PERFORMANCE MEASURE
2.2 Reduce manufacturing costs 2.3 Reduce external product failure rate	2.21	Review and reduce procurement costs for all products with life of more than 8 months. (Procurement/Materials manager)	* Cost reduction for product A, B, C of 15%
	2.22	Redesign and simplify product A and B. These products have an estimated life of 3 more years. (Engineering manager)	* Reduction in manufacturing costs for product A, B of 15%
	2.31	Improve manufacturing yields of current products. (Production manager)	* Reduce failures on lines by 50%
	2.32	Reduce failure rate of products B, C, & D (Engineering manager)	* Reduce top 5 failures by 50%
TARGET or GOAL	2.33	Etc.	
2.2 Cost reduction of 30% 2.3 Fail rate reduced from 4 to 1 %			

Figure 3.11   Preparation and deployment of a product division manager's plan.

Target/goal.   This is a broad indicator measuring accomplishment of an objective. It must be established for every objective and must be quantifiable.

Strategy.   This describes the procedure and method by which the targeted goal is to be accomplished.

Performance measure.   This is used to determine the progress or completion of a strategy. It consists of a statement and a number; the number indicates the target to be achieved.

Let us now examine some of these elements in detail, namely, objective, strategy, and performance measure. We start with the most crucial, the objective.

Objectives. Before objectives are prepared, develop an issue list. Refer to the Hoshin planning flowchart in Fig. 3.6. This starts with a review of the following:

- Company vision and the long-range (say) 5-year objectives
- Specific customer inputs and issues collected during the year
- The current economic situation
- Successes and failures in the previous year's plan at all levels in the organization

The last point above is extremely important—successes and failures of the previous year are reviewed and analyzed before a new plan is prepared. In the boxed example, we show an example of an issue list generated from the previous year.

These objectives will determine the direction in which your business is heading. As a guide, objectives should cover four important categories, abbreviated as QCDE:

Quality (Q) includes customer satisfaction issues and product and process quality.

Cost (C) includes all costs, such as administrative expenses, manufacturing costs, and productivity issues. The corollary to this is profits, and this is, of course, of paramount importance.

Delivery (D) includes new product design introductions, R&D and manufacturing product commitments, and delivery of products to the customer.

Education (E) covers human resource training and education as well as organizational issues.

The concept of QCDE is meant to ensure that nothing important is overlooked. *We do not suggest that you have four objectives—we suggest only that you review these four categories. In fact, we recommend that senior managers do not have more than two objectives— with several supporting strategies.* In actual practice, some of the objectives can have strategies that address more than one of these categories. For example, an objective to increase profits can address costs, delivery, and quality. Refer to the example in Fig. 3.11. Some companies that use Hoshin planning actually have prompts on their planning forms to ensure that managers have considered these four categories.

An Example of an Issue List from the Previous Year

In preparing a Hoshin plan, you must review the previous year's performance at the end of that year. What can you expect from an end-of-year review? It would be a list of problems, or lessons learned from a job well done. This list of issues can influence the following year's Hoshin plan. Here are some examples:

For a sales and marketing operation:

- Lost 35 percent of big deals when competing for a sale of product A1234. There is a need to understand what happened. Is it a product or marketing problem?

- Won 95 percent of big deals when competing for a sale of product B224. Why did we do well? Can we transfer the lessons learned to other new products and marketing?

- High inventory of demonstration units is now obsolete. What happened?

- Sales representatives are inexperienced due to extensive hiring during high growth. Need training next year.

- Both cross-functional projects did not meet target. What happened?

- High inventory of 2125 product. It seems market forecasts may be the cause— need to investigate.

For a design and manufacturing environment:

- Product C234 process yield fell due to purchase of inferior semiconductors. This created problems with customers. What happened?

- All cross-functional projects were accomplished. Can we expand the concept to other areas, especially sales?

- Product B224 was very successful, whereas product A1234 was not. Need to understand success and failure factors.

Before we go on, there are several points to be made about the concept of QCDE.

1. In total quality philosophy, the QCDE categories are considered essential for business success; progress in each category will help keep a company robust, healthy, and competitive. Complacency in any category may create problems. After generation of the key issues, which was discussed earlier, the issues can be sorted into the QCDE categories.

2. Ideally, the number of objectives should be limited to two or three maximum. This may be difficult at first but can be achieved if a manager focuses only on "breakthrough" objectives, that is, objectives that are essential for success. Experience has shown that an individual manager can only focus on two or three major objectives at a large organization. To try to do more may cause the manager to overstretch or lose focus.

3. In any year, not all the QCDE categories need to be addressed. For example, if in the previous years you had set up a comprehensive quality education and training program, you would not worry about it in the

current year because it is now a daily management item. The personnel or human resources department is managing this on a day-to-day basis.

4. Often, the QCDE categories will merge. For example, a profit objective could include strategies for quality, costs, and delivery, as shown in Fig. 3.11.

5. In setting the objectives, a hierarchy is important. So for the quality category, the general manager of the organization could set a top-level customer satisfaction objective with a broad overall customer satisfaction objective, say the result of an industrywide survey. But at the lower level, a product or service manager would have a specific objective that supports the top-level objective and goal, say an objective to improve a product or service that customers are unhappy about. A similar analogy will apply for the other QCDE categories.

6. The cost category is often overlooked in many organizations. This often becomes a concern only just before or after a company takeover, acquisition, management change, or sudden downturn in profits. When we mention costs, we do not mean just specific product or service costs. We mean *all costs*. And this includes overhead and other costs which have a tendency to creep up, especially during good times. Ignoring costs—that is, not managing them on a year-to-year basis and a daily basis—can result in the *self-feeding cycle of competitive decay*. One way of reducing costs is to improve quality, and this is addressed more specifically later. But other areas of equal importance are administrative and overhead costs, which must be managed on a routine basis, year in and year out. See the attached box for a detailed discussion of managing costs.

Managing Costs

A recent article in *Harvard Business Review,* "Vital Truths about Managing Your Costs," by Ames and Hlavecek[7], has an interesting discussion of managing costs. The authors state that there are several truisms that apply universally in the business world. Two of them are

1. Over the long term, it is absolutely essential to be a lower-cost supplier.
2. To stay competitive, inflation-adjusted costs of producing and supplying any product or service must continuously trend downward.

In particular, product costs including overhead or other costs such as designing, selling, delivering, and service must be managed. Especially dangerous is the fact that these tend to overaccumulate in good times when there is no pressure for tight performance and good sense. In particular, they stress that most costs are manageable, while only a few are truly fixed. The authors go on to discuss the self-feeding cycle of competitive decay, and we show a copy of their chart here, in Fig. 3.12. This viciously deteriorating cycle can work itself out into worsening conditions, and it must be prevented.

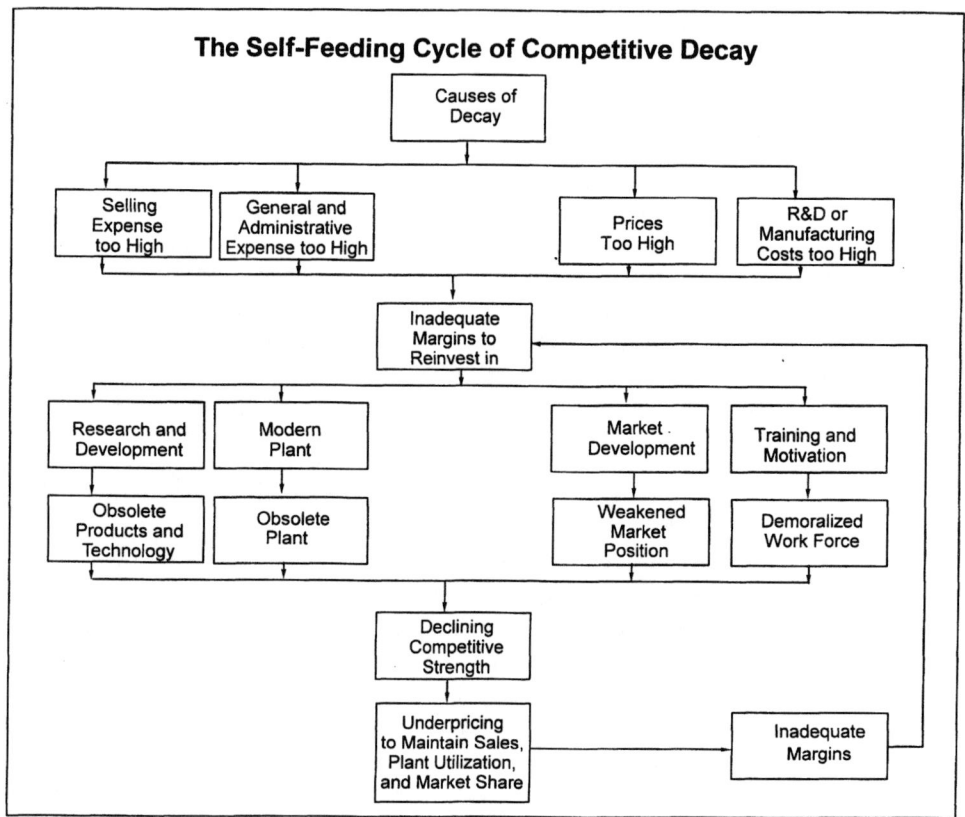

Figure 3.12 The self-feeding cycle of competitive decay. (*Copyright 1990 by the President and Fellows of Harvard College; all rights reserved.*)

Chief executive's objectives

We have given some guidelines in selecting and grouping objectives, basically using the concept of QCDE. If you are managing a complex organization—say, several divisions or several countries—then it may get more difficult to find a common denominator. In addition, you must be strategic and not have too many objectives.

How many is too many? Typically, if the chief executive has more than two objectives, it is too many. Experience has shown that one or two breakthrough objectives is about the maximum that an organization can manage.

Another good reason for the chief executive to have a few objectives is that this will allow lower-level managers to add objectives specific to their situation. Otherwise, a situation could exist in which the chief executive has, say, four objectives and the general manager adds a few more. Very soon the operation will be awash in plans and objectives, resulting in a loss of focus.

Our recommendation for the chief executive is to follow these steps before defining her or his objectives:

1. *Look for common issues or problems,* for example, in the area of quality and customer satisfaction, or profits.
2. *Review specific needs of a weak area in the organization,* for example, a weak product line, division, or country.
3. *Review future needs in your organization,* for example, start-ups in new countries, new markets to penetrate, new products to be introduced, or new technologies to be acquired.

With this guideline, the chief executive will be able to have one or two breakthrough, specific, and focused objectives for the organization.

Strategies and performance measures.   We have talked about objectives. Now we discuss strategies and performance measures.

Strategy.   A strategy describes the procedure and method by which the targeted goal is to be accomplished. An average of three to five strategies is recommended for each objective. However, the actual number can vary depending on the complexity of the objective. Too many strategies could result in loss of focus, and hence control.

Is there a technique for generating strategies? Yes. We recommend three methods that have been used successfully by many managers:

Generating strategies: Method 1

1. List each of the objectives with its goals.
2. Generate a list of strategies (by brainstorming or other means) by which the objective can be met. Many of them may be apparent, but you may need others.
3. Evaluate the choices by ranking each according to its contribution to fulfilling the objective, cost, feasibility, and other important limiting factors. If necessary, you can weigh each strategy as follows: contribution X cost X feasibility; but this is seldom necessary.
4. Establish ranking order of the choices.
5. Select the choices based on ranking.
6. Discuss the selected choices with managers to obtain endorsement. The discussion nodes shown in Fig. 3.6 are meant for this, and this is of crucial importance.

Generating strategies: Method 2.   The second method of determining appropriate strategies is a more scientific and data-driven

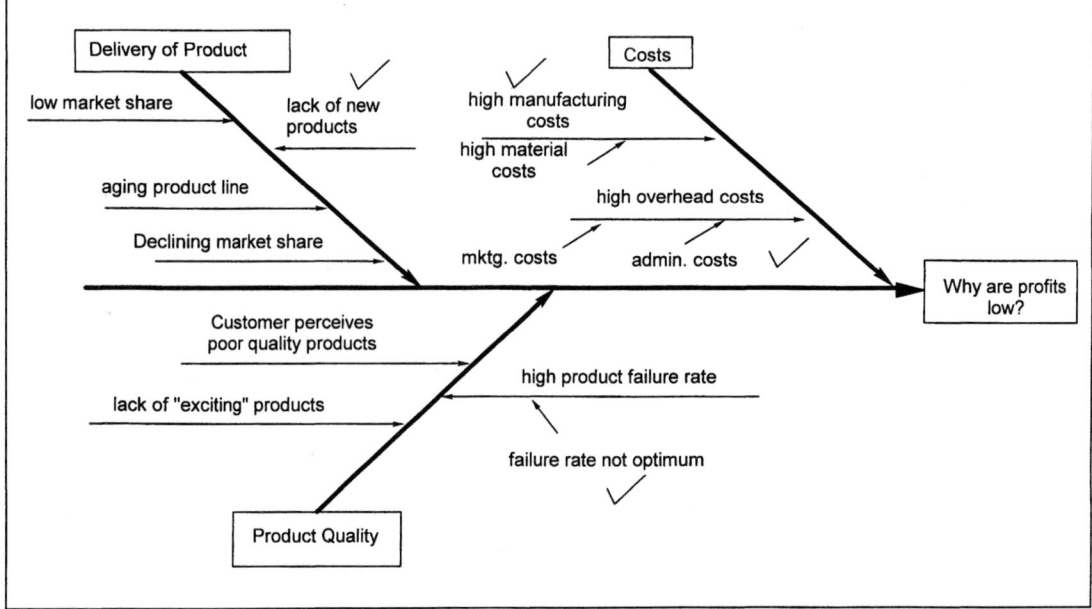

Figure 3.13   Cause-and-effect diagram for low profits.

method. As mentioned previously, the Hoshin planning process is a large PDCA cycle (see Chap. 4)—and preparing a Hoshin plan is the P (plan) stage of this cycle. Basically, this is what you do:

1. Once the Hoshin plan objective is determined, prepare a cause-and-effect diagram, where the effect is the selected objective. Many of the causes can then be brainstormed.

2. Determine the most likely causes that impact the objective, and verify them with data.

3. Convert the verified causes into strategies for the selected objective.

4. Ensure that the strategies, after implementation, succeed in meeting the objective and goal.

Let us review an example of how this is done. Suppose one of the selected objectives is "to increase profitability." Then a cause-and-effect diagram for "Why are profits low?" can be constructed. An example is shown in Fig. 3.13. Next, the most likely causes are selected, discussed, and verified with data. These have been marked with a check in the cause-and-effect diagram. The three most likely, and verified, causes could be

■ Lack of new products

- Product failure rate not optimum
- High manufacturing and administration costs

Appropriate strategies can now be prepared to reduce or eliminate these causes. In fact, we have done this in the Hoshin plan shown in Fig. 3.11. In the example, the strategies add up to the goal. This is not immediately obvious; but the situation statement at the top of the plan refers to a separate financial analysis. In the analysis, the specific performance measures were selected to ensure that the goal would be met if the performance measures were met.

This second method is very useful in formulating robust and effective strategies. Certainly, it is better than the standard strategy, these days, for improving profits—that of reducing workforce. This gives only short-term results, and worse still, it may not get to the root cause of the problem.

For more details on cause-and-effect diagrams, refer to the discussion in App. B on the seven tools. For more details on PDCA and verification of causes, refer to the P stage of the detailed discussion on the PDCA cycle in Chap. 4, "Managing Improvements and Breakthroughs."

Generating strategies: Method 3. The third method is a bottoms-up approach. First the necessary product or service strategies are generated, preferably from customer or market needs. Refer to the section on long-range planning in this chapter and Figs. 3.4 and 3.5. Then the strategies are grouped appropriately under a "super objective"—refer to Fig. 3.11, where strategies for new products, cost reduction, and reducing failures are grouped with an objective to increase profits. This powerful method is recommended because it gives freedom and creativity to generate strategies from customer and market needs.

Performance measures and types. A performance measure is used to determine the progress or completion of a strategy. It consists of a statement and a number; the number indicates the target to be achieved. There are two types of performance measures:

Result-oriented or end-of-process performance measure. This is a way to measure the outcome or desired result of the strategy, for example, an action plan, increased sales, or higher product line yield. It could also be a few substrategies which, if achieved, will result in completion of the main strategy. The owners of the substrategies need to be determined; it may be the owner of the main strategy, or there may be separate owners. Refer to the examples in the Hoshin plans in Figs. 3.10 and 3.11. Using substrategies can be a very powerful method as it allows a complex main strategy to be segmented into several substrategies, often with different owners. It is important, however, not to have too many substrategies. This method of having substrategies is only effective for a senior manager's Hoshin plan, because of the cascading process. It is not recommended for lower-level plans or the lowest-level Hoshin plan in an organization.

Process-oriented or in-process performance measure.   This is a way to measure the progress of that strategy, for example, phased results, interim action steps, or targets in various steps of the process.

Where possible, try to have one of each type of measure, but this is not always possible. In that case, do the following: At senior management level, a result-oriented measure will be sufficient; at lower levels, a result-oriented measure will be supplemented with an implementation plan, which will provide details on the various phases of the strategy.

Situational analysis.   In preparing objectives and strategies, a detailed review of the current situation is important. In the Hoshin flowchart, Fig. 3.6, we show the various items that are referred to prior to preparation of the plan. In the section "The Hoshin Planning Process," we have given some more details. Such a detailed analysis is crucial in selecting the few breakthrough objectives. In the analysis, you should also be able to determine the appropriate goals and performance measures to select. Such an analysis must be summarized and documented. In Fig. 3.11, we show a slightly different Hoshin plan form (different from Fig. 3.10). Here we included a brief situation analysis statement. A good analysis should show

- Why the objective was selected
- Why the accompanying strategies were selected
- How the goals and performance measures were determined

And, if necessary, refer the reader to a separate financial or root-cause analysis.

Deployment and cascading of objectives

As Hoshin plans cascade down an organization, this is what will happen: The senior manager's plan will be deployed to her or his managers. These next-level managers will select the strategies that are appropriate to them—these become their objectives, and the performance measures become their targets or goals; each of these objectives will generate a number of new strategies. The concept is illustrated in Fig. 3.14. Further discussions are given below and in Figs. 3.10 and 3.11.

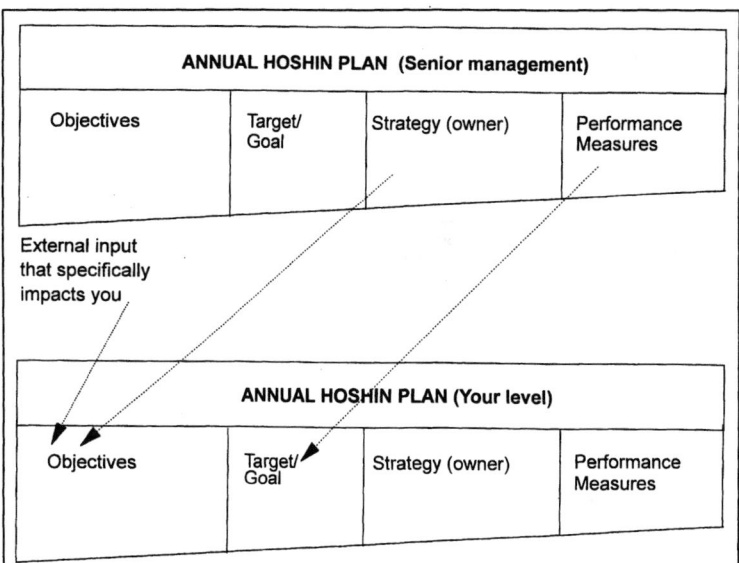

Figure 3.14 Deployment of objectives.

Deploying objectives in a large organization. Deploying objectives in a large organization is easy and effective with the Hoshin planning process. Certainly it is much easier than with any other planning process that we know of. We illustrate the concepts of deployment and cascading in Fig. 3.15. The figure shows how each strategy becomes the objective for the next-lower level. The cascading process continues downward until the last set of strategies is reached—these end up with implementation plans.

But not all strategies cascade down the organization. Some strategies will involve many departments and are assigned to a cross-functional team. Cross-functional teams are needed for strategies that require participation by many functions. For example, a strategy to introduce a new product will require the involvement of R&D, marketing, manufacturing, quality, and other functions. We illustrate all this in Fig. 3.15. There is more on cross-functional teams later in this section.

For this process to be effective, the chief executive's plan has to have only one or two breakthrough objectives—because at each lower level, the objectives will multiply. If done well, the result will be a very tightly knit plan, across the company, with everyone moving in the same direction.

Impact of cascading and how to prevent objectives from repeating. As Hoshin plans cascade down the organization, a number of things— good and bad—will happen. The number of objectives will quickly multiply, but if the plan is well made, there will be strong linkage

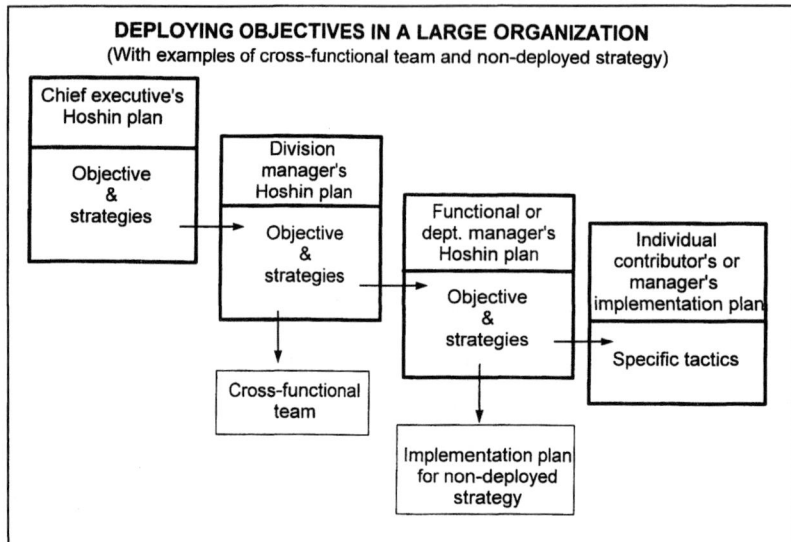

Figure 3.15   Deploying objectives in a large organization.

and tremendous synergy resulting in a sharply focused organization. This gives a strong argument for starting with fewer objectives at the top. Unfortunately, if top-level objectives are too tactical, then objectives will start to repeat. Or there can be other problems. Here are some guidelines for avoiding repeating objectives.

Problem.   Objectives repeat because they are too tactical. For example, if we have a five-star general or chief executive starting with a very tactical plan, such as "Take the hill" or "Hire five sales executives or 10 production operators," the cascading process will not work; the result will be objectives that repeat at many levels.

Solution.   Because of the nature of the cascading process, the top-level objective (and its supporting strategies) must be broad and as high-level as possible. If this is done, the lower-level objectives (and supporting strategies) can be more specific. For example,

- At the highest level you could have "Win the War," "Increase Customer Satisfaction," "Increase Market Share," or "Increase Profits."

- At the next level you could have "Destroy Xanadu," "Improve Product Quality," "Add a New Sales Force," or "Decrease Costs."

Problem.   Objectives (and supporting strategies) repeat at different levels if there are too many layers of management in an organization.

Solution.   This is more difficult to solve, but nevertheless needs to be done via a reduction in management layers.

Best method.   The best way to get good cascading of plans, which are compact and concise, is to stipulate only a few layers of Hoshin plans in an organization, say, two or three. This is then supported by implementation plans.

Use of a Hoshin plan deployment matrix and cross-functional teams. Look again at Fig. 3.11. We have shown the annual Hoshin plan of a general manager (of a division or plant). Her objective is to increase profits, and three specific strategies have been listed: introduce new products, decrease product failure rate, and reduce costs. The question that comes to mind is, To whom are these strategies deployed? One method is for the general manager to list the owner's name opposite each strategy—the owners are agreed upon during the discussion nodes shown in Fig. 3.6.

An elegant alternative is to use a Hoshin plan deployment matrix. This can be a very useful tool to ensure effective deployment of higher-level objectives. An example and a discussion of such a matrix are given in Fig. 3.16. Included in the matrix is a list of cross-functional teams necessary for the success of specific strategies. Also listed in the matrix are quality teams that will manage specific strategies.

Hoshin Deployment Matrix							Division: APEX		Page 2 of 2	
Objective No. 2 Increase profit	**Functions and Departments**							*Related Activity*		
	R & D	Market-ing	Finance & Admin.	Operation/Manufacturing					
				Procurement	Engineering	Production	Cross-functional team & leader	Project team per department
Strategies:									
2.1 Introduce new products per schedule	◎	◎	○	○	○	○	Yes, led by marketing manager	None
2.2 Reduce costs	△	○	○	◎	◎	◎	None	5 Teams, one each in all except R&D
2.3 Reduce external product failures	△	△	△	○	◎	◎	None	4 teams in Manufact-uring
Key: ◎ : High relationship, ○ :Medium relationship, △ : No relationship									

Figure 3.16 A Hoshin Deployment Matrix.   Such a matrix can be used to deploy an entity manager's objectives and strategies. The matrix can be used to determine who (function or department) picks up the various strategies that were shown in Fig. 3.11. This way there will be little confusion, and strategies and performance measures can be better formulated. Later in Chap. 4, we will show the deployment of strategies 2.2 and 2.3 to the operations manager's department. Note the matrix also lists which strategies require cross-functional teams, as well as which strategies will be managed by department-level teams.

These teams can present their progress during the quarterly Hoshin plan reviews.

It is crucial that the management team determine which strategies require cross-functional teams, consisting of representatives from different functions, such as R&D, marketing, manufacturing, and so on. Strategies that require multifunctional involvement will typically require such a team. In the example in Fig. 3.16, the strategy of introducing new products requires an entitywide team effort.

The features of a cross-functional team include the following:

- It is led by a senior manager and has representatives from all the functions that can impact the strategy.

- It agrees on allocation of responsibilities, tactics, and specific goals for individuals and departments.

- It meets regularly to measure progress, reset priorities, and request more resources when needed.

- It reports on progress to the senior management team. This should be done monthly because of the importance of the cross-functional strategies. The teams should also present progress reports during regular Hoshin plan reviews, usually every 3 months.

Implementation plan

Hoshin plans may cascade down several layers of management. At the last layer, however, there needs to be an implementation plan. The implementation plan lists detailed steps or tactics necessary to accomplish the strategies in the Hoshin plan. Hence, implementation plans are typically prepared at the lower management levels or by professionals.

The implementation plan format is shown in Fig. 3.17, with a discussion of each element. A completed example is given in Fig. 3.18.

Strategy with performance measure.   Each strategy from the Hoshin plan is listed here, with the corresponding performance measure.

IMPLEMENTATION PLAN				
No	Strategy with performance measures	Implementation Details	Timeline Jan-Dec	Who

Figure 3.17 Implementation plan.

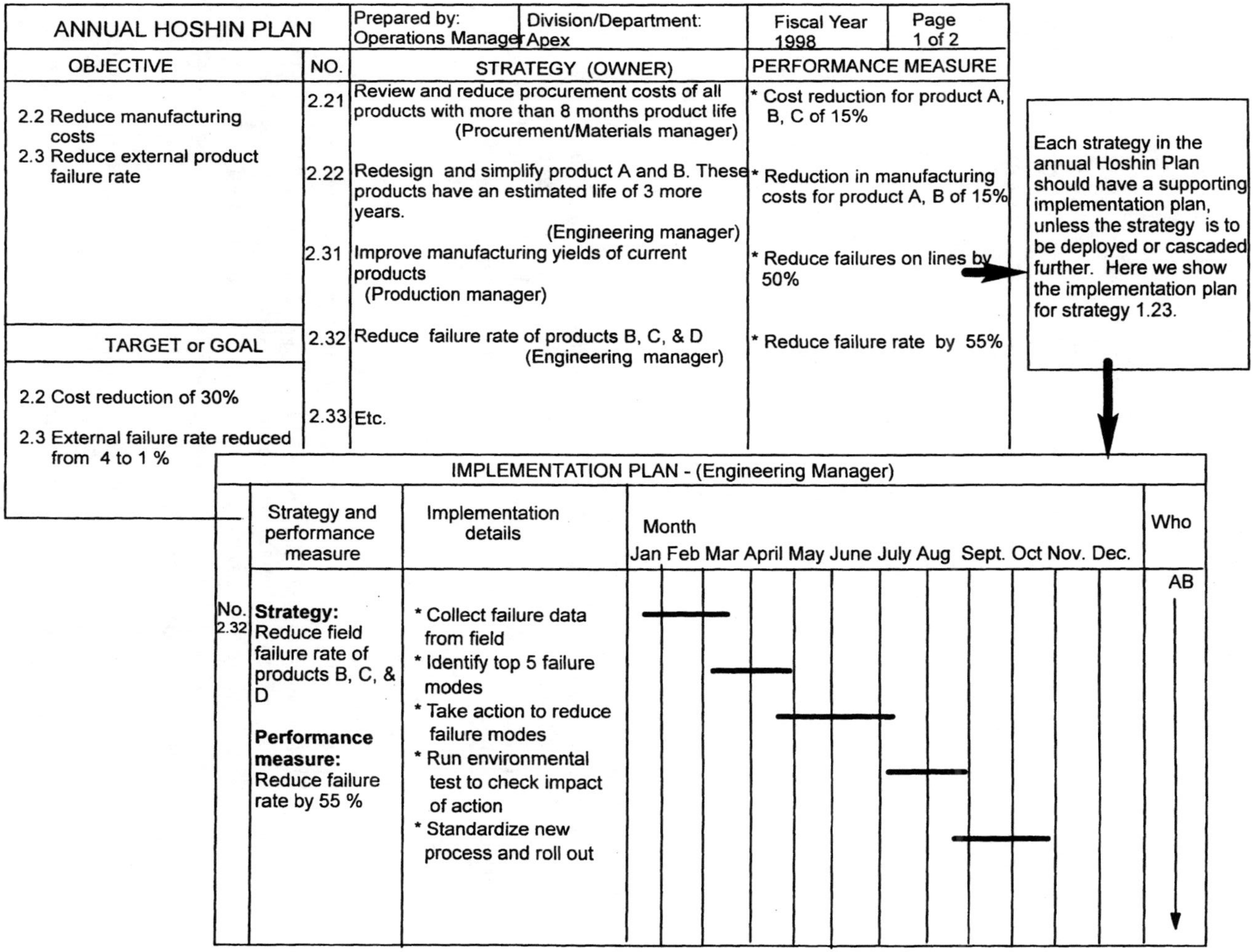
ANNUAL HOSHIN PLAN
Prepared by: Operations Manager
Division/Department: Apex
Fiscal Year 1998
Page 1 of 2
OBJECTIVE
2.2 Reduce manufacturing costs
2.3 Reduce external product failure rate
TARGET or GOAL
2.2 Cost reduction of 30%
2.3 External failure rate reduced from 4 to 1 %
NO.
STRATEGY (OWNER)
PERFORMANCE MEASURE
2.21
Review and reduce procurement costs of all products with more than 8 months product life (Procurement/Materials manager)
* Cost reduction for product A, B, C of 15%
2.22
Redesign and simplify product A and B. These products have an estimated life of 3 more years. (Engineering manager)
* Reduction in manufacturing costs for product A, B of 15%
2.31
Improve manufacturing yields of current products (Production manager)
* Reduce failures on lines by 50%
2.32
Reduce failure rate of products B, C, & D (Engineering manager)
* Reduce failure rate by 55%
2.33
Etc.
Each strategy in the annual Hoshin Plan should have a supporting implementation plan, unless the strategy is to be deployed or cascaded further. Here we show the implementation plan for strategy 1.23.
IMPLEMENTATION PLAN - (Engineering Manager)
Strategy and performance measure
Implementation details
Month
Jan Feb Mar April May June July Aug Sept. Oct Nov. Dec.
Who
No. 2.32
Strategy: Reduce field failure rate of products B, C, & D
Performance measure: Reduce failure rate by 55 %
* Collect failure data from field
* Identify top 5 failure modes
* Take action to reduce failure modes
* Run environmental test to check impact of action
* Standardize new process and roll out
AB

Figure 3.18 Developing an implementation plan from a Hoshin plan.

Implementation details and time line. Each strategy should be planned in detail. There are two choices:

- Gantt chart format: List all the steps for the strategy, and list dates in the time line.
- PDCA or *proactive* PDCA format: List all the steps in this cycle (see Chap. 4, "Managing Improvements and Breakthroughs"). In addition, list dates in the time line. This is the recommended format.

We stress the need for a detailed implementation plan; the benefits include these:

- Every Hoshin strategy in the organization eventually ends up in a detailed implementation plan, with an owner and a time line.
- Progress can be measured regularly by checking against the planned time lines, and deviations can be easily spotted.
- This provides an excellent format for reviewing an employee's plans and progress of those plans.

Ownership of responsibility must be clearly defined to avoid confusion.

Daily management, or business fundamentals, plan

As discussed earlier, the daily management plan focuses on keeping the house in order, that is, maintaining the performance of day-to-day, routine, or repetitive processes. No special effort other than establishing goals, control limits, and a monitoring system is required. The prerequisite is that these processes be well understood with a wealth of experience and knowledge, which is documented. *Daily management requires effective management of routine processes, discovering abnormalities or deviations, and preventing their recurrence.*

What are the processes that require routine management? Ideally, it should be the key processes in an entity; better still, they should be derived from a chart showing an entity's key processes. We discuss these soon, but first we review the recommended planning format, shown in Fig. 3.19, with an explanation of each element. A completed example is given in Fig. 3.20.

Item. List the item to be managed. The daily management plan has items that are well understood and documented. Little improvement is required for these items, and the expectation is that all customers (internal or external) are happy with the performance of this item. There is little high-level management involvement in these activities.

Goal and control limit. Each item will have a goal, and all goals must have control limits. When a process deviates from its control

DAILY MANAGEMENT/BUSINESS FUNDAMENTALS PLAN																	
Department		Fiscal year													Page		
No	Item	Goal/Action limit	When reviewed	Monitoring system or data source	Actual Performance												
					Jan	Feb	Mar	Apr	May	Jun	Jul	Aug	Sep	Oct	Nov	Dec

Figure 3.19 Daily Management/Business Fundamentals plan.

DAILY MANAGEMENT/BUSINESS FUNDAMENTALS PLAN																	
Department Manufacturing		Fiscal Year: 1998			Page 1 of 1												
No.	Item/Objective	Goal/Action Limit	When Reviewed	Monitoring System/ Data Source	Actual Performance												
					Jan	Feb	Mar	Apr	May	Jun	Jul	Aug	Sep	Oct	Nov	Dec
1.	Customer complaints resolution	* Resolution in 24 hrs/36hrs	Daily	Customer file													
2.	Product annual failure rate	* 0.5% / 0.7%	Monthly	Field report													
3.	Quality checkpoints conformance (refer Chap.5)	* Per chart	Daily	Checkpoints chart													
4.	Production loss, all lines (corollary of yield)	* < 5100 ppm	Daily	Line report													
5.	Product introduction postmortem completed	* Product intro plus 1 month	After intro.	Engineering													
6.	Expenses at target	* 100%, +0%, -15%	Monthly	Finance													
7.	Employee satisfaction score	* > 6.5 out of 10	6-monthly	Personnel													

Figure 3.20 A Daily Management/Business Fundamentals plan.   Shown here is the manufacturing/operations manager's daily management or business fundamentals plan. Note that this lists routine items that are done on a daily basis. These are items where there is a wealth of experience and knowledge—all these items should have standards on how they are to be performed. Hence, the goals and limits are based on experience and no implementation plans are required because there are established procedures.

limits, the deviation must be analyzed and understood, and recurrence prevented. There is more on this later.

When reviewed.   Here we list when this item is reviewed or checked.

Monitoring system or data source.   This lists where the information on performance is obtained for each item.

Owner.   Ownership of responsibility must be clearly defined to avoid confusion.

Guidelines for daily management plans. How many items should there be on a daily management plan? We recommend an average of 10 items for each department or manager. This is a guideline; the concern is that it will be difficult to monitor too many activities. Hence, priorities must be set. Some examples of items that are suitable are given below for design, manufacturing, sales, and service. These are processes that could have come from an entity's key process list or chart; we discuss these in greater detail in Chap. 5, "Process Management."

Items in sales and service entity

- The sales process. The planned sales for each month will be supported by the sales process, which can be monitored and managed. This requires monitoring the following (details are given in Chap. 5).

  Sales funnel to measure the health of current and future orders.

  Other key processes in sales and marketing, such as sales won/lost postmortems and marketing promotions.

- Capturing the customer's voice and needs during routine sales calls.

- Service contract management (planned calls, preventive maintenance calls, deviation from plans, etc.).

- Account management (management of major accounts).

- Management of customer satisfaction model for service. Refer to model in Chap. 2.

- Customer satisfaction (via surveys, customer complaint and feedback system, etc.).

- Expenses.

- Employee morale.

- Training and education.

Items in a design and manufacturing entity. Note that more details on the first five items are given in Chap. 5.

- The production plan for each month will be supported by the manufacturing process. This requires a process assembly and quality checkpoints chart.

- Managing the customer satisfaction model. Refer to Chap. 2.

- Capturing the customer's voice via various methodologies: QFD, routine visits, etc.

- Project postmortems (of newly released products).

- Product design for reliability (of new products).

- Expenses.
- Employee morale.
- Training and education.

Each of these items, with its goals and limits, will appear on the daily management plan.

Documentation of processes. Daily management is management of routine processes that are well understood and documented. What is good documentation? Managing a customer complaint system will require the type of documentation shown in Chap. 2. Customer satisfaction surveys will require a survey form and a flowchart of activities.

Other examples of appropriate documentation are given in Chap 5; there we discuss how to manage many of the key processes in a design, manufacturing, and sales company. These key processes will have performance measures that are captured and monitored on the daily management plan.

Managing abnormalities and deviations. Daily management also requires discovering of abnormalities or deviations and preventing their recurrence. This is important, and a system must be in place to allow for this. Good documentation helps. In addition, all deviations must be analyzed and understood, and recurrence prevented. There are two methods of doing this:

Use a Hoshin review table. This is illustrated and discussed in the section "The Review for Hoshin and Daily Management Plans." This is useful if deviations are reviewed every month or at some longer interval. Items requiring this method include expenses, sales quota, and customer satisfaction.

Use an out-of-control report. This is illustrated and discussed in Chap. 5 in the section on the quality assurance system. This is useful for items that require frequent reviews, for example, a manufacturing process.

Control limits

Each goal in the plan should have control limits defining upper and lower bounds, for example, $+10$ and -10 percent. Figure 3.21 illustrates this concept. As long as values of the measure fluctuate within the limits, no investigation is necessary. If the limits are exceeded, such as point A or B, causes for deviation should be uncovered and corrective action taken to bring the measure back within control in subsequent periods. *Remember that control limits are necessary for goals in the daily management plan.*

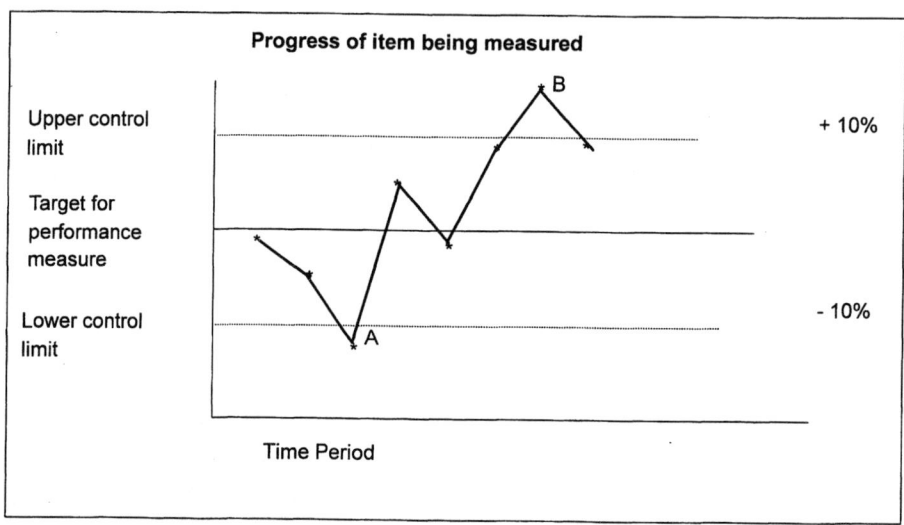

Figure 3.21   The concept of limits.

Ideally, control limits should be derived from data, which are used to construct a control chart. In such a case, we would have statistically computed upper and lower control limits. In an actual situation, however, this may be difficult, and the limits can be arbitrarily set. In this case, we would have limits that management set, based on experience or business needs.

Limits for daily management plans can be predicted because these are routine or repetitive activities. Setting limits has two purposes:

1. It is recognition that goals are difficult to meet exactly, because there will be fluctuation in real life. For example, can you meet your expenses 100 percent as planned every month? Probably not.

2. Management now has guidelines concerning when to intervene in a situation: As long as the daily management item runs within limits, management need not worry. This concept is similar to control limits in statistical process control theory. There is more on this in the discussion on control charts in App. C.

Setting numerical targets

Only breakthrough objectives—the key.   For an organization to succeed and move onward, it is imperative that the number of objectives be limited. Otherwise, there will be dilution of effort and a lack of control. The Hoshin concept requires the selection of only breakthrough objectives. Management must resist the inclusion of numerous, often stereotypical objectives, such as "perpetual virtues" or "motherhood and apple pie" or "100-year objectives"—that is, objectives that can be used year after year with little impact on the organization. The reasoning is that breakthrough objectives will move the organization to the right position in the marketplace, while the daily management plan ensures that the house is kept in order.

The numbers game. Numerical targets have to be set for both objectives and performance measures. When this is done, they must be scientifically derived from facts and data. When setting targets for any item, you must look at the following information:

- What was the achievement over the last few years for this item, and what success and difficulty have we had? What has been the average, and what has been the range observed (lower and upper control limits)?
- What are other departments or entities achieving for this item?
- What have our major competitors achieved for this item?
- What are the customer's needs or requirements for this item?
- What are the major trends for this item in the industry and the country?
- Do we wish to exceed our past performance or the competition for this item, and do we have the resources and technology to achieve the new target?
- What do our customers expect from us?
- What constitutes a breakthrough? Would it be a 10 percent change? Would it be a 30 percent change? Review the data to answer these questions.
- For guidelines on how to set targets for improvement, refer to the discussion in Chap. 4 in the box entitled "Rate of Improvement and Setting Targets."

With this information, you can set aggressive but realistic numerical targets. This gets you away from the numbers game, where numerical targets are set arbitrarily with dartboards or other equally ingenious means.

The review for Hoshin and daily management, or business fundamentals, plans

Basic methods of analysis. The annual planning process is not complete without a comparison of the actual results of the Hoshin plan with the targets. Any deviation between the actual and intended results must be analyzed in order to determine the cause. The assessment should be carried out using statistical analysis of the data.

Even if the results are satisfactory—that is, performance is much better than plan—the process must be checked. This means determining the appropriateness of strategies and performance measures that were established at the beginning of the fiscal year. It is important to understand the reasons for success and failure—only then can we learn to do better.

This review is extremely important for two reasons. First, it is important to understand why you have met, exceeded, or fallen short of targets and then to take corrective action. Second, you must use this information—the lessons learned—to influence and shape future plans.

The review format is shown in Fig. 3.22, with an explanation of each element. A completed example is shown in Fig. 3.23, with additional discussion. Note that the figure is segmented into four elements, which represent the PDCA cycle.

Objectives and strategies.   Each objective and its strategies from the Hoshin plan are listed and summarized here.

Actual.   Actual results are listed for comparison with goals (for objectives) and performance measures (for strategies).

Status flag.   This is not commonly used by companies that adopt Hoshin planning. We recommend the use of a status flag or symbol similar to traffic lights. These are shown in Fig. 3.23, and an explanation of the symbols follows:

- White circle (or on-track): Used when goal or performance measure has been met. If the goal or performance measure is planned to be accomplished at a later date, then this symbol indicates that we are proceeding according to the implementation plan.
- Crosshatched circle (or warning).   Used if there is a strong probability that a target or performance measure may not be achieved; or if the implementation plan is not being carried out as planned. Management intervention may be required.

HOSHIN REVIEW TABLE				
(Plan)	(Do)		(Check)	(Act)
Objectives & Strategies	Goal or Performance Measure vs. Actuals	Status Flags	Analysis of Deviation	Implications On Next Planning Period

Figure 3.22 Hoshin plan review table.

HOSHIN PLAN REVIEW TABLE - 3Q1998

STATUS: ○ ON TRACK　　◉ WARNING　　● OFF TRACK　　Ⓟ GOAL, STRATEGY OR PERFORMANCE MEASURE IS INAPPROPRIATE

Prepared by: Operation manager
Date: August 1998

Year: FY 1998　　Div: APEX　　Location: Manufacturing Operations

OBJECTIVE/STRATEGY (P)	ACTUAL PERFORMANCE (D)	STATUS FLAG	SUMMARY OF ANALYSIS OF DEVIATION (C)	IMPLICATIONS FOR FUTURE (A)
2.2 Reduce manufacturing costs 2.3 Reduce external product failure rate	Target 1: Cost reduction of 30 % Actual: Overall reduction of 16 % Target 2: Fail rate reduced from 4 to 1 % Actual: Currently at 2.5 %	○ ◉		
2.21 Review and reduce procurement costs for all products with life of more than 8 months.	**Perf. Measure:** Cost reduction for products A, B, C of 15% **Actual:** Achieved 19 %	○		
2.22 Redesign and simplify products A and B. These products have an estimated life of 3 more years.	**Perf. Measure:** Reduction in manufacturing costs for products A, B of 15% **Actual: 16%**	○		
2.23 Improve manufacturing yields of current products.	**Perf. Measure:** Reduce failures on lines by 50% **Actual:** 30%	◉	* Introduction of new low-cost parts and redesign have made it difficult to achieve our goal.	* Will continue to work on this item but we expect we cannot meet goal. Estimate is 35% at year end.
2.24 Reduce failure rate of products B, C, & D	**Perf. Measure:** Reduce top 5 failures by 50% **Actual:** Currently at 28%	◉	* All corrective action plans have been implemented. Results are not visible because it may take several months before the effect is seen in the field.	* No action planned. Expect to reach 40 % by 4Q98 and meet 50% by 2Q99.

Figure 3.23 A Hoshin plan review format.

- Black circle (or off-track). Indicates failure to meet target or performance measure. Management intervention is required.
- P within a circle (or inappropriate strategy or performance measure). We use this when we discover that our strategy or measure is inappropriate. This often requires a midcourse change during the planning year.

Analysis. Causes for the observed difference between the "plan" and "do" phases are looked into. When there is deviation, it is important to understand the root cause—ask why five times! Remember to analyze both successes and failures.

Implications. The outcome in this phase will influence the Hoshin plan for the next planning period—this could be the next quarter (if reviews are done every 3 months) or next year (if this is the last review of the planning year.)

Additional methods of analysis. Sometimes you may wish to provide a more detailed analysis than is provided with this format. This need may arise for technical problems, such as production yields. In such a case we suggest you use the out-of-control report that we show in Chap. 5.

Guidelines for conducting reviews. Here are some hints for conducting an effective review.

Frequency. For an entity or function we suggest quarterly (3-month) reviews for the Hoshin plan. For the implementation plan, a monthly review is recommended.

Format. We are proposing a standard format. The same standard format is used for quarterly and annual Hoshin plan reviews and for daily management, or business fundamentals, plan review. The format is shown in Figs. 3.22 and 3.23. Note the use of status flags to indicate if the goals and performance measures are on track, off track, and so on.

The symbols are very useful to give quick, visual feedback on the status of the plans. For example, if the objective's goal is not being met, there will be a black circle; and if some of the performance measures are white circles, it will be obvious that the objectives and strategies do not match.

Review procedure

- Discuss objectives and strategies.
- Review goals and performance measures. That is, compare actual performance against target.

- If there is deviation, state reasons for the deviation in the Analysis column and countermeasures in Implications column.

- If you are progressing to target or met target, then comment and give learning points in the Implication column. Remember, the review table is in PDCA cycle format, and the cycle must be completed and documented.

For fourth-quarter and annual review

- Use the same procedure and generate issues that impact next year's plans.

- Typically, the third-quarter review becomes the annual review. For this we suggest you do the following: Review the actual performance up to the third quarter, and provide a prediction for the fourth quarter. The prediction should be based on data and trends. This review will give you inputs for the next year's plan.

Duration of review.   Each manager could take up to 60 min for his or her presentation. Therefore, general managers and their staff will take about one day for a Hoshin review.

Some additional review pointers—what to look for

- The manager under review should be concise. In the Analysis column, get the facts and root causes. And in the Implications column indicate the corrective action that is planned.

- The manager under review should have backup data, especially because the review format requires a summary of data.

- Focus on failure to meet plan rather than success. But if there are good reasons for success, they should be highlighted, understood, and disseminated.

- If a target is scheduled for, say, the fourth quarter of the planning year, it is insufficient to say that you are on track during the early part of the year. The danger signals that may show up are as follows:

Comment during the first quarter: OK, on track.

Comment during the second quarter: OK, on track.

Comment during the third quarter: OK, expect to meet goal.

Comment during the fourth quarter: Oops! Problem. No resources. Missed goal.

Therefore you must check progress by reviewing the milestones, implementation plan, Gantt chart, etc. Check these; and if they are not available, there may be a problem. Be especially wary of comments such as "No problem" or "On track." Always check the facts, data, or milestones.

Changing objectives or strategies during the year. Often, an entity may discover that an objective or strategy is wrong or inappropriate. At other times, there may be external factors requiring changes in objectives or strategies. Making changes in the plan is therefore appropriate and recommended.

Reviews of implementation plans. For the implementation plan we recommend a monthly review. No special format is recommended. The review can be done by the owner of the implementation plan and the immediate supervisor. At this point, progress to plan is checked. Any deviation or slip should be analyzed and corrected. Notes or comments can be written directly on this plan. These plans can also be presented during a quarterly Hoshin review to support any statements or comments. For example, if you are on track with your plan, support the statement with your progress to the implementation plan.

Hoshin plan reviews are done bottom up. A final word on Hoshin reviews. They should be done bottom up. That is, implementation plans are reviewed at the lower level followed by reviews of higher-level plans.

Reviews of implementation plans. Implementation plans are reviewed, and then Hoshin plans that drove the implementation plans are reviewed; next Hoshin plans for the general manager's staff are reviewed; after that the chief executive can review the Hoshin plans for several product and/or sales divisions or general managers. This process is necessary—it should take less than 4 weeks—to ensure a thorough review with necessary corrections to keep the company going in the right direction.

Planning: Putting It All Together

Now we collect our thoughts and put together the ideas we discussed on the planning process. It begins with the company's or organization's vision and continues with the long-term plan, the annual Hoshin plan, the daily management plan, and the implementation plan and finishes with regular reviews of progress. The entire cycle then starts again. We illustrate this in Fig. 3.24.

Figure 3.24 summarizes all the requisite activities for a specific organization, entity, or department. A suggested time line is also given. The

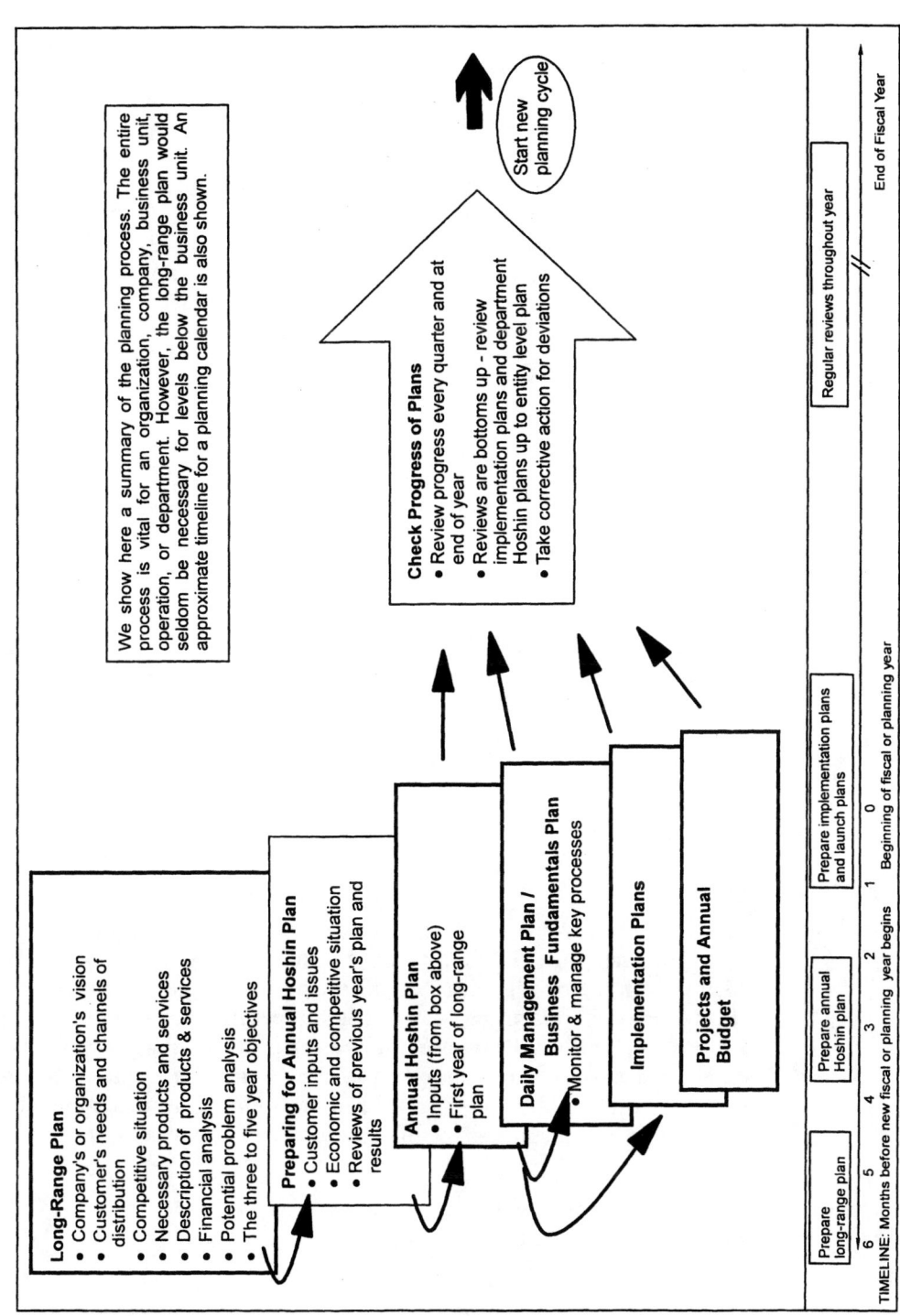

Long-Range Plan
- Company's or organization's vision
- Customer's needs and channels of distribution
- Competitive situation
- Necessary products and services
- Description of products & services
- Financial analysis
- Potential problem analysis
- The three to five year objectives

Preparing for Annual Hoshin Plan
- Customer inputs and issues
- Economic and competitive situation
- Reviews of previous year's plan and results

Annual Hoshin Plan
- Inputs (from box above)
- First year of long-range plan

Daily Management Plan / Business Fundamentals Plan
- Monitor & manage key processes

Implementation Plans

Projects and Annual Budget

Check Progress of Plans
- Review progress every quarter and at end of year
- Reviews are bottoms up - review implementation plans and department Hoshin plans up to entity level plan
- Take corrective action for deviations

Start new planning cycle

We show here a summary of the planning process. The entire process is vital for an organization, company, business unit, operation, or department. However, the long-range plan would seldom be necessary for levels below the business unit. An approximate timeline for a planning calendar is also shown.

Prepare long-range plan	Prepare annual Hoshin plan	Prepare implementation plans and launch plans		Regular reviews throughout year

TIMELINE: Months before new fiscal or planning year begins

6 5 4 3 2 1 0

Beginning of fiscal or planning year

End of Fiscal Year

Figure 3.24 Putting it all together: Summary of business planning.

most crucial and difficult portion will be the implementation plan. This is the point at which all the best laid plans must be implemented and culminate in results. Give special attention to this area.

Also, remember that all the various tactical plans or projects must be assigned to specific managers, professionals, or cross-functional teams. Where possible, this should be the only things they are working on in addition to their routine or daily assignments (daily management activity). This will ensure tight linkage from higher-level plans down to each individual person in the organization and will result in a single-minded approach by the entire team to achieve the objectives.

Planning calendar

In Fig. 3.24, we give a suggested time line for all the activities. It is important that you prepare a planning calendar for your entity for all the various activities. This should indicate activity along with the responsible person or department. It also sets expectations for the planning process each year.

Planning for special or R&D projects

You may wonder where any special, ad hoc, or R&D projects fit into this planning methodology. Since we feel that all activity must be planned, documented, and reviewed, we suggest the following: For critical projects or technology breakthroughs, the project obviously goes into the Hoshin plan; otherwise, the projects can be on the appropriate department's daily management, or business fundamentals, plan or on a separate project list.

Planning and budgeting

The annual budget and sales targets must be tied closely to the planning process. Once the long-term plan and annual plan are completed, the annual budget must be finalized, although preparation can start during the planning stage. It is important that the annual budget not be just a tweaking or simply an increase of the previous year's budget. Instead, it must support the upcoming year's plans. Hence, it is crucial that the budget be finalized after the plans are done; otherwise, plans and resources (people and dollars) may not match.

Questions and Answers on Long-Term, Hoshin, and Daily Management, or Business Fundamentals, Planning

Many organizations that start to use long-term or Hoshin planning do not reap its benefits immediately. There are several reasons for this and we will cover them through a question and answer session:

Question. What are the one or two most common problems seen when an organization starts to use Hoshin planning?

Answer. One of the most common problems is listing too many objectives at the higher levels of management. This will cause a high fan-out or cascading to lower levels. The result is a vast and unmanageable plan and a loss of focus for the organization. The key, then, is to have a few breakthrough objectives at the product division or entity manager's level. At the CEO or senior manager's level, *one or two breakthrough objectives are probably all that can be managed well.* In addition, senior managers need to avoid setting "100-year objectives."

Question. Hundred-year objectives—what are they?

Answer. That means an objective that can be used every year for the next 100 years. It is what many people in the United States call "motherhood and apple pie." Examples are objectives such as "understand our customers better," "achieve market success," or "work smarter and harder." These are statements that mean little. What you must have are specific objectives with measurable goals.

Question. Sometimes there is little choice, except to have too many objectives. For example, in a matrix-type organization, a high-level manager may have several bosses. Or just any manager may get several requests from superiors via the Hoshin plan or other inputs. How can you resolve this without generating five, six, or even 10 objectives?

Answer. Yes, this situation may occur. You have two choices to help minimize this problem. One is to group several requests into one or more "superobjective." For example, you may get requests to reduce costs, increase productivity, and reduce failure rates. These three items could become three strategies in one objective: "to increase profits." Refer to the example in Fig. 3.11. Thus you avoided getting three separate objectives plus maybe 15 strategies at your level. The trick is to be as strategic as possible.

The second choice is to delegate some issues or requests that appear at your level, especially if these are too tactical. Or you can try both choices. What all this means is that not every issue or request (from a higher level) needs to become an objective at your level. This way you are able to control the number of objectives at your level.

Question. What advice do you have on the use of daily management or business fundamentals plan?

Answer. There are four common mistakes to avoid:

- The list should be kept short. Senior managers should have a short list of perhaps 10 items. A longer list, several pages long, can result in a loss of focus and unnecessary generation of paper. In manufacturing or sales, at the lower levels, one item could be a list for managing the manufacturing or sales process.

- Keep Hoshin plans and daily management plans separate. A common problem is both types of items on a Hoshin plan. This mistake generates too many objectives, causing clutter, and can result in improper objectives and goals—specifically a lack of breakthrough items.

- A third mistake is the difficulty of setting goals for daily management plans. We recommend you set these goals based on experience. When you have no experience in a particular item, it may be wiser to make it a Hoshin plan strategy in the first year. For example, "set up and measure a process" may be the first-year strategy. In the second year, this process may now appear in the daily management plan, with a data-based goal.

- A fourth mistake is the lack of limits for all the goals (refer to the section "Control Limits" earlier in this chapter).

Question.   I have read that management consultant Tom Peters does not believe in planning. In fact, he suggests that you do not need plans and goals.[8] Can you comment?

Answer.   We have a tremendous amount of respect for Tom Peters. Typically, he brings fresh insight into management methods and thinking. We think he was against slow decision making, which is often cloaked in the guise of planning. His major concern, we believe, is corporate (or central planning) which is typically slow and removed from business needs. He is also probably concerned that when planning becomes routine, it becomes a mechanical exercise, hopeless, and nonstrategic. Refer to our discussion on pitfalls to avoid in long-range planning—this will help eliminate some of Peters's concerns.

On the other hand, in contrast to his numerous studies showing that planning is not helpful, we can quote studies indicating that planning helps tremendously. The successful Japanese companies are great planners—you only need to look at Matsushita, NEC, and Toyota. Nevertheless, there must be a balance. Too much planning is bad, and we have recommended short plans for both long-range and annual plans. In addition, when necessary, quick decisions are a must. Many successful products have been the result of quick management decision, in the absence of sufficient data.

Question.   How do you convince a skeptical manager to adopt the Hoshin planning methodology?

Answer.   There are two answers to this. First, we hope that your company adopts this as a standard to replace a weaker planning process. In such a case, our question to you is, Why is the manager not adopting the company standard? Second, analyze the current planning process. For example,

- What is the current process?
- How are current objectives deployed down the organization, and are they linked to the general manager's objectives?
- How are current objectives on the plan reviewed?
- What were last year's plans; what was achieved; what was not achieved? Were there problems?

If you discover enough weaknesses, it will be easier to convey the benefits of the Hoshin planning methodology.

Question. Must we have a long-term plan in addition to the annual Hoshin plan?

Answer. Not always. Most departments and operations will not need such a plan. But each organization—say, a product division or a sales division or headquarters—should have a long-term plan that addresses issues and strategies over a 3- to 5-year time frame.

Question. What about implementation plans?

Answer. Remember to have implementation plans. At a certain level in the organization, specifically the professional level (engineers, accountants, supervisors, sales representatives, etc.), you should have only implementation plans.

Question. I have heard that one of the best ways for a senior manager to be involved in total quality is via Hoshin planning. Any comments?

Answer. Yes, that is a valid statement. If you look at Fig. 3.24, you can see that the Hoshin plan requires addressing the entity's purpose and vision, customer needs and issues, quality, costs, process management, employee participation, and so on. These are all elements of total quality, and Hoshin planning ensures that you address them—if you follow the process correctly.

Question. Can the same item show up on the Hoshin plan and the daily management plan?

Answer. No. You should be either maintaining (or slightly improving) an item on the daily management plan or having a breakthrough improvement on the Hoshin plan. But on a year-to-year basis, an item can move from one plan to the other. For example, product reliability could be on the daily management plan in one year but due to a catastrophe, reliability could become a problem. Then in the following year, the operations manager could have a Hoshin plan objective to improve product reliability.

Question. Any other words of advice?

Answer. Yes. Do not get carried away with the long-range or Hoshin planning process. The long-range plan should be concise and should culminate in an annual Hoshin plan, supported by implementation plans. Senior managers should keep to two annual break-through objectives and a short—one-page—list of business funda-mentals or daily management items. That gives up to three or four pages in the format we have given. If your plans are longer, it may be appropriate; but we suggest that you look again at your priorities.

Summary: Business Planning

Planning is one of the most important processes in an organization—some would say the most important. It is what drives most other activities.

We have proposed the use of both long-range planning and annual planning methodologies. The combination of these plans, long-range and short-term, forms a powerful and planned approach to succeeding in the marketplace.

We have described a detailed long-range planning process that allows you to be strategic, to be better prepared to tackle the competition, and to meet customer needs. We discussed the power of a bold vision and how it can drive nations and organizations. We also advised on pitfalls to avoid in long-range planning.

For the annual planning process we have recommended the use of the Hoshin and daily management, or business fundamentals, plans. The Hoshin plan focuses on the organization's breakthrough objectives, while the daily management plan is used to manage the organization's day-to-day activities; more information on daily management is provided in Chap. 5. Hoshin planning and daily management are systematic and tightly coupled processes. They require effort and consensus and in return provide a focus, a single-minded approach by the entire management team. The process is designed to enhance the chances of success. During annual planning, we recommend that you review four crucial areas: quality, costs, delivery, and education. This will ensure that there is no neglect in these important, generic success factors. Also, during the annual planning process, you will discover many items that need improvement or breakthroughs. In the next chapter we discuss how to manage both these items.

Finally, a reminder that formal planning provides many benefits, including systematic thinking, better coordination, sharper objectives, improved performance standards, and management involvement. All these result in a planned approach to tackling the marketplace that eventually can end in higher sales and profits.

References

1. Charles F. Knight. "Emerson Electric: Consistent Profits, Consistently," *Harvard Business Review*, Jan./Feb. 1992, pp. 65–70.

2. Much of the Hoshin planning methodology mentioned here comes from the author's book, *TQC at Hewlett-Packard—The Asian Experience,* 2d ed., an internal publication. Additions and changes have been made based on the author's personal experience. The categories (objectives, goals, strategies, and performance measures) that we show are similar at companies that use Hoshin plan methodology in the United States and Japan. Their forms, however, may be different.

3. James C. Collins and Jerry I. Porras, "Building Your Company's Vision," *Harvard Business Review,* Sept./Oct. 1996, pp. 65–77.

4. Joseph B. White and Clay Chandler, "Toyota Seems to Relent amid Industry Woes," *Asian Wall Street Journal,* May 19, 1992.

5. Michael E. Porter, "What Is Strategy?" *Harvard Business Review,* Nov./Dec. 1996, pp. 61–78.

6. Henry Mintzberg, *The Rise and Fall of Strategic Planning* (New York: Free Press, 1994, p. 155).

7. B. Charles Ames and James D. Hlavecek, "Vital Truths about Managing Your Costs," *Harvard Business Review,* Jan./Feb. 1990, pp. 140–147.

8. Tom Peters, "Want to Get Ahead? Then Don't Plan, Do It," *San Jose Mercury News,* December 3, 1990.

4

Managing Improvements
and Breakthroughs

We need never ending improvement...
to establish better economy.
W. EDWARDS DEMING

Overview

One of the basic tenets of total quality is continuous improvement. The old adage "Don't fix it if it isn't broken" does not work anymore. In today's environment, if you do not fix it, your competitor will, and will take away your market share, too. Just look at the business that express mail companies have taken away from the U.S. Post Office; or look at what the Japanese electronics industry has done to its European and U.S. competitors.

We also need to improve things in a systematic way. There are numerous tools available that will help, such as the PDCA improvement cycle, to which we devote much time, design of experiments, and Taguchi methods. But before we go any further, here is an interesting incident:

> I visited an R&D lab recently and they showed me the various integrated circuit products that they design. There was one interesting driver assembly that they supplied to a display manufacturer. They told me that this current model was the third-version assembly. What was different between this and the first version? The first version had some failings, so they decided to improve it. The second "improved" version was short lived because it did not resolve adequately the failings of the first version, so they spent several more months developing the third version.
>
> But wait, why did the second version not solve the problems? Well, the R&D manager admitted, they thought they had the fix in the second version but they were wrong.
>
> Thought!...that's right, thought—they used gut feel and experience to arrive at the solution—little analysis had been done.

Alas, this is a common tale that I have seen on numerous occasions: the use of experience and genius to solve problems. This is little different from Russian roulette, except that it is the customer of your company who gets shot.

Every business must focus on both improvements and breakthroughs. This will require a focus on several activities, which will generate new business. Some ways to help generate new business are to

- Improve the current product or service: reduce costs, improve reliability, add new features

- Modify a current product or service with enough features to pass it off as a new product or service

- Design a breakthrough product or service that puts you ahead of the competition

In this chapter we discuss how to manage improvement and breakthrough projects successfully. In Chap. 3, we discussed business planning and how that can drive breakthroughs. We now discuss how to manage improvements, and we finish the chapter with a discussion of managing breakthroughs with a goal of achieving zero defects.

Managing Improvements—Selecting Items to Improve

How do we select items to improve? Every organization will have a poor-quality "iceberg" of visible and hidden problems. We need to know these problems. Figure 4.1 shows a poor-quality iceberg contributed by Mike Ward of Hewlett-Packard. In it we list some of the obvious and hidden problems that could arise in a large company.

It is extremely dangerous to ignore or hide these problems. Consider the following statement: "I am sick and tired of visiting plants to hear nothing but great things about quality and cycle time—and then to visit customers who tell me of problems."[1] Sound familiar? The solution is to have a system in place that captures and solves these problems. Throughout this text we provide methodologies for identifying problems. Briefly these include

Chapter 2, "Customer Obsession." The various customer satisfaction surveys that are conducted, competitive analysis and benchmarking, the customer complaint and feedback system, the product satisfaction model, and of course product and service failure information systems are discussed.

Chapter 3, "Business Planning." Items from here include all the issues raised during the preparation of the long-term and annual

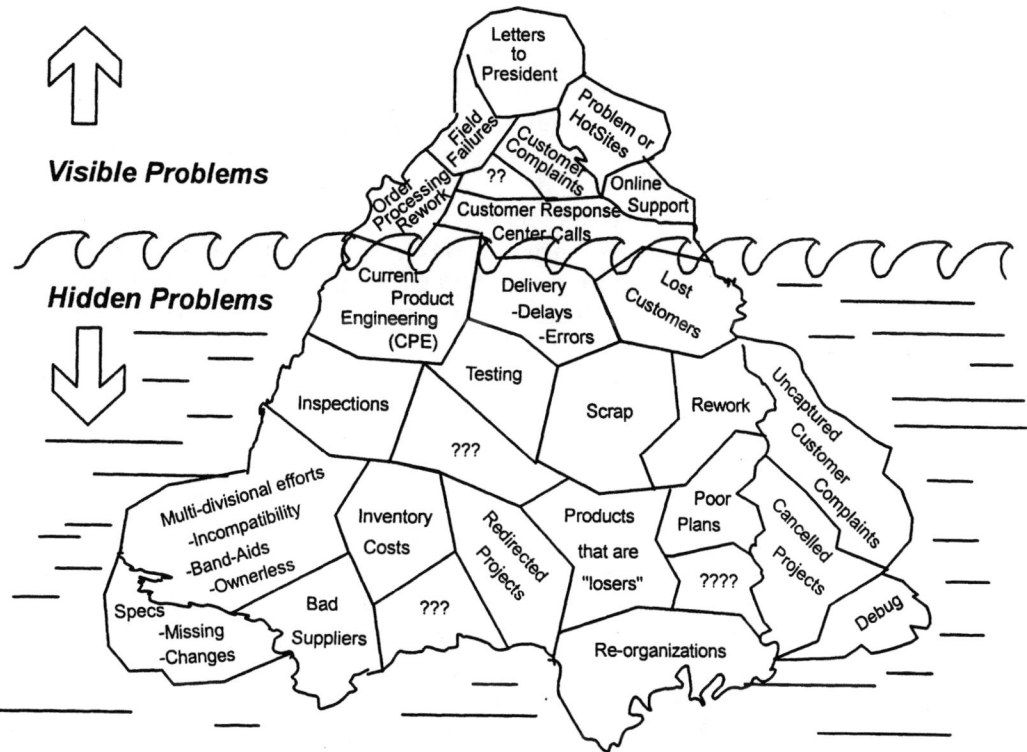

Figure 4.1 The poor quality iceberg. This figure shows the iceberg of problems that may exist at an organization. Typically, we may be attacking the obvious problems that exist, not realizing the numerous hidden problems or opportunities for improvement. These opportunities, if left unresolved, can lead to waste, higher costs, and customer dissatisfaction, resulting in lost business. A well-managed organization should have a small iceberg.

Hoshin plans as well as items discovered during Hoshin and daily management plan reviews.

Chapter 5, "Process Management." All the processes that need to be managed and improved upon are addressed.

Chapter 6, Employee Development and Participation, and Leadership." All the issues raised by employees are discussed.

The above listed methodologies will provide fertile ground for identifying items for improvement. Before we start improving, priorities must be set; otherwise, we may have too many things to do. This is done routinely during Hoshin planning, discussed in Chap. 3, and via the quality assurance system, discussed in Chap. 5.

Cost of Poor Quality

The various items shown in the poor-quality iceberg can be listed, grouped, and converted to dollars wasted. This is a popular method used by many consultants in the United States and Europe. It was popularized by Feigenbaum[2] and Joseph Juran[3] as quality costs. Others call it the *cost of quality,* which is not appropriate—a better description is the *cost of poor quality.*

Juran talks about three types of costs: internal failure costs, external failure costs, and prevention costs. He then goes on to describe a model that gives "optimum quality costs." He recommends that these costs be made to decline on a continual basis. Certainly, this is one way to get management's attention to start a quality improvement program—that is, by focusing on dollars wasted. In many companies this is the only way to get management's attention. This is very different from the approach adopted by Japan and the newly industrialized countries. Here quality is improved because there is a strong drive toward perfection and customer satisfaction, resulting in increased market share and profits.

The notion of managing quality costs is disputed by Dr. Hitoshi Kume.[4] He contends that western companies "are so concerned about identifying quality costs that there seems to be an impression that quality control activities cannot exist without a formal quality cost system." He goes on to say that he has tried and failed to introduce this concept in Japan—partly due to the fact that it is impossible to include all of a company's quality information in quality costs. He gives several examples to support his argument. For example, a mature product (such as a TV set) will have low quality costs and low profits—because of high competition. Yet a new innovative product could have high quality costs and yet have high profits—because of no competition and a high retail price. He goes on to say: "The goal of business management is to increase profit, not to reduce cost." Therefore cost increases are acceptable, as long as a company can obtain more profit to offset the additional cost.

The most important thing that management needs to do is to ensure that the design, production, marketing, and product meet the customers' needs. Hence, although cost should be lowered on a continual basis, lower quality costs are not necessarily a sign of successful management. If all costs are equal with those of other firms, success can only come from continuous development and introduction of new products that meet customer needs—because the largest loss is probably the loss of market share, and quality costs do not measure this. In fact, in the same vein Dr. Deming has said: "The most important costs are unknown and unknowable."

In summary, we do not recommend the laborious collection of cost-of-poor-quality data. Nevertheless, this may be useful in a specific department that has very high poor-quality costs. Instead, we suggest using the guidelines given in the section "Selecting Items to Improve." A good management system will automatically and continuously focus on improvements. Concurrently the focus must be on developing products and services that meet and exceed customer needs.

Problem-Solving Styles

How many ways can you solve a problem? Several. On the basis of experience, we show three styles below. In all three styles we show the reaction to a problem, how the root cause is determined, and the solution implemented. Our humblest apologies for picking on specific careers or personalities—although we have stereotyped some careers, we realize that there are many exceptions to our observations. Our intention is to show a cluster of styles that have evolved over the centuries.

The shotgun approach

Method. The cause of a problem is determined from gut feel, intuition, experience, or intelligence. The solution, to eliminate the cause, is then implemented. This method is also called the *genius approach.*

What are the pros and cons of this method? The solution can be implemented quickly. The effectiveness of the solution depends on an individual's or team's intelligence or experience level. We have observed

brilliant individuals who have come up with very effective solutions. In most cases, however, the solution is often wrong because it misses the root cause. Hence, the problem may not go away, and will need to be dealt with again, via a second improvement cycle, or a third. The result could be wasted effort, frustration, and time.

Who uses this method?   Many people. Typically, this method is used by many politicians, some chief executives and senior managers, and nonanalytical people. We have observed it in individuals who are impatient and want a quick solution, or those who want a political solution that provides the least resistance and makes everybody happy. Sounds familiar? It is definitely not recommended, except for very experienced or brilliant individuals.

The explorative approach

Method.   The problem is explored in detail, and the individual or team goes through a winding, tortuous, journey through many ideas, data, and alternate causes and solutions. One of the solutions is then selected from the alternatives, fine-tuned, and implemented.

What are the pros and cons of this method?   There are many good aspects to this method, including the gaining of group consensus and generation of creative solutions. This method takes medium to long

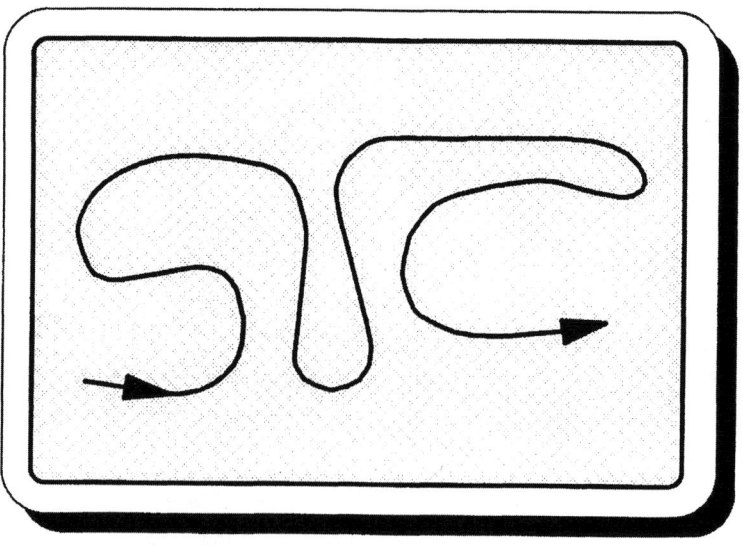

time, but there is no systematic approach on how to proceed. Hence, the wrong solution could be implemented because the root cause is not understood.

Who uses this method?　This method is used by intellectuals and creative people. We have also observed that many Europeans use this method. Why? We suspect the reason is that Europe has a tradition of generating ideas that dates back to the Renaissance period. Europe has been and still is a source of artistic and unique ideas, albeit often expensive ones.

The analytical approach

Method.　The individual or team members go through an analysis and background of the current situation. Then they search for and validate the root causes of the problem. Finally, solutions are identified and implemented.

What are the pros and cons of this method?　This method generates a powerful solution that gets to the root cause of the problem, hence there is little possibility of a recurrence. This method can be taught and documented. It takes medium to a long time, but many people find it frustrating to proceed in such a systematic fashion, which requires discipline to stay on track.

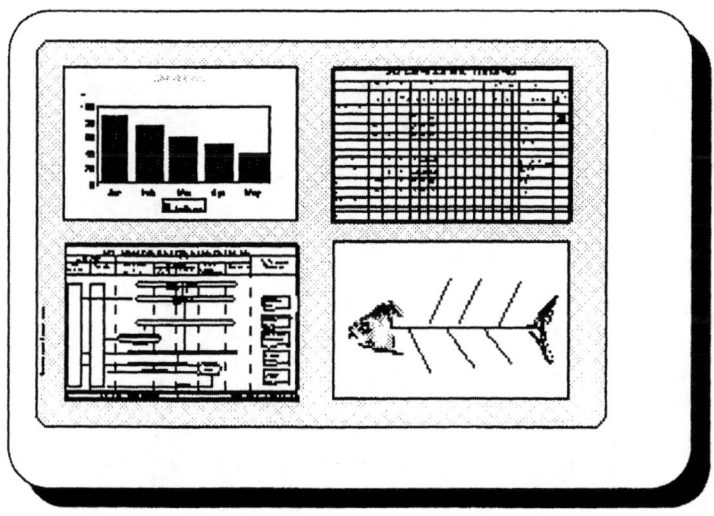

Who uses this method?   It is used by analytical people and those trained in quality improvement techniques. Many Japanese managers have been observed to use it, probably because of Japan's history and culture of aggressive quality improvement. It is also popular in quality-oriented companies and organizations. We now discuss the analytical approach, known as the PDCA improvement cycle, in detail.

PDCA Cycle

At a seminar I once attended, a participant asked the legendary Kaoru Ishikawa, "What is the most important tenet of quality?" He replied without hesitation, "The PDCA cycle."

The PDCA (plan, do, check, act) cycle was originally developed by Walter Shewhart, the originator of statistical quality control. It was popularized by Edwards Deming[5] and is often called the Deming cycle. It gained wide popularity in Japan, through the efforts of Deming. Since then, there have been numerous versions called the TQC story, the QC story, CA-PDCA, etc. Very briefly, here is an explanation of some of the versions.

The Shewhart and Deming cycle

A cycle is designed to help improve a process. It is also meant to be used as a procedure for finding the root cause through statistical analysis. It is divided into four steps:

1. What is to be accomplished? What data are available? Are new observations needed? If yes, plan and decide how to get more data.

2. Carry out the change to be accomplished, preferably on a small scale.

3. Observe the effects of the change.

4. Study the results; what can we learn or predict?

The PDCA cycle

The PDCA cycle is very similar to the Deming cycle. The four words describe the stages very nicely and are more explicitly stated as follows:

1. *Plan.*   Determine goals and methods to reach the goals.

2. *Do.*   Educate employees and implement the change.

3. *Check.*   Check the effects of the change. Have the goals been achieved? If not, return to the plan stage.

4. *Act.*   Take appropriate action to institutionalize the change.

The CA-PDCA cycle

The thinking behind this is that you need to check or analyze the current situation before you start to plan, do, check, and act. The logic behind this is correct, but why not just add a step in the plan that requires analysis? That was Shewhart's original intent. Doing this will allow the original PDCA cycle to be retained.

The QC story

This attempts to cut through the confusion of the various improvement cycles and provides a sequence of activities similar to those in the CA-PDCA cycle, without using the words *plan, do, check,* and *act.* A word of caution on the QC story: Many people have the impression that the QC story is only meant to document a project when it is completed. *This is incorrect.* It is meant to be used as a step-by-step guide for solving a problem, and as a procedure to document a completed project. The same concept applies to the PDCA cycle that we now discuss in detail.

Modified and improved PDCA cycle

Figure 4.2 shows a modified PDCA cycle, which retains the original intent of the cycle but includes the various improvements of the other versions. This is the cycle we recommend.

The PDCA cycle is often depicted as a wheel, as shown in the center of Fig. 4.2. This is an important concept, because one turn of the wheel represents one improvement cycle, which brings us to the beginning of the next cycle. When one cycle is completed, there are two alternatives that can be pursued: control the improved process or go through another improvement cycle.

Relationship between Improvement and Control

Let us look at the relationship between improving and controlling something. As previously stated, at the end of an improvement cycle we have two choices: Put the improved process under control, or start another improvement cycle after you put it under control. We illustrate this concept in Fig. 4.3. The choice is influenced by the nature of the current project and other priorities. The purpose of putting it under control is to maintain the improvements that have been made—because it is very easy to slip into old habits and lose the gains. Hence, proper training and documentation are crucial to help retain the gains. In the next chapter, we discuss process control and management of processes.

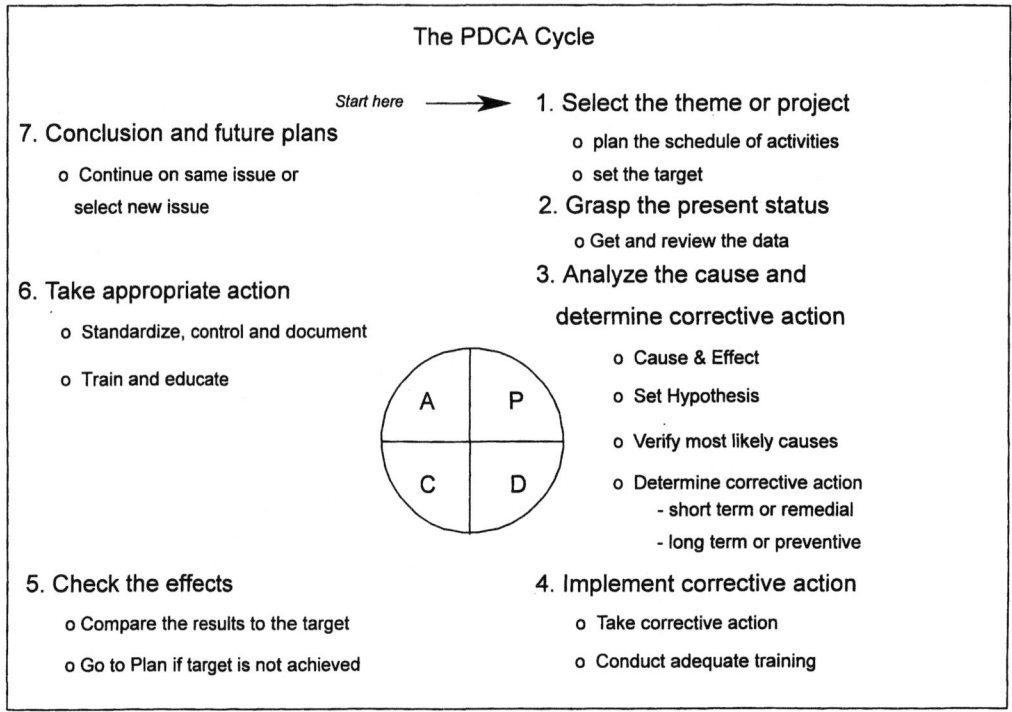

Figure 4.2 The PDCA cycle.

The alternative is to go into another cycle of improvement, after the process is put under control—if not now, later. At that point, good documentation of the current project—the analysis, the validation, the choices made, the gains, and what remains to be improved—is very important. Having such information will make the next improvement cycle easier and faster.

Benefits of the PDCA Improvement Cycle

Here are the main benefits of the PDCA improvement cycle:

- It is a systematic, problem-solving process that provides the quickest route to an effective solution.
- It ensures an agreed upon schedule for project completion.
- It ensures an agreed upon goal or target, usually set with data.
- It ensures a detailed analysis of the failure modes.
- It ensures verification and elimination of the most likely failure modes.

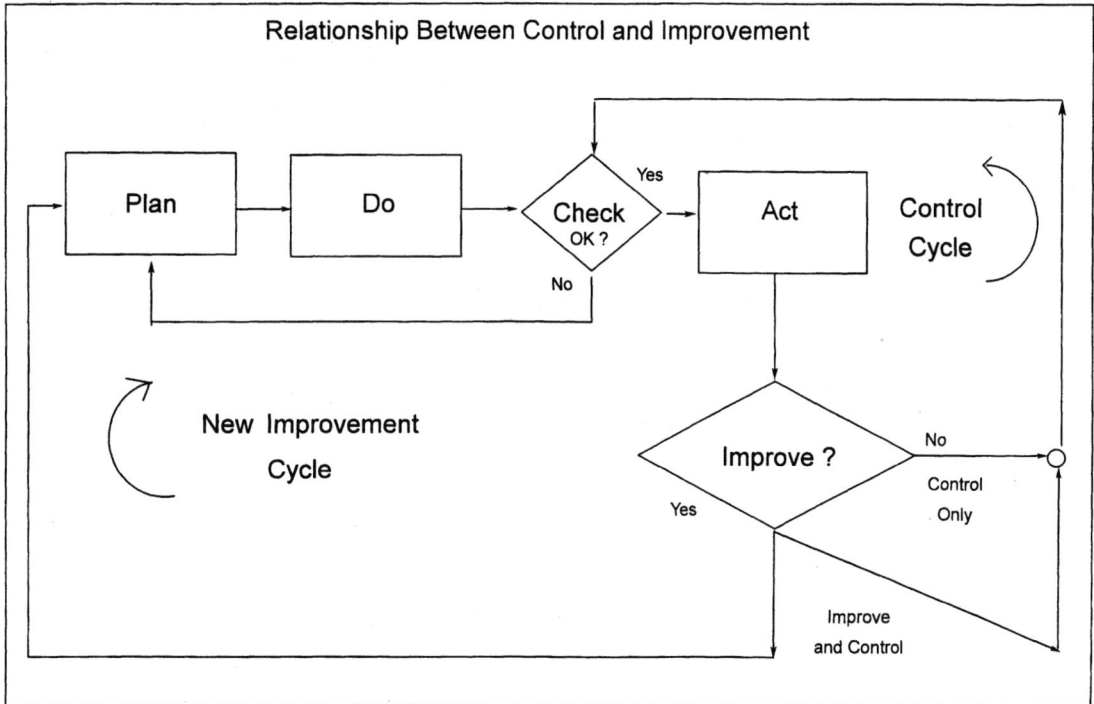

Figure 4.3 The relationship between control and improvement.

- It requires implementation of controls to monitor and manage the new improved process.

- It requires training in and documentation of the new process. It requires documentation of before-and-after failure data. This will be useful for the next improvement cycle.

- It will ensure no recurrence of the problem, thus ensuring continuous improvement. This is achieved through standardization of the new improved process.

- Managers and supervisors may come and go, but if PDCA improvement cycle is institutionalized and mandatory, then employees will always be systematic and analytical in eliminating root causes of problem areas.

The last point just stated is important. We paraphrase it as follows: People may come and go, but processes stay. The PDCA cycle is part of a process—the improvement process. Over the years, we have seen numerous improvement processes, each customized to meet an individual's needs or pet theory. This causes an excessive amount of relearning and inefficiency.

We suggest you minimize the use of other improvement processes. Instead, standardize on the use of the PDCA process shown here. This will result in a common language and facilitate communication of improvements in an organization. Many companies have adopted this as a standard, including most Japanese companies and Florida Power & Light and Hewlett-Packard in the United States.

Detailed PDCA cycle

Shown next is a detailed PDCA cycle. You will find it useful as a reference and a training tool. Each of the seven steps in Fig. 4.2 is explained in greater detail. In addition, we have listed some quality control (QC) tools that could be used in each step. To illustrate the steps in the PDCA cycle, we give a true example. The example comes from Hewlett-Packard Malaysia.

Plan Stage

Step 1: Select the theme or project

- *Objective of this step.*   To define clearly the problem to be resolved.
- *Discussion.*   Here we define the project, understand its background, set a target, and prepare a schedule of activities. The following are substeps of step 1.

   Step 1a: Project background and reasons for selection.   The project can be selected from the department objectives or customer complaints, or it can be a continuation of a previous project. And of course it should be within the improvement team's control.

   Step 1b: Set a target.   This should include a statement of the item, a numerical number to be achieved, and a time frame for project completion. For example, improve product yield to 85 percent by May 1991. Set reasonable and realistic targets; otherwise, it will be difficult to achieve them. The target can also be set after step 3, when more data are available. The situation will vary with each project, but an initial target may be appropriately set here. The target can be set according to these guidelines: Use current data to set a breakthrough goal, use competitive data to equal or better the competition, or use the rule of thumb of reducing defects by 50 percent every 12 months. Refer to the rule of thumb given later in this chapter in the box "Rate of Improvement and Setting Targets."

   Step 1c: Prepare a schedule of activities.   This lists the seven steps in the PDCA cycle and the expected time frame for each step. For our first project we may estimate, and with experience this will become easier. We recommend the implementation plan format, shown in Chap. 3, to record the schedule. This is very important, as it sets outer bounds on the project.

- *QC tools that can be useful:* Pareto diagram and trend charts.

Example of Step 1: Project Theme

Step 1a: Project background and reasons for selection

To improve first-pass yield at final test for product QDSP-6666. This product is a small display, used in cars and computers.
1. Low first-pass yield at final test (86.2 percent).
2. Too much rework at final test.
3. Shipment timeliness adversely affected.

Step 1b: Set a target

To achieve 96 percent first-pass yield at final test by July 1988. This is equivalent to reducing failures from 13.8 to 4 percent, and is in line with the department goal.

Step 1c: Prepare a schedule of activities.

This is not shown here.

Step 2: Grasp the current status

- *Objective of this step.* To understand the problem area and to highlight specific problems.

- *Discussion.* Here we study the effects of the problem, by reviewing the available data. Our study should be approached from several facets, such as time, location, and type. Suppose we want to reduce the percentage of failures in a facsimile machine. Then for time, we can look at the failures between the day and night shifts and over a period of time. For location, we can look at which part of the machine has the greatest number of failures, for example, top, bottom, side center. For type, we can determine which aspect of the machine is causing failures, for example, printed-circuit board, power supply, or plastic case. The available data can be presented in graphs and Pareto charts. We should get a process flowchart of the product or process being studied. If it does not exist, we must prepare a chart.

- *QC tools that can be useful.* Process flowcharts, Pareto charts, trend charts, control charts, histograms, process capability indices.

Example of Step 2: Current Status

In Fig. 4.4, we show a graph and Pareto diagram. The graph gives the final test yield data for the display assembly, and the Pareto diagram gives a breakdown of the causes of failures. A process flowchart is available but not shown here. Note the following for the Pareto diagram categories: Open digit means any digit that does not light up. Wrong sequence means that the displays fail to exhibit the correct sequence per the product specifications.

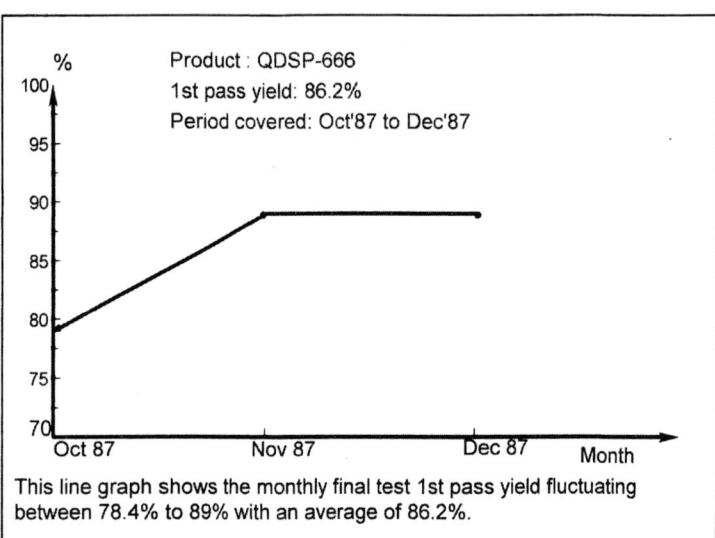

This line graph shows the monthly final test 1st pass yield fluctuating between 78.4% to 89% with an average of 86.2%.

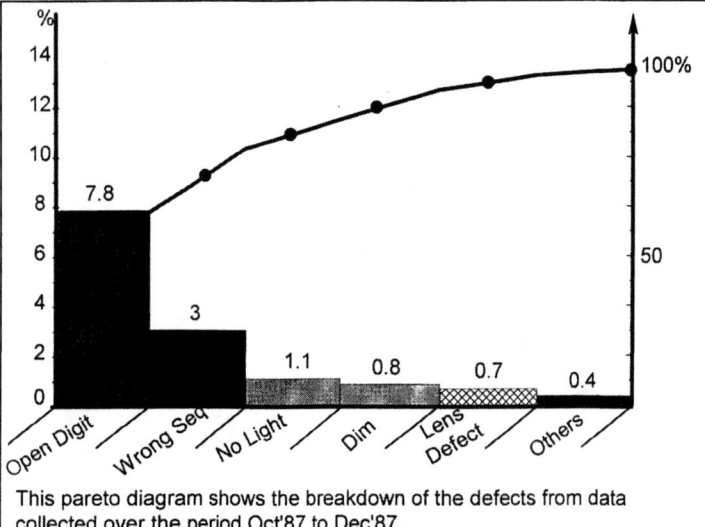

This pareto diagram shows the breakdown of the defects from data collected over the period Oct'87 to Dec'87.

Figure 4.4 The current status of display assembly failures.

Example of Step 3: Analyze the Cause and Determine Corrective Action

Step 3a: Prepare cause-and-effect diagram

The two highest fail categories in the Pareto diagram—open digits and wrong sequence—were selected, and a cause-and-effect diagram was prepared for each item. The group members then brainstormed for all possible causes of each item, by using the questioning techniques of the four Ws (what, where, when, why) and one H (how). The cause-and-effect diagrams are shown in Fig. 4.5.

Step 3: Analyze the cause and determine corrective action

- *Objective of this step.* To find out the root cause of the problem and plan for corrective action.

- *Discussion.* In this step we examine the causes of the problem, isolate the root causes, and determine corrective action. The detailed substeps (3a, 3b, and 3c) follow:

  Step 3a: Prepare cause-and-effect diagram. The item to examine is selected. This can be the first or second bar in a Pareto diagram of defects, or it can be the first two or three bars. At other times we may select a specific item we want to improve. A cause-and-effect diagram is then prepared. All the causes in the cause-and-effect diagram can be derived through a brainstorming session. Possible and impossible causes should be listed. Now use data obtained in step 2, to eliminate unlikely causes. The cause-and-effect diagram can now be simplified and redrawn.

- *QC tools that can be useful.* Pareto diagram, cause-and-effect diagram.

  Step 3b: Prepare a hypothesis and verify most likely cause
- *Discussion.* We prepare a hypothesis by selecting the most likely causes of the cause-and-effect diagram; this can be done by using the group's experience or voting. This list of most likely causes must now be verified with data. But we should use new data to determine whether there is a relationship between the selected causes and the effect; this may require us to conduct experiments. We will now have a short list of verified causes—the root causes of the problem we want to reduce or eliminate.

- *QC tools that can be useful.* Check sheet, stratification, statistical design of experiments.

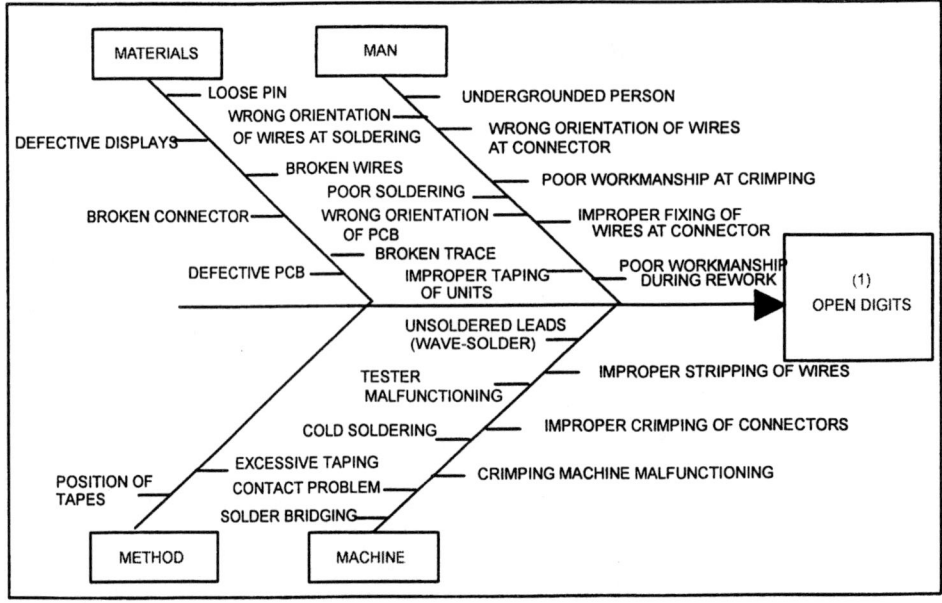

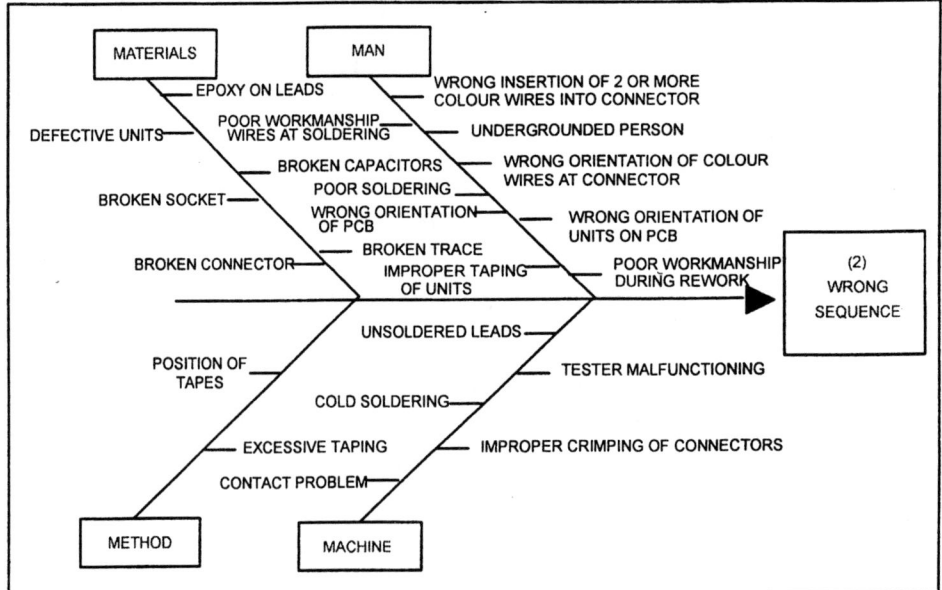

Figure 4.5 Cause-and-effect diagrams for open digits and wrong sequence.

Example of Step 3b:
Prepare a Hypothesis and Verify Most Likely Cause

Members deliberated on the possible causes and then voted for the most probable causes. The basis for the selection was the operators' job experience and knowledge. They are as follows:

1. Open digits
   a. Wrong orientation of wires inserted into connector
   b. Improper fixing of wires into connector
   c. Broken wires
   d. Unsoldered leads
   e. Defective units
2. Wrong sequence
   a. Wrong orientation of wires inserted into connector
   b. Wrong orientation of wires at soldering
   c. Improper crimping of connectors

Verification:

Additional failure data were collected in order to verify which of the possible causes were the most likely causes. The analysis findings were then summarized in Fig. 4.6. Members also noticed that most of the causes are workmanship-related. The most probable causes were verified to be

1. Wrong orientation of wires inserted into connector
2. Wrong orientation of wires at soldering
3. Improper fixing of wires to connector
4. Unsoldered leads

TYPE OF DEFECT	Open Digit	Wrong Sequence	Open Digit & Wrong Seq	Total
Wrong orientation of wires at connector	16 (53.3%)	13 (76.5%)	4 (100%)	33 (64.7%)
Wrong orientation of wires at soldering	4 (13.5%)	4 (23.5%)		8 (15.7%)
Improper fixing of wires to connector (Wires came out loose)	3 (10.0%)			3 (5.9%)
Unsoldered leads	4 (13.3%)			4 (7.8%)
Broken wires	1 (3.3%)			1 (1.96%)
Defective unit	1 (3.3%)			1 (1.96%)
Improper crimping of connector	1 (3.3%)			1 (1.96)
QUANTITY ANALYZED	30 Sets	17 Sets	4 Sets	51 Sets

Figure 4.6 Verification of most likely causes.

Step 3c: Determine corrective action

- *Discussion.* We decide on the corrective action. Sometimes the corrective action is obvious. If not, we need to decide on the action. Creative alternatives should be generated by using brainstorming or cause-and-effect diagrams. There will usually be two types of corrective action:

  A quick fix or remedial action. This could include inspection for the defect or repair of the defect.

  A long-term fix or preventive action. This could include elimination of root cause, hence preventing the problem from recurring. This is more important, but due to constraints the quick fix may be implemented first.

  It may be necessary to conduct a trial of the proposed action, to determine that it works. Only then should it be proposed.

- *QC tools that can be useful.* Check sheet, checklist.

Example of Step 3c: Determine Corrective Action

Since the four major causes were workmanship-related, the members attempted to list the snags and difficulties in the processes that are related to each of the causes. Members then suggested and discussed what could be good solutions or preventive actions for each. The discussion is summarized in Fig. 4.7.

Members then discussed the requirements for each of the work holders and soldering block, and had a volunteer draft out the preliminary drawings. A final meeting was held with the vendor, after which the vendor finalized the drawing, selected the material, and fabricated the respective fixtures.

Do stage

Step 4: Implement corrective action

- *Objective of this step.* To implement the plan and eliminate the root causes of the problem.
- *Discussion.* Employees who execute the correction must understand the corrective action. Good communication and training will be necessary. The following are recommended substeps:

  Step 4a: Prepare instructions and flowcharts for procedures that are complicated.

  Step 4b: Adequate training must be provided.

  Step 4c: Follow the plan exactly.

	CAUSES		SNAGS AND DIFFICULTIES		RECOMMENDED ACTIONS
1.	Wrong orientation of wires inserted into connector	a)	Insertion of the 16 wires in their correct sequence into the connector was done with the aid of the reference chart and by memory.	a)	Design a workholder for the connector with the appropriate colour of the 16 wires painted on it to act as guide for correct insertion of the wires.
2.	Wrong orientation of wires at soldering onto PCs.	a) b)	Plastic bags were used to store the 16 different colours. Soldering of the 16 wires in their correct sequence into the PCB was done with the aid of the reference chart and by memory.	a) b)	Design an appropriate workholder at soldering for the storage of the 16 wire types that arranges them in the appropriate sequence. Modify the soldering block to include colour guides for the 16 wires.
3.	Improper fixing of wires into the connector (wires came out)	a)	Each crimped wire should be fully inserted into the connector until there is a "click" sound.	a)	Implement a "pull test" at wire insertion for every wire to ensure that each crimp is correctly seated in the respective sockets.
4.	Unsoldered leads.	a)	There is limited space on the PCB for tape adhesion when taping the display components for wave solder.	a)	Implement 100% inspection on all taped units before wave soldering to ensure that all the corner leads are not covered with tapes so that all leads will be soldered during wave soldering.

Figure 4.7   Determine corrective action.

Step 4d: Record any deviations from plan and collect data on results.

- *QC tools that can be useful.*   Checklist, check sheet, trend charts.

Example of Step 4: Implement Corrective Action

A training session was held for all production staff members to educate them on how to use the fabricated work holder and the soldering block. The recommended action was then implemented in workweek 22, and results were tracked everyday. The results were tracked on the graph shown in Fig. 4.8.

Check stage

Step 5: Check the effect of corrective action

- *Objective of this step.*   To check the effectiveness of the corrective action.

- *Discussion.*   We now check the effect of the corrective action. There are several substeps that must be followed:

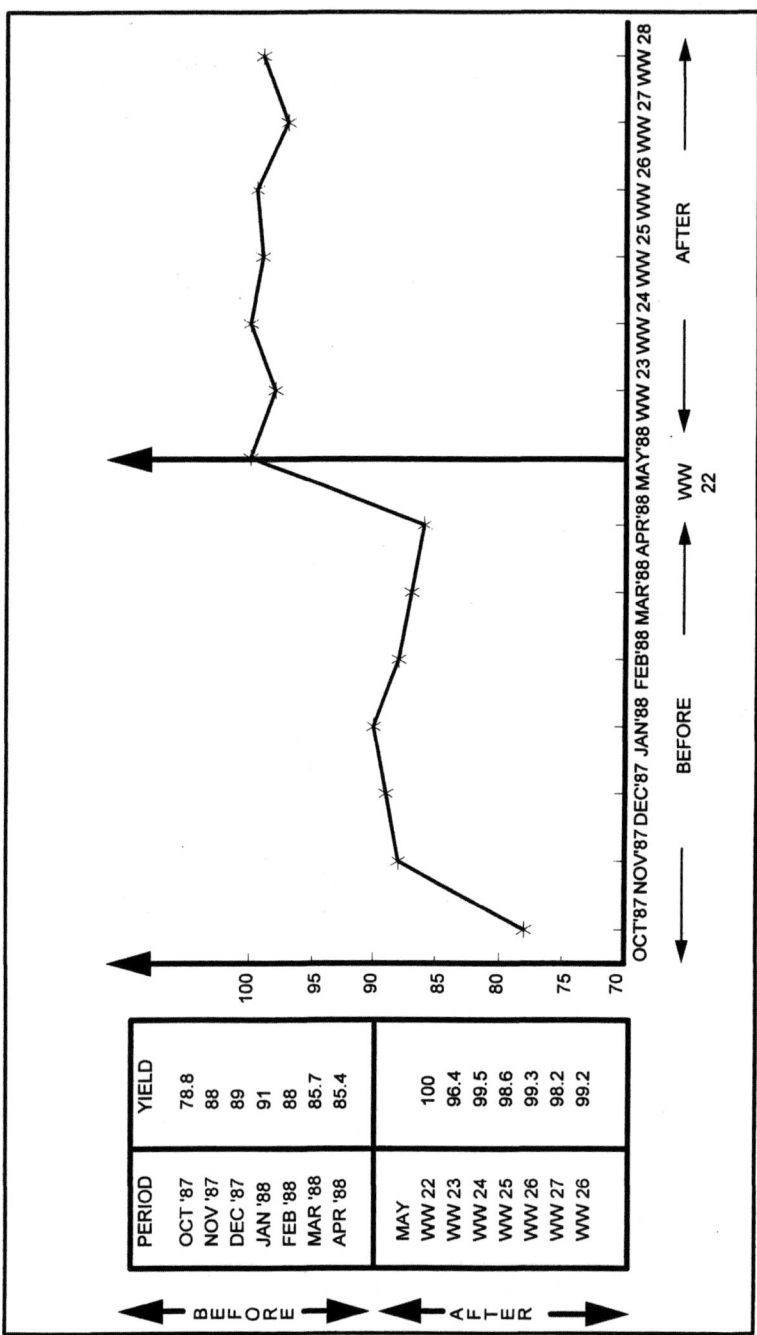

Figure 4.8 Production line yield for display assemblies.

Step 5a: Compare overall result. Here we review the overall results. We must also review improvements on a paired Pareto diagram in order to compare before-and-after performance. Before-and-after results should be compared on all other items selected for study in step 2; use the same tools for making the comparison, such as bar graphs, paired Pareto diagram, trend chart, control charts, histograms, and process capability indices.

Step 5b: Failure to meet results. If failure is due to improper implementation, then we must go back to step 4, implementation. Otherwise, we go back to step 3, analysis. If we fail to meet our goals, it is very likely that we missed the root causes, and further analysis will be required.

Step 5c: Results have been achieved, goal has been met. If the overall result is equal to or better than the target set in step 1, we review the before-and-after data—especially the Pareto diagrams—and check that there are no side effects, that is, that there is no increase in the other categories of failures.

- *QC tools that can be useful.*   Paired Pareto diagram, trend charts, control charts, histograms, and process capability indices.

Example of Step 5: Check Effects of Corrective Action

A detailed analysis was done. The data were then plotted on a paired Pareto diagram. The results show a marked improvement in open digits. A check was made for unwanted side effects. There were none—in fact, some of the other Pareto bars also decreased (see Fig. 4.9).

Act stage

Step 6: Take appropriate action

- *Objective of this step.*   To ensure that the improved level of performance is maintained.

- *Discussion.*   The corrective action that has been successful in improving performance must be documented in current operating procedures. There are several substeps.

Step 6a: Documentation, standardization, and control. The corrective action (implemented in step 4) that has been successful in improving the performance level should be documented in current operating procedures or standards. Poor documentation can result in problem recurrence in the future. It is very important to convey this information to other parts of the organization, which may have generated the root cause of this problem. Refer to the section "Standards Update Request Format" later in this chapter.

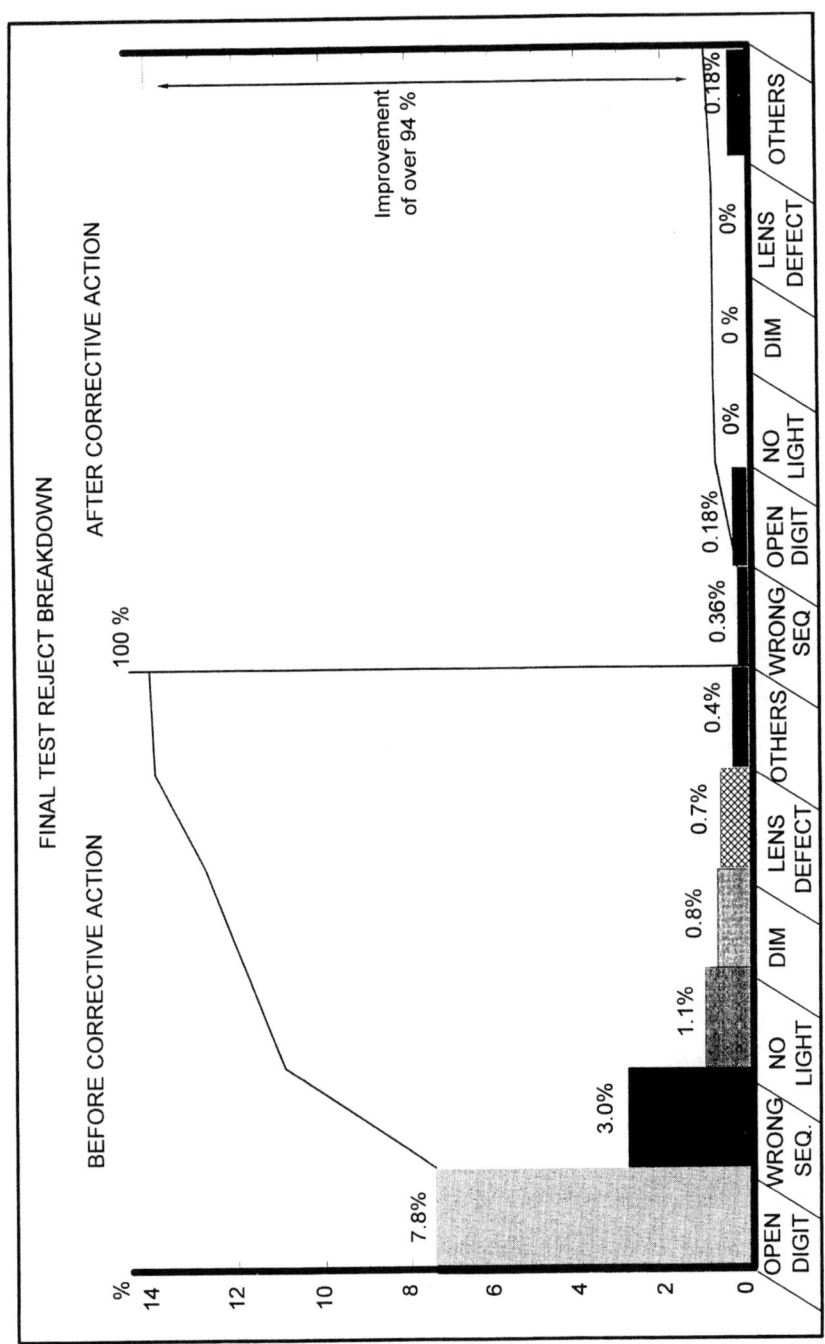

Figure 4.9 Paired Pareto diagram of before and after corrective action.

Example of Step 6: Take Appropriate Action

The team then recommended to the process engineer that use of the respective fixture and the other corrective action steps be incorporated into the process specification. The results were conveyed to all employees, and the importance of the new procedures was emphasized.

It is also important to identify critical process parameters to control. The idea is to monitor the appropriate parameter and to detect any deviation from the new standards. Deviations should be promptly analyzed and eliminated.

Step 6b: Training. Ensure appropriate training in the new methods and standards. Employees must fully understand the changes that have taken place and the new procedures.

- *QC tools that can be useful.* Trend chart, control chart, check sheet.

Step 7: Decide on future plans

- *Objective of this step.* To use the experience gained for future projects.
- *Discussion.* The best area to look for a new project is from the results in step 5. If our new trend chart or Pareto diagram has sharp peaks, we must eliminate them. If the bars in our new Pareto diagram have even heights, we must change our stratification base and then decide what to eliminate. If we have created side effects, then we must work on eliminating them. We may also start afresh with a new breakthrough activity. But, first, we make sure that the entire process we have been through is documented according to the seven steps listed here. We make the recommendation because the documented project will be a good learning tool for new employees, and we provide a historical record of improvements. The decision to continue on the current project or to select a new one has to be based on priorities and resources.

Example of Step 7: Future Plans

The team's future plans are:
1. To continue monitoring of the first-pass yield of QDSP-6666
2. To move onto the third project
3. To achieve a 100 percent direct labor participation in QCC activities for our line (present status is 90 percent participation)
4. To extend the corrective action of this project to all cable assembly-type devices in the production line

---

Seven Quality Control Tools and Other Methodologies

The seven tools of quality control are relatively well known:

- Data collection check sheets and checklists
- Pareto diagram
- Cause-and-effect diagram
- Stratification
- Graphs and histograms
- Scatter diagrams
- Control charts

More details on each of these tools are given in App. B. In Japan, a heavy emphasis is placed on education in and use of the seven tools. Often an analogy is made between these seven quality control tools and the seven tools used by the Samurai warriors of ancient Japan.

The Samurai warrior had seven tools, a sword, a helmet, bow and arrow, and so on; he never ventured anywhere without these tools, which he needed for protection and success. In a similar vein, the seven quality control tools are essential for today's workers, engineers, professionals, and managers. They are needed for extracting information from data, conducting a proper analysis, and making correct decisions.

According to Kaoru Ishikawa, about 95 percent of the problems in the workplace can be solved using these tools. We have done an analysis at one Deming Prize–winning company—Yokogawa Hewlett-Packard—and are able to verify Ishikawa's statement. For the remaining, more difficult problems, you would use the seven new management tools, design of experiments, Taguchi methods, and so on. But you would use them within the context of the PDCA cycle. Some details of the seven quality control tools and the seven new management tools are given in App. B. For others, the recommended readings in the Bibliography provide a guide.

Education of Employees

One of the important steps in the PDCA cycle is that of educating employees in the new improved process. This is crucial; otherwise the improvements cannot be maintained or the process controlled. This must be done quickly and efficiently every time.

We read with astonishment a remark made by the president of Boeing Aircraft:[6]

Rate of Improvement and Setting Targets

What is a good rate of improvement? How aggressive should improvement targets be? For over 10 years we have come across several quality teams that use a rule of thumb of a 50 percent reduction in defects for each project, but we have never been able to trace the source of this rule.

A recent article—"Setting Quality Goals" by Schneiderman in *Quality Progress,* April 1988—sheds some new light on this question. Empirical evidence suggests that most improvements (more specifically reduction in defect levels) can be made at the rate of 50 percent in a very narrow time range of 6 to 9 months. Reduction of defects in an autonomous environment such as manufacturing or administration takes a little less time. Reduction of defects in a complex environment such as between a factory and its supplier may take longer. The data show that the rate of improvement is independent of cumulative volume—the learning curve—but is dependent only on time. The reason for this phenomenon is simple. Typically a few causes, the first few items on the Pareto diagram of defects, are responsible for a majority of the defects. Hence eliminating the first few items will quickly reduce defects by 50 percent or more. This cycle can be repeated several times.

As an initial rule of thumb, plan to set a target of 50 percent reduction in defects every 6 to 9 months. By defects we mean waste, scrap, inventory levels, time to do something, and most other undesirable items. Once you set a target based on this simple rule, plan to deviate from it only if data suggest you do so.

After an unfortunate air crash, the error was traced to some workers' mistakes after a repair....When the president of Boeing's Seattle plant was asked, "How long will it take after re-education has begun before the technological strength (of your company) will begin bearing fruit?" his answer was seven years. Seven years! How can we ride around in jumbo jets for seven years not knowing what types of defects they might have?"

We hope this is an error. The comments here seem to reflect the different attitudes, between U.S. and Japanese managers, toward problem solving. In one case a more leisurely pace seems possible; in the other, immediate attention and no recurrence seem to be the rule.

As we get continuous improvement, the process gets changed and is redocumented, and the workforce needs reeducation. Due to growth or attrition, the workforce needs continuous training—otherwise defects may increase. See the attached box for a discussion of inadequate worker education and work standards.

Source of Errors and Defects

What causes errors and defects? "The root cause is incomplete knowledge or imperfect work," says Hitoshi Kume. This causes problems everywhere, in every company, every country, and individuals. Errors and defects will always occur. It is your job to minimize this occur-

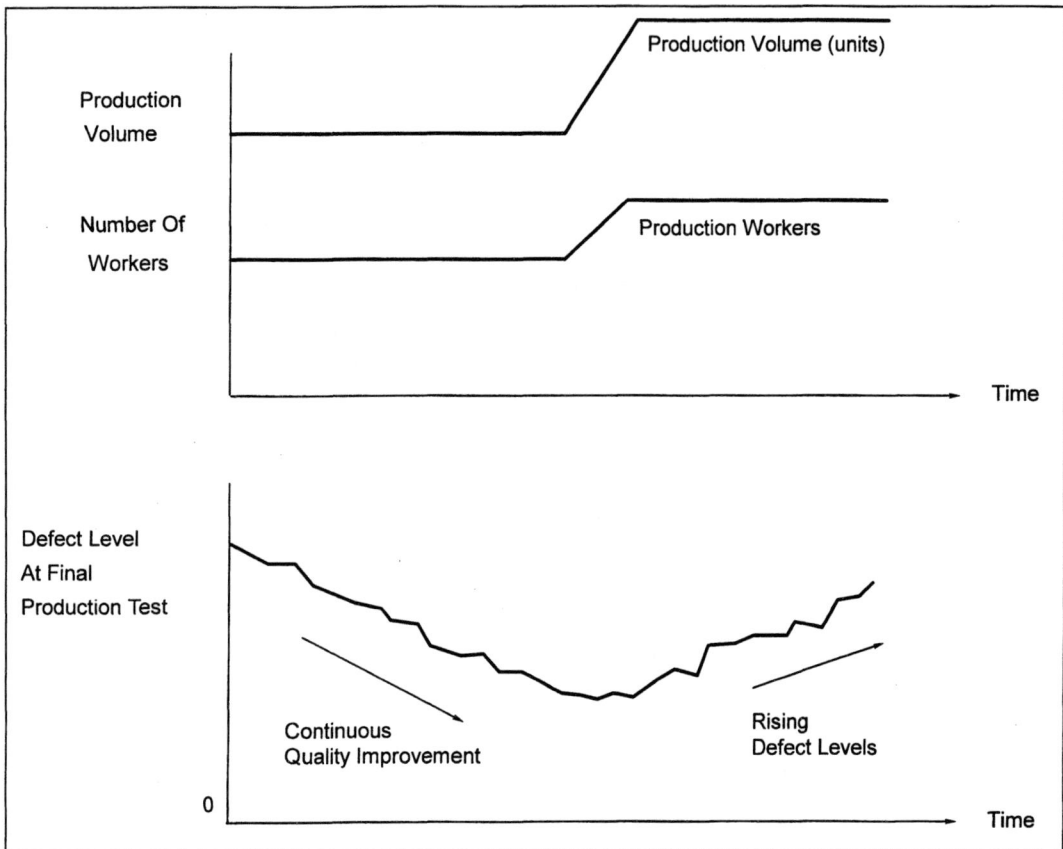

Figure 4.10   The effect of poor training and documentation on defect levels when production increases.

Inadequate Worker Education and Work Standards

What happens when workers are poorly educated or trained? Or if work standards or procedures are inadequate?

Figure 4.10 illustrates a real case of what happened when production volume picked up and new workers had to be hired. The new workers were inadequately trained or unaware of informal assembly procedures. The effect, in this case, was that all the results of continuous quality improvement were destroyed due to inadequate work standards and training. Hence, what is required is "living" documentation—documentation that changes as the process changes—and continuous training of workers and new workers. The same situation will occur in a service environment. Inadequate training will cause problems when customer transaction volume increases or employee turnover occurs.

rence. The introduction of standards will help reduce errors. This is the focus of the discussion in this chapter and the next, "Process Management."

Human errors

As you keep improving a process, each successive improvement becomes more difficult. One of the fruitful areas to work on is the reduction of human errors. An effective technique to reduce human error is called *foolproofing*—this ensures that smart machines which reject human misuse are employed. But you cannot always resort to foolproofing. You need to check causes of human error and determine whether it was lack of training, poor training, boredom, and so on. A good way to proceed is to use a human error troubleshooting tree to determine the source of the error. Figure 4.11, which was prepared by Katsu Yoshimoto of Yokogawa Hewlett-Packard,[7] provides some help. You will find it extremely useful as you attempt to get the last iota of improvement from a process.

One of the keys is to set up a foolproof system. Many Japanese feel that the term *foolproofing* could be offensive to some, hence Shigeo Shingo of Toyota Company coined the term *poka-yoke,* which means *mistake-proofing* or *fail-safeing.*[8] Implementing poka-yoke is an effective way to reduce human error and move toward the goal of zero defects.

Where Can You Use the PDCA Cycle?

You can use it for all problem-solving and improvement projects. Examples include improving your own work, improving product quality, improving service, improving marketing and sales activity, and reducing costs. The list is endless. Professor Iizuka of the University of Tokyo[9] has suggested four types of problem classification:

1. Reduction of existing defects.

2. Improvement of a system, such as a quality assurance system or manufacturing process.

3. Acquisition of new knowledge, such as optimal conditions in research and development or design improvement.

4. Introduction or construction of a new system or drastic improvement of an existing system.

For an entirely new system—a product or service breakthrough— we will propose a proactive PDCA, which is discussed later in this chapter.

PROBLEM SYMPTOM/CAUSE	POSSIBLE COUNTER MEASURES	
	TENTATIVE FIX	PERMANENT FIX OF THE SYSTEM
HUMAN ERROR OCCURS		
THERE IS NO DOCUMENTED PROCEDURE		
THERE IS LITTLE MANAGEMENT ATTENTION	HELP THE MANAGERS GET ATTENTION	TRAIN THE MANAGERS ... ESTABLISH/REVISE TRAINING MECHANISM
THIS IS LOW PRIORITY FOR MANAGEMENT	ANALYZE DATA AND RE-SET PRIORITIES	ADDRESS IN HOSHIN PLAN } ESTABLISH/REVISE DIVISION-WIDE QUALITY SYSTEM
INCONSISTENT PROCESS	GET CONSISTENT PROCESS	
THERE IS A DOCUMENTED PROCEDURE		
THE WORKER DID NOT FOLLOW THE PROCEDURE		
WORKER DID NOT KNOW THERE WAS A PROCEDURE		
WORKER WAS NOT TRAINED IN THE PROCEDURE	TRAIN THE WORKER	PLAN PERIODICAL ... ESTABLISH/REVISE TRAINING MECHANISM TRAINING
WORKER FORGOT THE PROCEDURE	RE-TRAIN THE WORKER	
WORKER KNEW THERE WAS A PROCEDURE		
WORKER DID NOT LIKE TO FOLLOW THE PROCEDURE		
THE PROCEDURE WAS DIFFICULT TO FOLLOW		
IT TOOK TOO MUCH TIME	RE-TRAIN THE WORKER	REVISE THE STANDARD FOR GENERATING PROCEDURES
IT REQUIRED DIFFICULT WORK		
IT REQUIRED DIFFICULT JUDGMENT	TEACH WORKER IMPORTANCE OF THE PROCEDURE	
PSYCHOLOGICAL PROBLEM AFFECTED WORKER	MANAGE THE WORKER	SET UP FOOL PROOF.... REVISE DESIGN-REVIEW DOCUMENTS AND DESIGN STANDARD DOCUMENTS
WORKER MISUNDERSTOOD THE PROCEDURE		
THE PROCEDURE WAS NOT EASY TO UNDERSTAND		
IT HAD TOO MANY WORDS, COMPLEX DRAWINGS ...		
IT HAD MIXTURE OF SENTENCES, DRAWINGS	RE-TRAIN THE WORKER	REVISE THE STANDARD FOR GENERATING PROCEDURES
IT HAD TOO SMALL, CROWDED CHARACTERS }		
PSYCHOLOGICAL STIMULUS AFFECTED WORKER	MANAGE THE WORKER	SET UP FOOL PROOF.... REVISE DESIGN-REVIEW DOCUMENTS AND DESIGN STANDARD DOCUMENTS
THE WORKER FOLLOWED THE PROCEDURE		
THE PROCEDURE WAS INCOMPLETE		
NECESSARY CONTROL ITEM WAS NOT INCLUDED	REVISE THE PROCEDURE	REVISE THE STANDARD FOR GENERATING PROCEDURES
CONTROL LIMIT WAS NOT ADEQUATE }		
THE WORKER MISUNDERSTOOD IDENTIFICATION		
IDENTIFICATION WAS NOT EASY TO UNDERSTAND	REVISE THE IDENTIFICATION	REVISE THE STANDARD FOR IDENTIFICATION
*IDENTIFICATION ON WORK-ORDER, MATERIAL LOCATION, ETC.		
PSYCHOLOGICAL STIMULUS AFFECTED WORKER	MANAGE THE WORKER	SET UP FOOL PROOF.... REVISE DESIGN-REVIEW DOCUMENTS AND DESIGN STANDARD DOCUMENTS

Figure 4.11 Problem solution model for human error.

We add a fifth item:

5. Managing an ongoing system, such as the Hoshin planning cycle. In the Hoshin planning cycle we need to plan, implement strategies and tactics, check progress, and take corrective action when problems occur. These are steps in the PDCA cycle. Look at Fig. 3.6; it is in PDCA format.

When you construct a new system (item 4 above), there is no need to have too precise an analysis. Since the goal is to have a new system or to drastically improve a current system, according to Professor Iizuka, you should recognize existing deficiencies and agree on what you wish to achieve. To use the PDCA cycle for new ideas or systems—a product or service breakthrough—we will propose a proactive PDCA cycle, later in this chapter.

Let us elaborate on the PDCA cycle for managing an ongoing system (item 5 above). In Fig. 3.11 we show the Hoshin objective to increase profits. That objective was deployed (in part) to the operation (or manufacturing) manager. In turn, that was deployed to the operation manager's staff with the objective of reducing costs and failure rates. All that activity forms a vast PDCA cycle that needs good management; we illustrate that in Fig. 4.12.

On the left of the figure the steps of the PDCA cycle are listed as they apply to the objective, which was to increase profits. The operation manager adopted two strategies for achieving this; the strategies were to decrease failures and reduce costs. Figure 4.12 shows how these strategies were deployed in the manufacturing operation and how results were monitored. The targets and the current status for each department are displayed on bar charts; and progress is measured regularly.

This is a modification of Komatsu's flag system,[10] which is used to manage their cost reduction programs. This is the detailed implementation plan of a Hoshin plan strategy and is a very powerful tool, giving impressive results. Komatsu has used this methodology, charts and all, as an effective management tool. The term *flag system* comes from the target versus actual bars which are displayed on a large chart or flag. These flags with corresponding plans are displayed and reviewed regularly. This helps to raise awareness and commitment in achieving Komatsu's cost reduction program.

Example in PDCA format

During our discussions on the detailed PDCA cycle, we showed an example from a manufacturing environment. In App. A, we provide an example from a sales environment—a follow-up to a customer satisfaction survey. Both examples follow the PDCA cycle steps discussed earlier. Study them and note how systematic the process is—

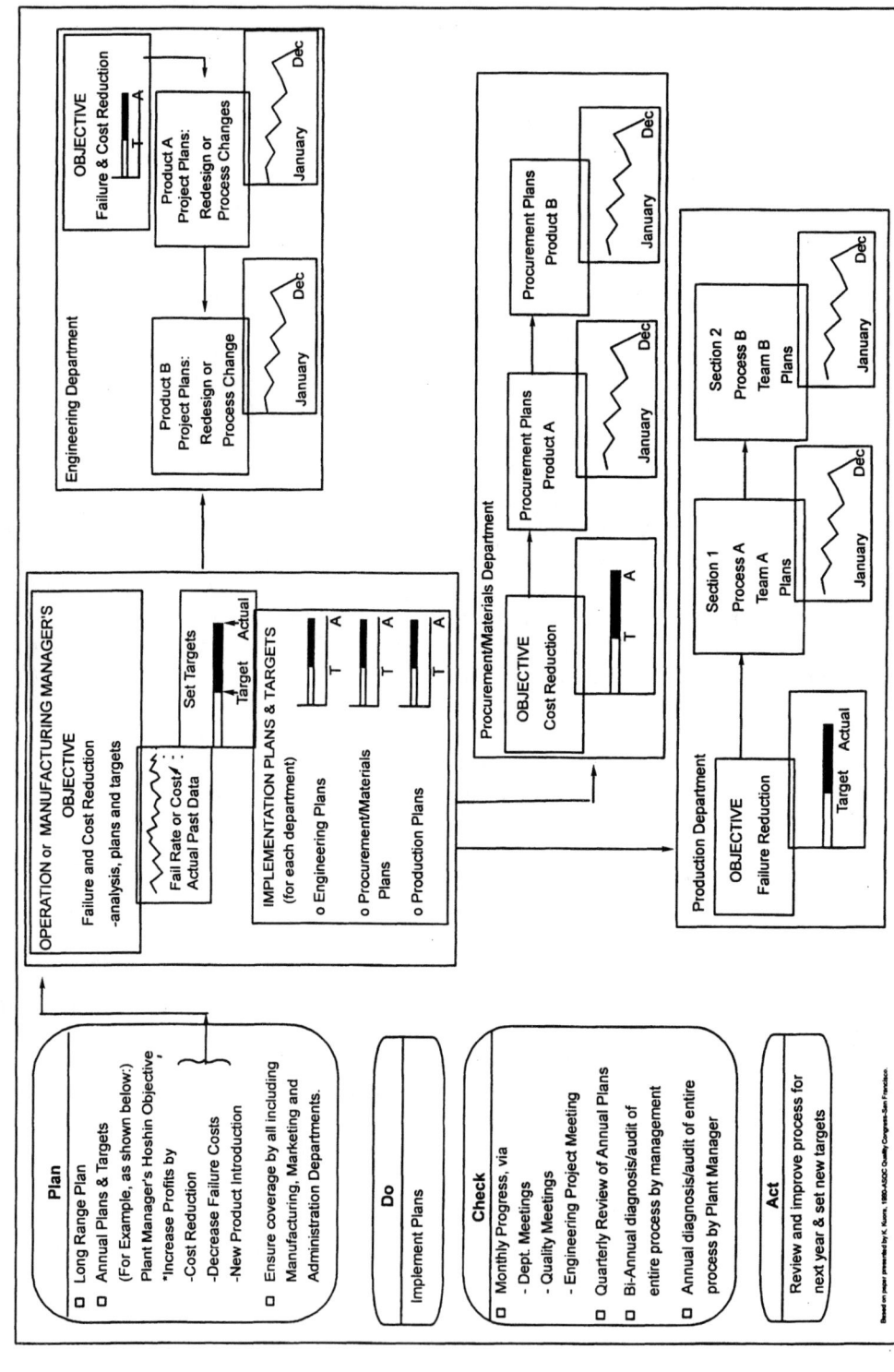

Figure 4.12 Using the PDCA cycle for management of an improvement plan.

how the current situation is analyzed, how the hypothesis is generated, and then how the hypothesis is validated. Only then is a solution prepared. There is no fire fighting, no jumping to conclusions, no intuitive or genius approach.

We show in Fig. 4.13 the results of a major improvement project. This is a summary of the Yokogawa Hewlett-Packard (YHP) solder wave project.[11] This project has become a legend in the quality community, and it has been shown countless times within and without the Hewlett-Packard Company. There are two interesting points to this project, quoted by Kenzo Sasaoka, president of Yokogawa Hewlett-Packard:

- Because of the success, "the soldering process became the catalyst of the quality movement" at YHP.

- The soldering machine used was not state of the art, but instead "a jalopy" transferred from another plant more than a decade before, "which never fails to astound" every visitor.

Notice that YHP went through three major improvement cycles and was not satisfied until it reached a failure rate of 4 parts per million. That is equivalent to finding 4000 specific people in China's population of 1 billion. Think about that. More important, when would you have been satisfied? After the first or second cycle? Or would you have gone on and on as YHP did?

Standards

We discussed the need for having a standard improvement methodology—the PDCA cycle. Everybody should use the same improvement methodology in the organization. This concept is similar to that of having standards in all areas of commerce and industry—standard electric sockets, standard material for building a house, and so on. This makes shopping, planning, and work so much easier. Similarly, we need standards in the quality environment.

Standards represent proven best practices that are institutionalized in an organization. All employees must be trained to understand and use the standards that are relevant to them. Good standards will save time as employees will not need to reinvent the wheel in other parts of the organization. Instead they can use their creative abilities to invent things in new, unexplored areas. This will result in optimum use of your most valuable resource—all your employees.

Benefits of standards

Let us restate the benefits of standards or proven best practices:

- Standards minimize the need for new resources and ideas—just use a standard form or a standard procedure across an entire organization.

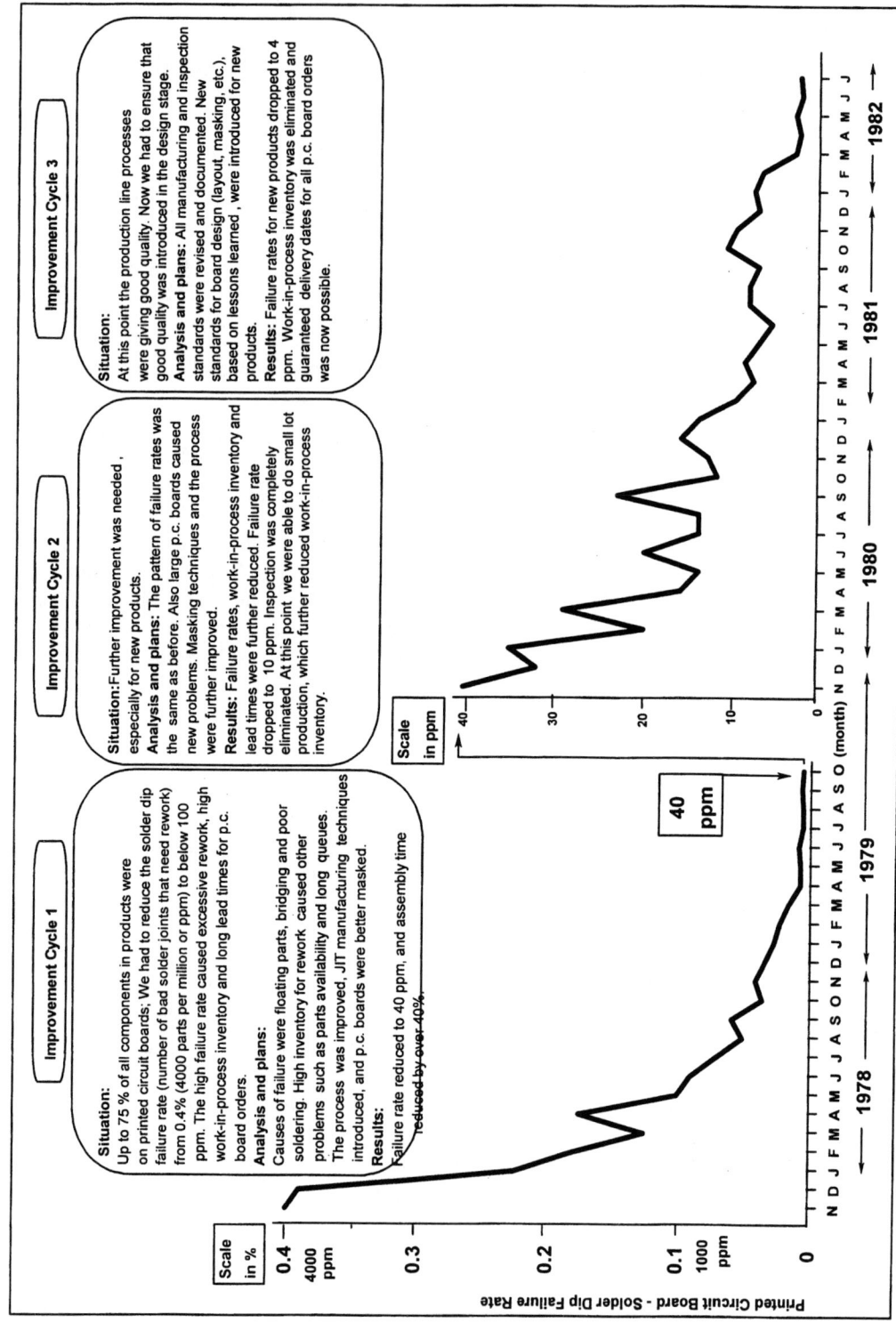

Figure 4.13 Improvement of printed circuit production process.

- Standards minimize human errors and other types of defects, thus saving time and reducing rework.
- Standards make it easier to communicate ideas and information.
- Standards, when established, allow time for creativity in other new areas; otherwise time is spent in:

Slower problem resolution, if the standardized PDCA improvement cycle is not used

Redesigning where there are no design standards, for example, component or reliability standards

Retraining where training is poor and not documented, for example, operator or designer training

Longer time to learn when managers or employees move to different locations in an organization using different standards for planning, problem solving, designing, and so on

Important points for creating standards

The following pointers will be useful when standards are created and maintained:

- Create standards which can be followed.
- Revise standards which are difficult to follow.
- When training workers to follow standards or procedures, consider their receptivity and ability to follow the standards.
- Observe whether standards are being followed, and understand why workers do not follow them.
- The standards must be easily accessible by designers, engineers, buyers, and other company employees. There is more on this when we discuss standards update requests in this section.

Preventing recurrence

One of the steps in the PDCA cycle is to standardize, control, and document the new improved process. How important is this? Extremely. Let us illustrate the importance with a true story:

Recently, I was reviewing a manufacturing operation by conducting a Total Quality Review. One of the improvement projects they showed me was entitled "Reduction of scratches in display modules." They manufactured and shipped display modules for instruments and aircraft cockpit displays. These got scratched during packaging, when they were inserted in shipping tubes. As a result of improvements they were able to reduce scratched displays to zero. Very impressive.

But wait a moment. I have a long memory; and I remarked that many years ago—maybe 6—I had reviewed a similar project for a different product. Then, as now, there was a project to reduce scratched displays, caused by shipping tubes. Why had they not learned from their errors? Why had they not conveyed the lessons learned to the designers of the shipping tubes? I told them: "If you don't learn from your errors, if you don't convey the lessons learned into new standards, then you are doomed forever to have projects to reduce scratches in display modules."

Does this sound familiar? Every company has a similar situation. The outcome is problems that keep recurring throughout the company, and improvement projects that are repeated, resulting in lower productivity and customer dissatisfaction.

The solution is to convey the lessons learned into standards or to upgrade existing standards. How do you upgrade existing standards? We propose the use of a standards update request.

Standards update request format.   Often, there will be a need to update standards—especially design standards. This must be done via a formal process rather than a verbal request or complaint. In the case of the customer complaint and feedback system, we suggested the use of a corrective-action request (CAR) to formally drive the change in the organization. Here we propose the use of a standards update request (SUR) format that can be used to drive the change. The format used can be on paper (in the beginning) or via an electronically transmitted medium. We show a recommended and completed format in Fig. 4.14.

In the figure, the proposal to update the standard is conveyed to the standards owner in the company. This could be an R&D standards manager, a specification control manager in the purchasing or materials department, and so on. The manager must ensure that the standard is updated by modifying design software or changing drawings and specifications. In all cases the new standard must be documented and filed—preferably in a computer file. The use of a keyword—as recommended in the SUR format—will allow easy access for reviewing design standards by designers, buyers, and other company employees.

Sources for standard updates are as follows:

The improvement cycle.   When an improvement project is completed and if an improvement in a process or design is proposed.

Customer complaints and feedback.   When customer complaints are received, problems are analyzed and an improvement in design is made. We show an example of this in the attached and completed SUR format in Fig. 4.14.

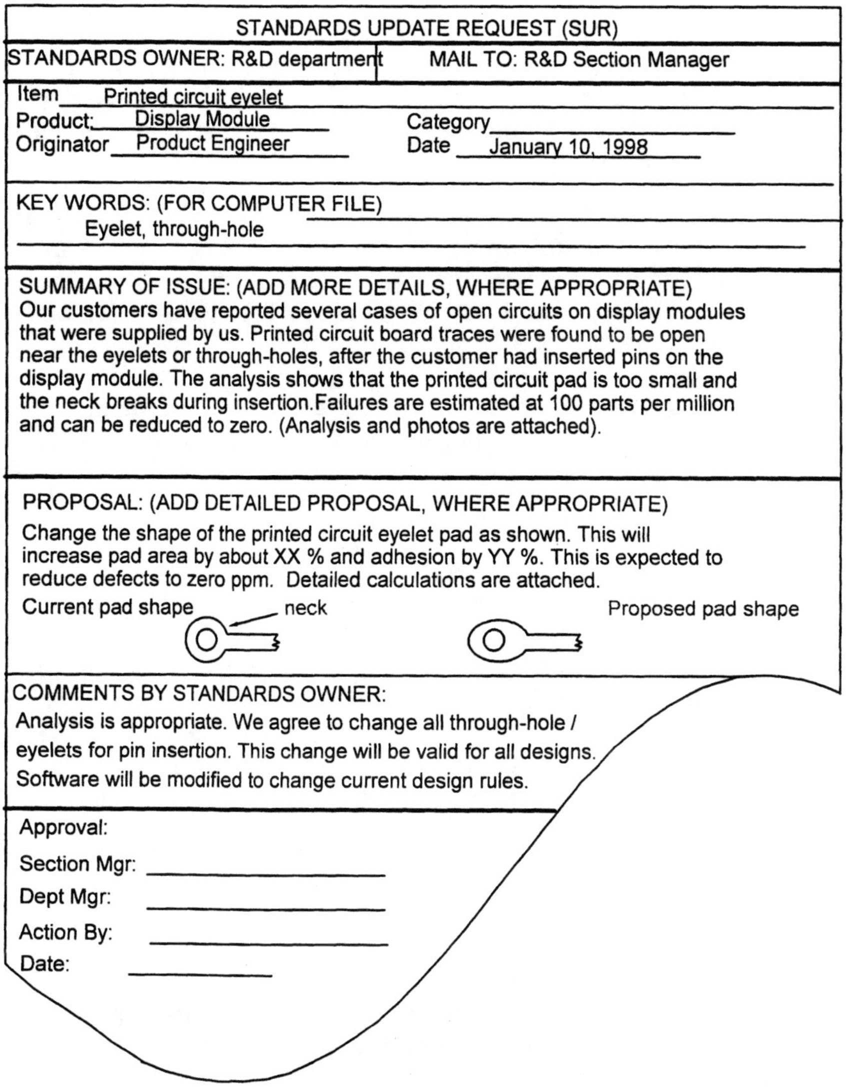

STANDARDS UPDATE REQUEST (SUR)

STANDARDS OWNER: R&D department MAIL TO: R&D Section Manager

Item Printed circuit eyelet

Product: Display Module Category

Originator Product Engineer Date January 10, 1998

KEY WORDS: (FOR COMPUTER FILE)

Eyelet, through-hole

SUMMARY OF ISSUE: (ADD MORE DETAILS, WHERE APPROPRIATE)

Our customers have reported several cases of open circuits on display modules that were supplied by us. Printed circuit board traces were found to be open near the eyelets or through-holes, after the customer had inserted pins on the display module. The analysis shows that the printed circuit pad is too small and the neck breaks during insertion.Failures are estimated at 100 parts per million and can be reduced to zero. (Analysis and photos are attached).

PROPOSAL: (ADD DETAILED PROPOSAL, WHERE APPROPRIATE)

Change the shape of the printed circuit eyelet pad as shown. This will increase pad area by about XX % and adhesion by YY %. This is expected to reduce defects to zero ppm. Detailed calculations are attached.

Current pad shape neck Proposed pad shape

COMMENTS BY STANDARDS OWNER:

Analysis is appropriate. We agree to change all through-hole / eyelets for pin insertion. This change will be valid for all designs. Software will be modified to change current design rules.

Approval:

Section Mgr:

Dept Mgr:

Action By:

Date:

Figure 4.14 A standards update request format.

Daily process management. As processes are managed, problems and errors may be discovered. For example, during a project post-mortem, weaknesses in the original design may be discussed and a SUR generated.

This concept of standards will be revisited in the next chapter, "Process Management." There we focus on standard processes—process that are well understood and documented, processes that stay even as managers and employees come and go.

Problem-Solving Hierarchy

As we have mentioned, problems and errors will always occur. Our goal, then, is to minimize problem and error occurrence. As we decrease our errors and eliminate problems, we go through the following phases or hierarchy:

1. *A fire-fighting mode.*   Here problems occur helter-skelter throughout the organization, and we fix them on an ad hoc, nonsystematic basis. Everything is disorganized. Often, the bearer of bad news is punished. The person who fixes the problem and douses the fire is rewarded and considered a hero; but no attempt is made to understand why the fire occurred, or why there are so many fires.

2. *Systematic problem-solving mode.*   Here we are in control, fixing problems systematically, eliminating root causes, and operating in an environment of continuous improvement. The PDCA cycle is used constantly and consistently; systems are in place to detect problems and solve them quickly.

3. *Prevention of problems by prediction mode.*   Here we go beyond systematic problem solving; instead, we predict and prevent potential problems from occurring. We are able to understand customer needs before we design a product or service—this will prevent low customer acceptance and subsequent design changes. Also, before the release of a product or service, we are able to predict potential design problems and eliminate them. Some tools that will allow us to do this are the *proactive PDCA* cycle, *quality function deployment* (QFD), and *failure mode and effects analysis* (FMEA). The proactive PDCA cycle is discussed next, while the other tools are discussed in App. C. This third stage is crucial if an organization wishes to move toward a zero-defect mentality and goal.

Clearly, this is the phase we want to be in—always. What follows is a discussion of how we can be in this phase—preventing problems, aiming for zero defects, and providing hassle-free and exciting products and services.

Managing Breakthroughs—Aiming for Zero Defects with the Proactive PDCA Cycle

In any organization, we need both improvements and breakthroughs. In Chap. 3, we discussed business planning and how that can drive breakthroughs. We have discussed managing improvements with the PDCA cycle. We now discuss how to introduce and manage breakthroughs with the *proactive PDCA cycle*— for small to medium-size products and any new service. For large or massive projects, more complex tools will be required. These are beyond the scope of this text; nevertheless, the concepts used here should be understood and applied in large projects.

The proactive PDCA cycle

The PDCA improvement cycle is very powerful for managing improvement projects, where we have a good history and data. But what process can we use for new ideas, when we are designing a breakthrough product or service and where we want a successful introduction? We propose a *proactive PDCA* cycle. What is the purpose of the proactive PDCA cycle?

> The purpose of the proactive PDCA cycle is to help manage the design of breakthrough products and services, with zero defects and maximum market acceptance and customer satisfaction.

Figure 4.15 shows the proactive PDCA cycle. This is a modified PDCA cycle, in which we have made some changes. Instead of looking backward (history), we look sideways (at competition) and forward at new

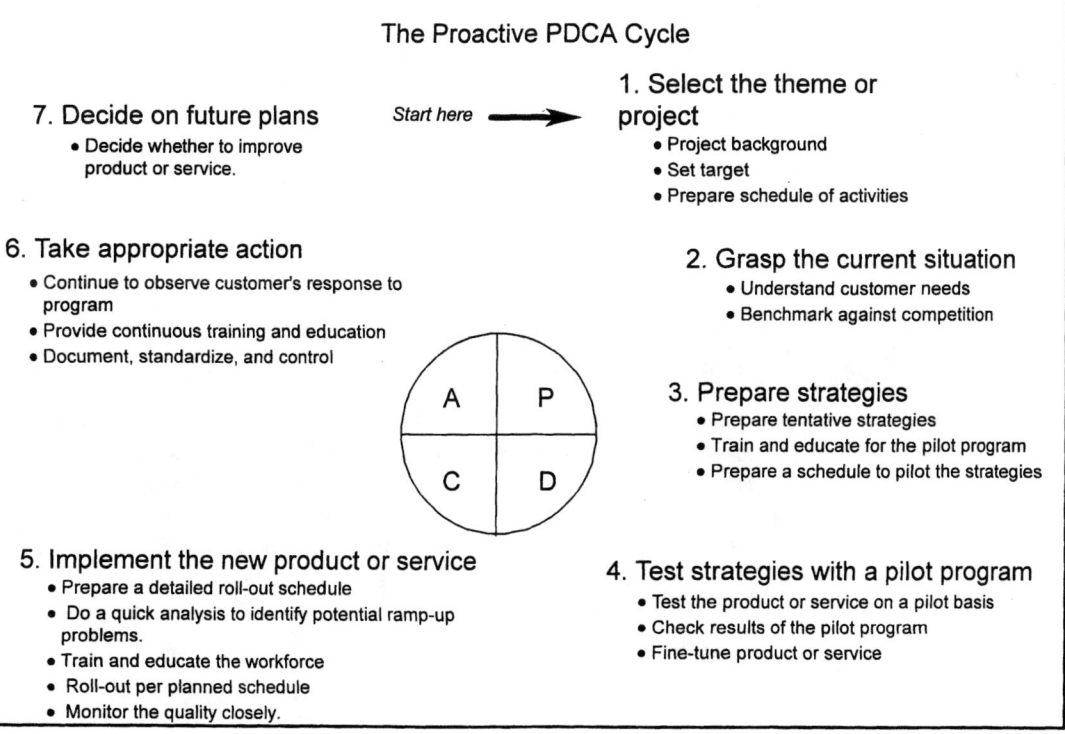

Figure 4.15 The proactive PDCA cycle.

or future customer needs. Our desire is to have a successful product or service that is well accepted in the marketplace. At the same time, we want to predict and prevent problems and aim for a goal of zero defects.

You will find the proactive PDCA cycle useful in managing breakthroughs and as a training tool. Each of the seven steps in Fig. 4.15 is explained in greater detail. In addition, we have listed some tools that could be used in each of the steps.

Step 1: Select the theme or project

- *Objective of this step.*   To clearly define the new product or service to be implemented.

- *Discussion.*   Here we define the product or service to be implemented, understand its background, set a target, and prepare a schedule of activities. The following are substeps.

  Step 1a: Project background.   The project can be selected from the department objectives or customer needs gathered from a survey or feedback.

  Step 1b: Set a target.   This should include a statement of the item, a numerical number to be achieved, and a time frame for project completion, for example, a product or service definition, quantity to be produced or market share/position to be achieved, and time frame to achieve the result. The target can be set according to these guidelines: Use current data to set a breakthrough goal, use competitive data to better the competition, or plan to move out into the next zone—preferably the outermost—of the total product concept chart for your product or service. Refer to the discussion on the total product concept in Chap 2.

- *Tools that can be useful.*   Pareto diagram, trend charts, total product concept chart.

  Step 1c: Prepare a schedule of activities.   This lists the seven steps in the proactive PDCA cycle and the expected time frame for each step. We recommend a reasonable time frame of 6 to 12 months. If you take more time, either your project is very massive or you are taking too long; hence you need to relook at your project size or plans. We recommend the implementation plan format, shown in Chap. 3, to record the schedule. This is very important, as it sets outer bounds on the project.

- *Tools that can be useful.*   Pareto diagram, trend charts, Hoshin implementation plan.

Step 2: Grasp the current situation

- *Objective of this step.* To look at customer needs, benchmark the existing offering, and understand competitive strategies.

- *Discussion.* Here we do an analysis of what customers want, attempt to benchmark ourselves against the competition, and understand the direction of competitive strategies. The detailed substeps follow:

  Step 2a: We need to understand what the customer's expectations are. These are constantly changing, becoming higher expectations. We can interview customers and get their needs. But often the customer does not know what he or she wants, and we need to guess. Wherever possible, try to stratify the customer's expectation as follows:

- What is the basic need? Are we meeting it? Refer to the discussion on must-be quality in Chap. 2. You must get closure on this item before moving to the next item.

- What is the spoken need? Are we meeting it? Refer to the discussion of attractive quality in Chap. 2. Again get closure on this item before moving on.

- Is there an unspoken need?

  Often the unspoken need can be understood by looking at current problems and complaints. For example, if customers complain about poor service or poor entertainment on an airline service, this gives us our cue for a pressing customer need.

  Unspoken needs could also be things that are the unexpected surprises or delightful items. It helps if we start to complete the total product concept chart, discussed in Chap. 2. This will help provoke our creative energies and provide new ideas.

- *Tools that can be useful.* The total product concept, check sheets, Pareto charts, trend charts, customer complaint and feedback system, customer meetings, and focus groups.

  Step 2b: We need to benchmark our offerings and understand competitive strategies. It is important to understand what our strongest competitor is doing and how customers are responding to competitive offerings. Refer to the benchmarking section in Chap. 2.

- *Tools that can be useful.* Competitive benchmarking, check sheets, process flowcharts, Pareto charts, trend charts.

Step 3: Prepare strategies

- *Objective of this step.* To create tentative strategies for products or services.

- *Discussion.* Having understood customer needs and competitive strategies, we are now ready to generate tentative strategies to meet customer needs. After preparing these strategies, we will need to train and educate the workforce and prepare for a focused rollout. There are three substeps:

Step 3a: Prepare tentative strategies. Remember, we are looking for a breakthrough and not continuous improvement; hence our plan should be to be better than the competition. Here is a list of questions you can ask that will help move you in the right direction:

What new features can we provide, with new technology to ensure that products and services are hassle-free?

What new customer segments can we reach out to? Refer to this discussion in Chap. 3, "Business Planning." What products or services can we offer them?

What must we do to create a total experience? To ensure repurchase by customers?

What new ideas will stretch and move us from the current zone to the next zone, preferably the outer zone, in the total product concept? Refer to Chap. 2, "Customer Obsession."

For a product, the product is now designed and manufactured. The product should be designed to be trouble-free and to meet customer needs, using the predict-and-prevent process we discussed earlier, that is, FMEA and QFD. These are elaborated upon in App. C. For a service, you can also use the QFD concept during the design stage and use a simplified FMEA table to identify potential problem areas. An example of a service designed with the QFD concept was given in Chap. 3. The detailed process should be documented, with performance measures at critical points in the process. Some information on this is given in Chap. 5.

- *Tools that can be useful.* Brainstorming, QFD, FMEA, total product concept chart, flowchart, Pareto chart, check sheet, computer simulation of product or service.

Step 3b: Train and educate a selected workforce for launching the pilot program. If we have a new product, the workforce needs to understand how to manufacture and sell it. If it is a service, the workforce needs to be trained in providing the service. Because we will be launching a pilot program, we need only to train the workforce involved.

Step 3c: Prepare a schedule to roll out pilot program. We should plan for a focused rollout of the program in a selected market or location. Refer to comments in step 4.

Step 4: Test strategies with a pilot program

- *Objective of this step.* To test the product or service and ensure it is trouble-free.

- *Discussion.* We need to test the product or service to ensure it is trouble-free and well received by customers, check the results of the pilot program, and fine-tune the product or service to ensure it meets the original objective. There are several substeps.

Step 4a: Test the product or service. There are several things we must do to test products or service:

- For a product, the product should have been designed to meet customer needs and should be trouble-free. The product should be tested using the standard environmental tests, such as stress, burn-in, and temperate cycling. Where possible, the product should also be tested in a specific market to understand its receptivity by customers. Japanese electronic companies routinely test-market their products in Tokyo before rolling them out elsewhere—many never go further, but the test market provides a controlled environment that can help demonstrate both product quality and receptivity.

- For a service, it will be beneficial to pilot the service in one location or market to understand the quality of the service and receptivity by customers. When Walt Disney started its Disneyland operation in Tokyo, it reportedly had a perfect launch, with no failures in any of the shows or services on the first day. Japanese companies were astonished at such quality, because although they are very good at ensuring product quality, they are weak in service strategies—new or ongoing. The key to Walt Disney's success was good design of every service, workforce training, computer simulation and engineering of every operation, and thorough testing to weed out potential failures. If Walt Disney can do this for a complex service introduction, such as Disneyland, it clearly can be done for any service or product.

- Whether you test a product or service, remember to have a large enough sample size to measure the success of the program.

Step 4b: Check results of the pilot program. We must now check the results of the pilot program by collecting customer and employee feedback, measuring actual performance to design specifications, and reviewing any failures in our product or service. There are several outcomes that could occur:

Results have been achieved, target has been met.   If we have done a thorough job of understanding customer needs and provided a robust design, the customer feedback should be positive. In this case we should fine-tune and improve the product or service, with information gathered from the pilot program, and implement it on a wider scale.

Failure to achieve results.   If the product or service is poorly received and failure is due to improper implementation, we must go back to step 4a, training and education. Otherwise, we need to go back to step 3, prepare strategies. If we fail to meet our goals, it is very likely that we missed customer needs, we proposed something that may have been too progressive, or the design may have been bad. We must make the decision to continue or to abandon the project. The decision to continue the current project or to select a new one has to be based on priorities and resources. Sometimes if the failure is marginal, we may release the product or service, but with a containment plan to fix the cause of failure after release. We would do this if the product or service were far ahead of the competition.

Step 4c: Fine-tune product or service.   The information gathered from the previous step is used to fine-tune and improve the program. We need to eliminate any design or process failures detected in the pilot program to ensure we can meet our target.

Step 5: Implement the new product or service

- *Objective of this step.*   To launch the product or service in all selected markets or locations.
- *Discussion.*   We are now ready to launch the new product or service. There are several substeps here.

Step 5a: Prepare a detailed rollout schedule.   We must prepare a detailed rollout schedule for our new product or service for all selected markets or locations.

Step 5b: Do a quick analysis to identify potential ramp-up problems.   This is important if the planned product or service ramp-up is steep. If possible use FMEA. Potential problems could include machine capacity, training program for many people, and rollout of a product or service to many locations, hence requiring complex monitoring and control.

Step 5c: Train and educate the workforce. This was discussed in step 3b. This time, however, because the rollout is more extensive, all the workforce involved needs to be trained.

Step 5d: Roll out new product or service per schedule. The new product or service is now implemented. It is important to ensure that all involved have been trained and that the rollout is according to plan.

Step 5e: Monitor the quality closely. This is to ensure that expectations are met.

- *Tools that can be useful.* Trend charts, control charts, process quality checkpoints, and FMEA.

Step 6: Take appropriate action

- *Objective of this step.* To ensure that the performance of the new product or process is maintained.

- *Discussion.* We must now continue to monitor performance, train and educate new employees not familiar with the new product or service, and complete final documentation of the new product or process. There are several substeps.

Step 6a: Monitor and observe customer's response to the new product or service. The new product or service should be well received because we tested it before launch. Nevertheless, we must continue to monitor customer feedback and responsiveness to the program.

Step 6b: Provide continuous training and education. Ensure continuous training in the new methods or standards for new employees. This is important to ensure a high level of performance.

Step 6c: Documentation, standardization, and control. The new strategy (implemented in step 5) should be documented in current operating procedures or standards. It is also important to identify critical performance measures or process parameters to control—in product manufacturing or service implementation. The idea is to monitor the appropriate parameters and to detect any deviation from the new standards. Any deviations should be promptly analyzed and eliminated. Finally, when improvements are made in the offering, update all documentation and retrain all staff.

- *Tools that can be useful.* Checklists, control charts, process quality checkpoints.

Step 7: Decide on future plans

- *Objective of this step.* To decide whether to further improve new product or service or to use the experience gained for future projects.

- *Discussion.* In the spirit of total quality creation, we must decide on the next step. We should determine if we achieved zero defects in the new product or service at introduction, and if not, why not? The lessons learned will help to move the next project toward zero defects at introduction. We could work on making the current product or service better or select the next project. This decision will have to be based on department priorities and objectives.

Note: When preparing a schedule of activities, plan for an overlap between steps 4 and 5, since some items can be done simultaneously. Also observe that steps 4 and 5 have both a do and a check stage.

Benefits of the proactive PDCA cycle

There are several benefits in using the proactive PDCA cycle.

- It is very useful for planning for breakthrough product or services. We recommend it for small products and any new service.
- It requires you to look at customer needs and the competition before proposing a new product or service.
- It requires you to test a product or service in a limited market, location, or customer base and to make adjustments so that the product will be well accepted by customers.
- It takes you into the prevention of problems by prediction mode— our objective is to design a breakthrough product or service, which is thoroughly tested and achieves zero defects in the marketplace or customer site.
- The above points can result in a product or service that is successful and well received in the market.

Questions and Answers on Managing Improvements and Breakthroughs

Let's review some frequently asked questions on managing improvements and breakthroughs.

Question. Why should we restrict ourselves to one specific improvement methodology (that is, PDCA cycle)? This will restrict our creativity.

Answer. If there is a rigorous, proven, and successful improvement methodology, you should adopt it instead of inventing your own. There are two advantages to doing this. If there is a standard methodology in your company and everybody is trained to use it, then you have a common problem-solving language in your company.

There is no need for retraining as employees move across the company. Creativity is very important. But it should be used in new unexplored areas or in providing creative solutions that thrill your customers. Hence you will not stifle creativity; instead, you are pointing it in the right direction.

Question. How do we ensure that we improve the right things?

Answer. Priorities need to be set. In the beginning of this chapter, we discussed how you can select items to improve. This will form your initial list. From that you develop priorities, which are influenced by the following:

- *Linkage to the annual Hoshin plan.* That is, is this an item that shows up in or supports a Hoshin plan strategy?
- *Customer complaints.* Frequent customer complaints must be given top priority.

Question. Must we document every completed project in the PDCA format?

Answer. No. The important point is that for every project the PDCA framework is followed. Completed projects should be filed and sorted in PDCA categories. That is, you can access the project targets, schedule, analysis, implementation plan, check data, and so on. However, one or two projects should be documented in the PDCA format, examples of which are in the Appendixes. These documented projects can be used for training, promotion, publicity, and presentations.

Question. You quoted Kaouru Ishikawa as saying that the PDCA cycle is the most important thing in total quality. Does that mean that improvements are the most important part of total quality? If so, what about new product design and planning for the future—are they not important?

Answer. They are very important. Continuous improvement is a basic tenet of total quality; and the PDCA cycle plays an important part. But the PDCA cycle can be used for managing other activities as well: the long-term plan, the annual Hoshin plan, projects in R&D, and education programs. The PDCA cycle can also be viewed from a macro perspective. For example, for a design and manufacturing company, P (plan) is the design stage, D (do) is the manufacturing stage, C (check) is the selling and getting customer acceptance stage, and A (act) is the customer feedback stage. That brings us back to P or the next design. We have also proposed a proactive PDCA cycle for designing new products and services.

Question. I am managing a business and need to plan for the future. How can the PDCA and proactive PDCA cycles help me?

Answer. In any business you need both improvements and

breakthroughs. For example, most car manufacturers introduce a new model every few years. In between new-model introductions, they focus on continuous improvements before the next-generation model is introduced with great fanfare. Hence every business will focus on several activities to generate new business:

- Improve the current product or service: reduce costs, improve reliability, add new features.
- Modify a current product or service with enough features to pass it off as a new product or service.
- Design a breakthrough product or service that puts you ahead of the competition.

We propose managing improvements with the PDCA improvement cycle. But for breakthroughs you need a different outlook. The proactive PDCA cycle helps give you an aggressive outlook toward generating a breakthrough product or service, with minimum or zero defects and maximum market acceptance. Remember the proactive PDCA cycle is only a guide to ensure you do not overlook anything, such as training, competitive benchmarking, and reliability. The creativity will have to come from the project team. So, with the two different PDCA cycles, you have a model to help generate new business.

Summary: Managing Improvements and Breakthroughs

In any business you need both improvements and breakthroughs. This allows you to generate a stream of products or services. Hence every business will focus on several activities to generate new business: improve the current product or service by reducing costs, improve reliability, and add new features; modify a current product or service with enough features to pass it off as a new product or service; design a breakthrough product or service that puts you ahead of the competition.

To help this activity, we have proposed the PDCA improvement cycle and the proactive PDCA cycle for designing breakthrough products and services.

We recommend the use of the PDCA improvement cycle for all improvement projects, because it provides a systematic improvement process. This cycle is especially useful because it can be used to manage each improvement cycle as a project with schedules and targets. In addition, it helps to identify the root cause of problems and to remove them. Within this cycle you can use the numerous statistical tools such as the seven tools, the seven new tools, design of experiments, and Taguchi methods. We realize that the PDCA cycle is very structured, but nevertheless we recommend that you adopt it and use your creativity in unknown areas where there is no experience or knowledge, for example, providing creative solutions or alternatives for your customers.

Finally, to manage breakthrough products or services, we have proposed a proactive PDCA cycle. The proactive PDCA cycle can be used to manage breakthroughs for small to medium-size products and any new service. The proactive PDCA cycle sets up a framework for designing products and services to meet customer needs and minimize possible failures, problems, or poor receptivity by customers. It also helps you achieve zero-defect products or services. This moves you into the zone of providing potentially exciting and hassle-free products and services.

References

1. John Akers, *Newsweek,* June 10, 1991.
2. A. V. Feigenbaum, *Total Quality Control* (Singapore: McGraw-Hill, 1986).
3. J. M. Juran, Frank M. Gyrna, and R. S. Bingam, *Quality Control Handbook* (New York: McGraw-Hill, 1951).
4. "Business Management and Quality Cost, The Japanese View by Hitoshi Kume," *Quality Progress,* April 1985.
5. W. Edwards Deming, *Out of the Crisis* (Cambridge, MA: MIT Press, 1982).
6. Comments taken from "The Japan that Can Say No" by Akio Morita and Shintaro Ishihara. Anonymous English translation.
7. Prepared by Katsu Yoshimoto of Yokogawa Hewlett-Packard. Adapted from an article "Principles of New Product Development" by Yoshinori Iizuka, 1987 International Convention of Quality Circles, Tokyo.
8. *Poka-Yoke: Improving Product Quality by Preventing Defects,* edited by Nikkan Kogyo Shimbun Ltd. (Cambridge, MA: Productivity Press, 1988).
9. "Key Points for Success in Problem Solving," 1990 ASQC Quality Congress Transaction, San Francisco.
10. Kozo Kuora, "Survey and Research in Japan concerning Policy Management," 1990 ASQC Quality Congress, San Francisco.
11. Extracted from an article by Kenzo Sasaoka, of Yokogawa Hewlett-Packard, "A Challenge to Revitalize Management," Quality month text No. 155, published by JUSE in Japanese.

5

Process Management

People come and go but processes stay.
ANONYMOUS

Overview

In Chap. 3 we discussed daily management, or business fundamentals, plans. The daily management plan lists items or key processes that are managed on a day-to-day basis. Progress is measured by selecting appropriate performance measures, setting goals, and reviewing them regularly—often daily. In this chapter we review in detail how key processes in the design, manufacturing, and sales environment can be managed or controlled.

In Chap. 4 we discussed the relationship between improvement and control. All the key processes in an organization must be controlled. The following incident illustrates the point.

> My friend and mentor Noriaki Kano, a member of the Deming prize committee in Japan, was once reviewing a manufacturing operation. He had been impressed by everything the managers had shown him. As we passed a production operator working on a bonding machine, he stopped to talk to her.
>
> He asked her if she knew the reject rate of her bonding machine. "Oh yes," she replied, "it is 0.15 percent." What did she do if for any reason the rejects exceeded that number? She replied that she had been told to call the supervisor whenever rejects exceeded 0.20 percent. Kano decided to check the records to verify her statement. He flipped through her records and found a day when the lot reject percentage was 0.24 percent—so he asked her, "What did you do?" She replied that 0.24 was so close to 0.20 that she did nothing. "Fair enough," he said and looked at the records again and found another lot about 1 week before with a reject rate of 0.30 percent. "What did you do then?" he asked. She replied that she had indeed informed her supervisor.

The supervisor happened to be standing by and informed us that she had told the engineer. Kano remarked, "It's been a week, what has the engineer done?" The engineer was nowhere to be seen. I suggested looking for the engineer and sent someone to find him. Meantime, the production manager was getting a little nervous and suggested we move on. Kano stood his ground and kept asking to meet the engineer. We finally found the engineer. He had been informed but what had he done? He had done nothing, no documentation or analysis was available; he just never got around to doing anything, but he planned to do something—when he had time.

The learning points of this episode are as follows:

1. A good job of characterizing the process had been done according to the textbooks.

2. When the process went out of control, the company never had the discipline to analyze and take corrective action.

3. Eventually the process or the bonding machine could have given more rejects, gone out of control, but nobody would have noticed—until it was too late.

After this experience, Kano commented, "The workers are awake, the engineers are just waking up, but management is asleep." His point was that the workers were doing their best and alerting management to potential problems; the engineers were busy and stretched; but management was oblivious of the real problems. In our example, there was a possibility that if the current neglect continued, problems would multiply and the process would run out of control. The outcome could have been product problems and customer dissatisfaction. Here is where we can apply the concept of process management.

In this chapter we discuss the basics of process management and how to get started, provide guidelines, illustrate the concepts with examples, give a short case study, and finally discuss some common, or generic, processes for a design, manufacturing, and sales environment.

The Process Concept—and Its Contribution to the Organization's Value Delivery System

Before we define process management, let us discuss the process concept. Every activity is part of a process. This applies to the manufacturing and service sectors—in fact, any activity at work or home. Figure 5.1 shows a simple process flow. In the figure, we have converted raw materials to food or a product. *Hence, a process is an activity to which we have added some value. A process is therefore part of a value chain.* Every organization will have a multitude of processes— some managed by individuals and others managed cross-functionally.

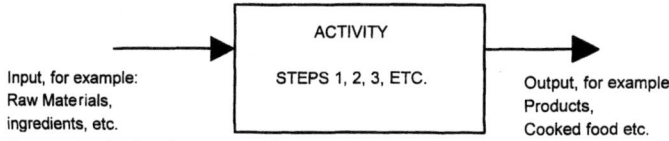

Input, for example:
Raw Materials,
ingredients, etc.

ACTIVITY

STEPS 1, 2, 3, ETC.

Output, for example:
Products,
Cooked food etc.

Figure 5.1 A simple process flow.

These processes add up the value delivery chain or system of the organization, which provides the organization's product or services. If this value delivery system is managed well—via its processes—we get customer satisfaction, repeat purchases, and increased business or revenue.

To improve the quality of a product or service, the value chain or the processes within that value chain must be improved by changing or modifying them. We discussed improvements in Chap. 4 and showed how improvements and control are linked. We mentioned that every process must be controlled or managed. In part, we manage a process in order to retain the gains after improvement; but we also manage a process to better control its output or results.

Why manage processes?

Achieving good results is paramount in any company—because they indicate commendable performance, which is what management strives to have. Good results, however, are a *lagging indicator* of performance; only when results are achieved do we know that we have performed well. But we must be able to predict results, and for this we need a *leading indicator*. A well-managed process, monitored with properly selected performance measures, can be a leading indicator and will give predictable results. For example, if a prize-winning recipe is followed diligently, it will give an outstanding dish. Similarly, good design, manufacturing, and sales processes will give predictable and outstanding results in any company.

Processes also drive the value delivery system of an organization. So if we select the correct set of processes and manage them well, we can ensure good control of the value delivery system and can provide the highest-value deliverables to customers.

To reiterate, *a well-managed process can be a leading indicator of good results in any company.* In this chapter we discuss how to control or manage a process. Given that every activity is a process, we identify key or important processes in the design, manufacturing, and sales environment that must be managed. We also discuss how to manage processes on a day-to-day basis, hence our use of the term *process management.*

Process Management

Process management means identifying and monitoring a process, ensuring it meets a target (within limits), discovering abnormalities, and preventing their recurrence.

Before we go any further, we want to stress several important concepts about processes:

- People, and managers, come and go (in any organization), but processes stay. Hence there must be a system and understanding to ensure processes are well managed and not people-dependent. *Otherwise processes may change with people changes, or people will try to compensate for poor processes by tampering with them.* We do not want that.

- Processes may be poorly or well managed. Our goal, of course, is to manage them well and to have predictable results from each process. We show two examples of what this means.

- Processes can be managed by individuals (for example, an assembly line operator or receptionist), or they can be cross-functional (for example, a product definition and design process). Cross-functional processes are more complex because they span an organization and there could be poor management at the boundaries between departments of such processes—these are often called the *white spaces* of an organization. We must have a method to manage such cross-functional processes.

Requirements for process management

What are the requirements of a good process? For a start, the process should be documented, everybody should be trained, and the performance measures to monitor performance of the process should be well understood. If the process goes out of control, corrective action should be taken to bring it back into control.

Process management at a factory. Let us look at an example at a facsimile (fax) machine factory. In it there could be a printed-circuit board assembly line and at the end of the line a test station. One of the performance measures for the process could be the failure rate at the test station. Figure 5.2 shows what to expect with an unmanaged process, a well-managed process, and an improved, well-managed process. The lower the failure rate, the better the quality of the printed-circuit boards coming off the line. The result is less rework, higher productivity, and a predictable output from the production line.

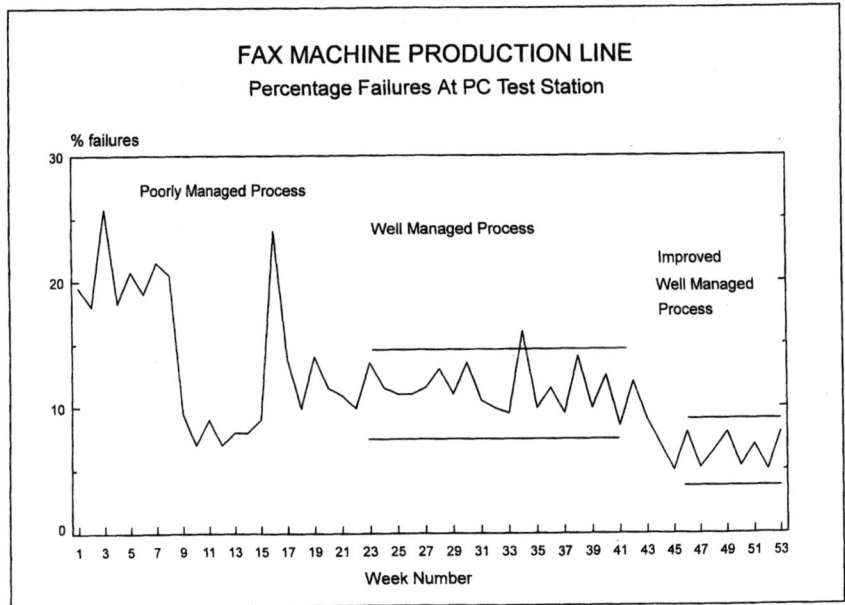

FAX MACHINE PRODUCTION LINE
Percentage Failures At PC Test Station

Figure 5.2 Measuring and managing failure rate.

Process management in a marketing and sales operation. The same concept applies to the sales process in a marketing and sales operation, as shown in Fig. 5.3. Assume one of the performance measures in the operation is the success rate, defined as

$$\text{Success} = \frac{\text{number of qualified prospects who purchase}}{\text{number of qualified prospects}}$$

Doing well here means that the marketing and sales people have done a good job of selecting, screening, and qualifying prospects during the preselling or marketing stage. The result is that the chances of selling are now higher and more predictable. Less energy is expended, and productivity is higher.

Our purpose is to aim for a well-managed and predictable process by documenting, monitoring, and controlling the process. When any key process becomes stable and well managed, it becomes part of the daily management, or business fundamentals, plan as an entity. Managers can then use their creative energy for other problem processes or new unexplored areas. See the comments in Chap. 4 about standardization.

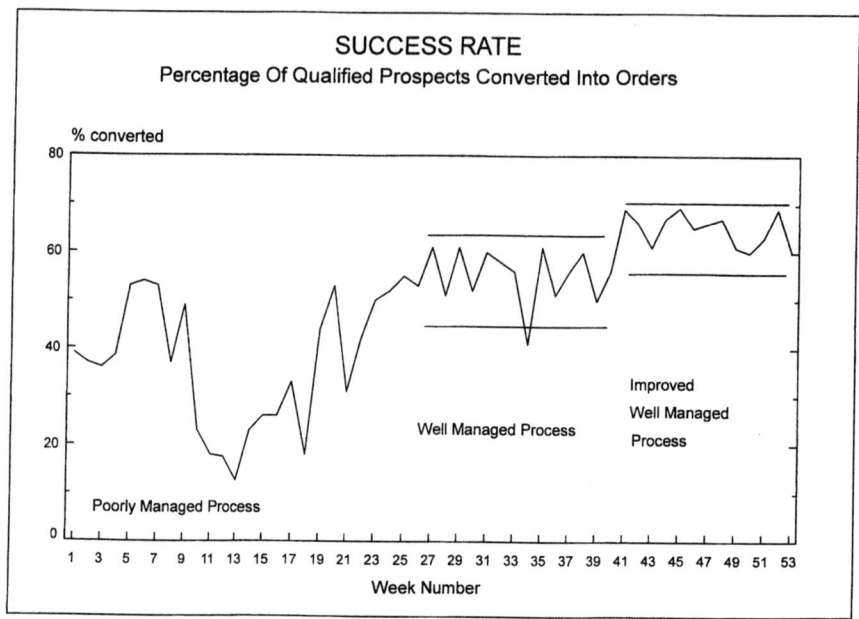

Figure 5.3   Measuring and managing success rate.

Prerequisites for good process management

How do you ensure that a process is well managed? What are some of the necessary prerequisites? Here is a short list that is a good guideline.

1. *The process must be documented and must include a process flowchart.* The documentation includes key steps with performance measures for each step. The process could be stand-alone or a cross-functional process, in which case the linkages across the organization must be shown.

2. *There must be process performance measures, with targets.* Refer to Fig. 5.1. Each of the steps (1, 2, and 3) could have a target which must be consistently reached. So, for example, in the facsimile machine factory discussed earlier, step 4 could be the test station. The performance measure could be the board failure rate, with a specific target of, say, 2 percent. If the other steps are well defined, understood, and managed as step 4, we have a well-managed and predictable process in the factory.

3. *There must be a method to manage out-of-control conditions.* In managing a process, the concept of statistical quality control must be understood and applied. We discussed the concept of limits in Chap. 3, "Business Planning." If you are still unfamiliar with this concept, refer to App. B and the Bibliography.

Let us go back to Figs. 5.2 or 5.3. In the left of both figures we have a very unpredictable, wildly gyrating, or poorly managed process. Using the concept of statistical quality control to eliminate assignable causes (such as poor training or inferior machines) will result in a well-managed process. We can see from the charts that the process is fluctuating between two statistically identifiable limits—the upper and lower control limits. When the process is contained within these limits, we have a managed process. We can further improve the process by understanding the reason for failure; for example, we need better raw material for the printed-circuit boards or a better way of qualifying prospects, perhaps by understanding their needs better. If we are able to do this, we will get an improved, well-managed process, as shown on the right of Figs. 5.2 and 5.3.

We will find that occasionally the process will deviate and cross the control limits. It is important that, when this occurs, the reason for the deviation be analyzed and corrective action be taken to prevent recurrence. Otherwise, this may be the start of a deteriorating cycle. Failure to understand and correct a deviation can also give the wrong signal to employees—that is, if there is a deviation, we can ignore it. There is more on deviation analysis later in this chapter, when we discuss out-of-control reports.

4. *There must be a training plan.*   All employees must be trained in the appropriate process steps that are applicable to them. And if the process is changed or improved, the changes must be documented and employees trained in the improved process. This is one of the steps in the PDCA cycle.

5. *The process must be continuously improved.*   An aggressive management will continuously set new and tougher targets for each important performance measure in the process. Thus not only must a process be managed, but also it must be continuously improved. Here lies the secret of success for any company.

Identifying and managing key processes

Getting started on process management. So how do we get started, and how do we apply the concepts mentioned above? How do we apply these concepts to an organization, company, or department? More specifically, how should a general manager, senior manager, or department manager use these concepts to manage better? We will answer these questions and give some examples for both the manufacturing and sales environment. Most importantly, we will show how senior mangers can manage these processes to create an organization that runs like clockwork—this will require linking key processes to business planning.

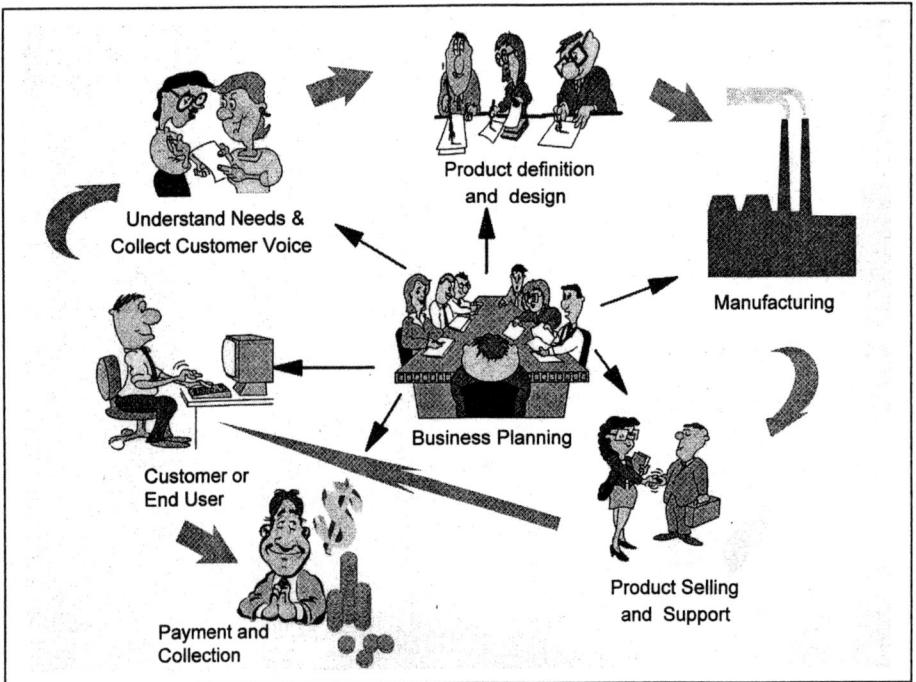

Product definition
and design

Understand Needs &
Collect Customer Voice

Manufacturing

Business Planning

Customer or
End User

Product Selling
and Support

Payment and
Collection

Figure 5.4 Sketch of key business processes in a company that designs, manufactures, and sells products. These processes represent the value chain or value delivery system of the company.

Sketching the process—by visualizing the value delivery system of the organization. In any organization there is a value delivery system. This system consists of processes beginning with customer needs and ending by meeting those needs. So, in a company that designs, manufactures, and sells products, we can start by doing a quick sketch of the key processes in the company, as shown in Fig. 5.4. (I am indebted to Koh Hwee Miem of Hewlett-Packard for this sketch, which is actually used for training.) Next, we write down the activities below each picture, as shown in the sketch. We start with the customers and their needs on the left; then we move to product definition, manufacturing, sales, delivery, and product consumption by the customer. At the center of the figure is the business planning process, which seems to drive, or is driven by, the various items.

This sketch looks cute and is a great conversation piece to show the kids what you do at work, but we need more details. For example, we need to list the owners and success factors (or performance measures) for the processes. So we get a blank sheet of paper and draw some columns, each representing the major functions in the company, and then insert the processes shown earlier. The major functions are mar-

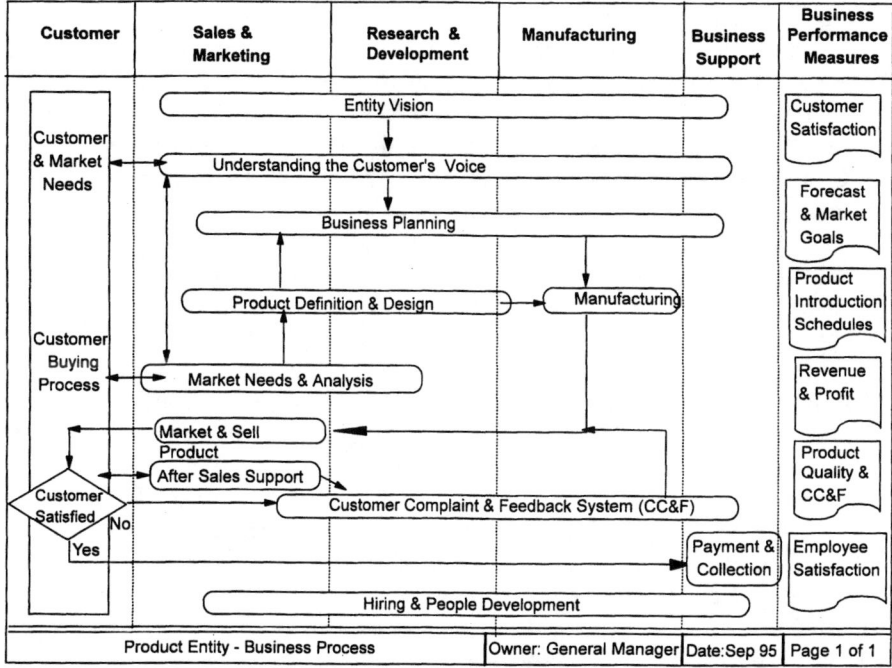

Figure 5.5 Top-level process map of a business unit that designs, manufactures, and sells products. These products also reflect and measure the value delivery system of the business unit.

keting and sales, research and development (R&D), manufacturing, business support, and, of course, the customer. We show this in Fig. 5.5.

Figure 5.5 represents the *value delivery system* of the company, or, as we call it, the top- or entity-level process map. This is the top-level processes map of a business unit, or company, that designs, manufactures, and sells products. In the figure we also inserted, on the right side, a list of performance measures of the key processes. We now review some of the items in Fig. 5.5.

Each box in the figure represents a key process, for example, product definition and design, and product manufacturing. Ownership will be within a department (for product manufacturing) or cross-functional (for product definition and design). Across the top are the entity vision and business planning processes—these are entitywide activities, which we discussed in detail in Chap. 3. We also show the process of understanding the customer's voice, which again is an entitywide process.

The business planning process is very important, as it influences everything else in the entity's process map, including what products to introduce and what processes to improve and manage. *This is crucial—processes cannot be improved or managed in a vacuum, instead they must be part of the bigger scheme of things and must fit into business planning and management's priorities.*

Cross-functional processes.   Several of the processes shown are cross-functional. We have already discussed business planning. Another is product definition and design—this has multiple owners, specifically marketing, R&D, and manufacturing. For this to work well, the various functions must work as a team. There are specific tools that can ensure good teamwork, for example, QFD and the total product concept, discussed in Chap. 2.

Business performance measures.   On the extreme right of Fig. 5.5 is a list of business performance measures. Typically, there will be one performance measure for each process. The measures should be as high-level as possible. Why? Because they represent the company's key processes. Hence, the performance measures should represent what is important to customers and shareholders. When you are preparing a next-level process map, that is, for one of the functions, such as sales and marketing, you will again have the performance measures as high-level as possible, and they will represent the function's key deliverables or outputs of the entity or its value delivery system. We give more examples later.

Process management and business process performance measures versus daily management, or business fundamentals, plans.   All the key processes should have performance measures, with targets and limits. The specific targets are not shown in Fig. 5.5 because we wish to keep the process chart generic. The targets for performance measures will change with time, but because we keep our process map generic, we can keep it unchanged. So, where do we show and measure the performance measures? In Fig 5.6 we list the performance measures (from Fig. 5.5); these are then entered into a daily management, or business fundamentals, plan, shown on the right side of Fig. 5.6. Here we have listed the targets and limits.

Notice that the figure is a daily management, or business fundamentals, plan for the product division's general manager. *And there you have it! We have connected the entity's business plan (which will consist of an annual Hoshin plan and a daily management, or business fundamentals, plan) to the entity's key processes. The daily management, or business fundamentals, plan becomes the business scorecard for managing the fundamental processes of the business.*

The business scorecard.   The daily management plan, or business fundamentals plan, is the business scorecard that keeps track of the key business processes. There should be one at every level in an organization, from the chief executive or general manager to the engineer or supervisor. Note that in Fig. 3.24, we showed the big picture—the business planning process, which includes monitoring and managing key process-

Daily Management or Business Fundamentals Plan					
DATE: 26 FEB 97	**Prepared by: Product Division General Manager**				
				Actual Performance	
Item	Goal	Limit	Data Source	Nov	Dec, etc.
CUSTOMER RELATED					
● Customer Satisfaction			Marketing		
- survey score	8.5	7.5			
● Product Quality			Quality Dept		
-annual failure rate	2%	4%			
● Customer Complaint & Feedback system			Quality Dept		
- complaints in ppm	1000 ppm	1200 ppm			
- time to resolve 90%	4 days	6 days			
● Market Share			Marketing		
- overall, all printers	70%	65%			
INTERNAL/OPERATIONAL					
● Product Introduction	Per Target	none	Marketing & R&D		
- schedule					
● Forecast accuracy	+/- 30%	+/- 45%	Marketing		
FINANCIAL					
● Revenue	Target	--10 %	Business Support		
● Operating Profit	35%	30%			
● Net Profit	8%	6%			
PEOPLE RELATED					
● Employee Satisfaction					
- survey score	6.5	5.5			

Business Process Measures
- ● Customer Satisfaction
- ● Product Quality
- ● Customer Complaints & Concerns
- ● Product Introduction
- ● Forecast Accuracy
- ● Market Share
- ● Revenue
- ● Operating Profit
- ● Net Profit
- ● Employee Satisfaction

Business performance measures from Fig. 5-5.

Figure 5.6 Business process measures for a business unit that designs, manufactures, and sells products. These are the performance measures for the entity discussed in Fig. 5.5.

es in any organization, company, business unit, operation, or department. Here, we have shown how this concept is implemented. Also, note that not every department in an organization will have an annual Hoshin plan; but all departments should have an implementation plan handed down from above, and they should have a daily management plan or business scorecard to manage their key processes or activities.

Linkage between top-level (business unit) process map and next level. Should there be a linkage between the business unit or top-level process map, the next level, and the next level? Yes. Look at Fig. 5.5, the company's (top-level) process map of the general manager. In it, there are both stand-alone processes and cross-functional processes. All have owners, as indicated by the functions across the top of the figure. These functions will manage the next level of these processes. Let us give some examples.

In Fig 5.5, the product definition and design process is owned by marketing and R&D. The general manager should then request that the R&D and marketing managers show it in their process maps and have it on their daily management, or business fundamentals, plans as joint owners. In the case of the manufacturing process, the owner will be the manufacturing manager, who runs the manufacturing operations and will manage it via his or her daily management plan. In the case of the business planning process, however, there is no need to go any lower, as the general manager should be the owner. She or he may have a facilitator to help guide the process. The entity vision is definitely managed at the top level; refer to Chap. 3 for more details.

In Fig. 5.7, we show conceptually how the levels will link up. At the top is the top-level or entity process map, with its daily management, or business fundamentals, plan. The sales and marketing functions in the entity can be exploded into several functions. In our example, they would break into the entity marketing function and several sales functions representing different sales regions in the world. We show here the break into one function—a regional sales operation, with its own regional marketing arm. The detailed process map for the sales region is shown in Fig. 5.8.

How many levels down should you take this linkage? Three, maybe four. If you go any farther, you may have too complex an organization, with too many layers. You could end up creating a bureaucratic monster, so watch out for and avoid such situations!

Multilayer processes. In every function or department, each process within the process map can be exploded to a more detailed level, with a series of processes; and each process within this next level can be exploded yet again. We have seen processes explode to about four to five levels. Our advice to you is: do not be too concerned about this. Where it makes sense, keep to one layer; but where the process is very complex, go down another level. If you have to go to a third level, question the need—is it really necessary? Basically it is all right to delegate to the next-level department, if resources permit. But be wary about exploding the process to more than the second layer in a department. This may be overkill, unproductive, and unnecessary.

Process documentation, training, and improvement. Earlier, we gave several prerequisites for good process management. They include training, documentation, and improvements. Who does this? It should be the process owners. In the case of the top-level process map, shown earlier, the owners are the functions listed across the top. So, for the product design and definition process, it is jointly owned by R&D and marketing, with manufacturing involvement; for product manufacturing it is the manufacturing operations. For product selling and support processes, it is within the sales operations. These functions will manage the training, documentation, and improvements. The cue for improvements will come from participants in training classes, actual performance, and, of course, customer feedback.

In the process map, we also show business planning, which is an entitywide process. This is driven and owned by the general manager.

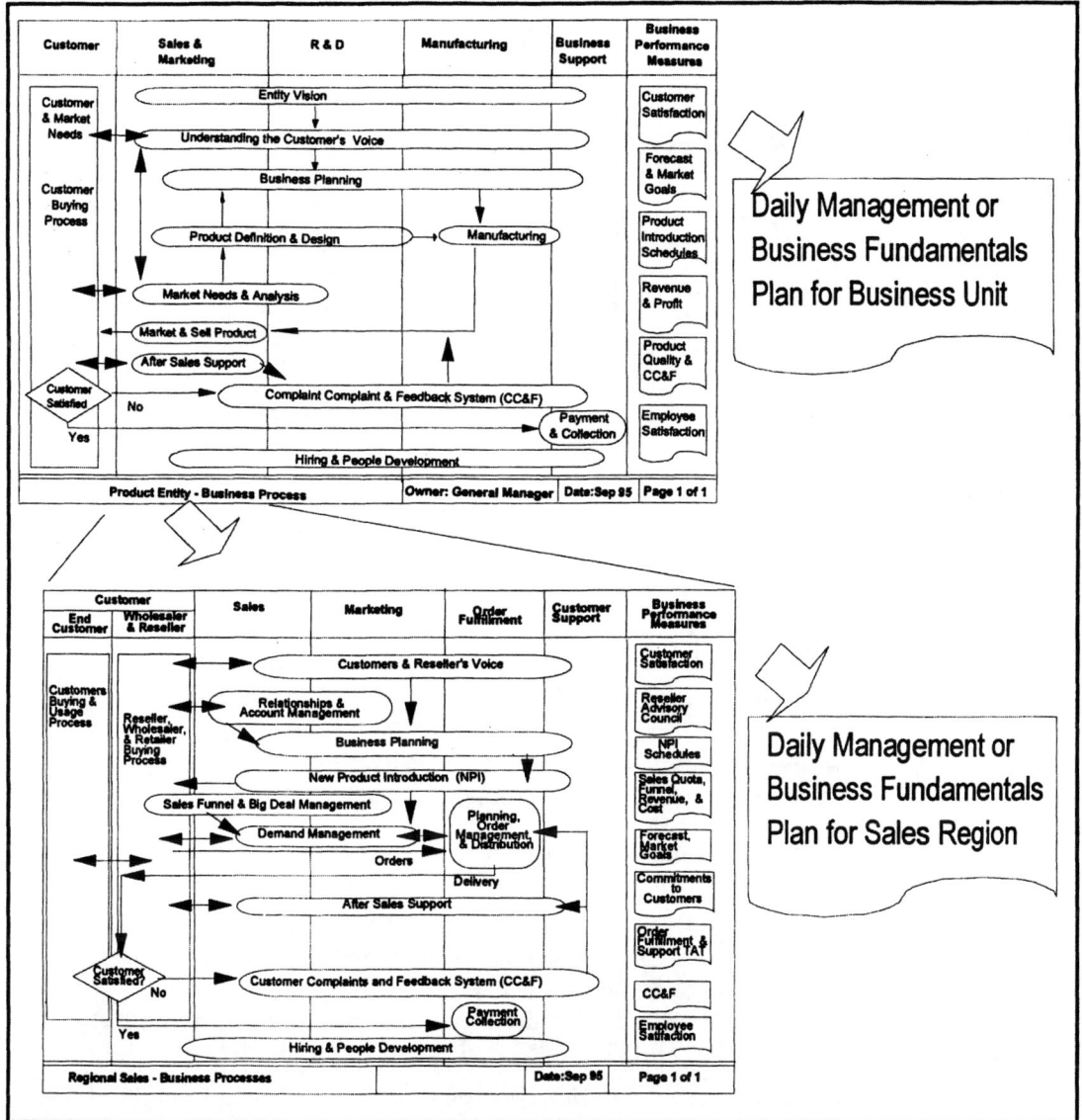

Figure 5.7 Deriving a sales region process map from the entity, or business unit, process map. The entity, or business unit, map is shown in Fig. 5.5 and the sales region map is shown in Fig. 5.8.

The documentation for business planning could be a manual that exists in the company, similar to what we show in Chap. 3. In fact, this author uses Chap. 3 as the formal documentation for business planning in his entity. Training for new managers consists of reading and using Chap. 3. A word of advice—do not make it too complicated.

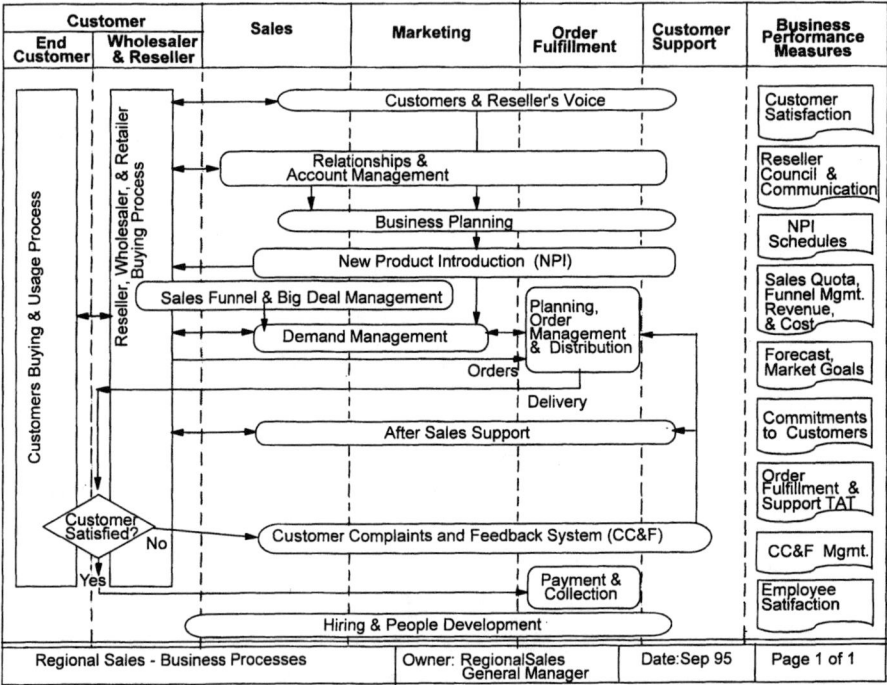

Figure 5.8 Process map of a regional sales operation.

Guidelines for preparing a key process map. Let us discuss some guidelines for preparing a key process map:

- The key functions or departments (of the operation being mapped) should be shown across the top (or across one side).

- The key processes, the flow of activity, and the linkage between functions should be shown. Note that some processes are confined to a function (for example, product manufacturing), while others (for example, product definition and design or planning) are cross-functional and will be shared across functions.

- Plan to show only key processes. It is easy to get carried away in preparing a process map. The way to keep focused is to keep each process map to one page. Be tough!

- Keep in mind that the key process map attempts to show the value delivery system of your company; that is, it shows what processes deliver the value or key deliverables that the customer wishes to get. With that in mind, it will be easier to list the key processes.

- Business performance measures for the processes are shown in the column on the right. The performance measures indicate the output or expectations of each process. All performance measures will have a target, which will go into a performance measure list or daily management plan—*this becomes your business scorecard that measures the pulse of your business.*

- When you have prepared the list of performance measures, do a sanity check with your staff, peers, and customers. Ask if this is the way in which you should be measured.
- The performance measures will include customer measures. Check that these are meaningful to the customers—ask customers how they measure your performance to them and ensure those measures are included.
- We recommend that you list the performance measures but not their targets. This allows you to keep the process map generic, while routinely adopting more aggressive targets. More importantly, some processes may be weak or new in the entity and cannot have a performance target. In this case, the plan of action would be to set up the process. How? By planning for it in the annual plan. We discuss this in the case study.
- Do not worry about getting the process map right the first time—it will take time. The process map will improve as your business planning process matures. It may take a few iterations, and it will get better with time. The key is to get started and to monitor results routinely. In some cases, do this daily.

Case study of an order fulfillment operation.   Let us review a case study of an operation implementing process management. We reviewed processes in a regional sales operation in Fig. 5.8. One of the functions we showed was order fulfillment. We show now the next-level process map for the order fulfillment operation. This operation operates in the Asia-Pacific region, manages the supply chain from the factories and other suppliers, is responsible for all aspects of order management with resellers and dealers, ships to customers, and manages credit and collection.

The process map for the order fulfillment operation is shown in Fig. 5.9. This was refined over several years to the level shown. Over the years, new activities were added, for example, credit and collection. Hence, the process map must keep pace with the changing business environment. Some key points to note:

- In Fig. 5.10 we show two items—the performance measure listing on the left and the daily management plan on the right. Look at the performance measure listing on the left, and note that the measures have been sorted into four categories. This is an arbitrary but convenient segmentation. Let us quickly review them.

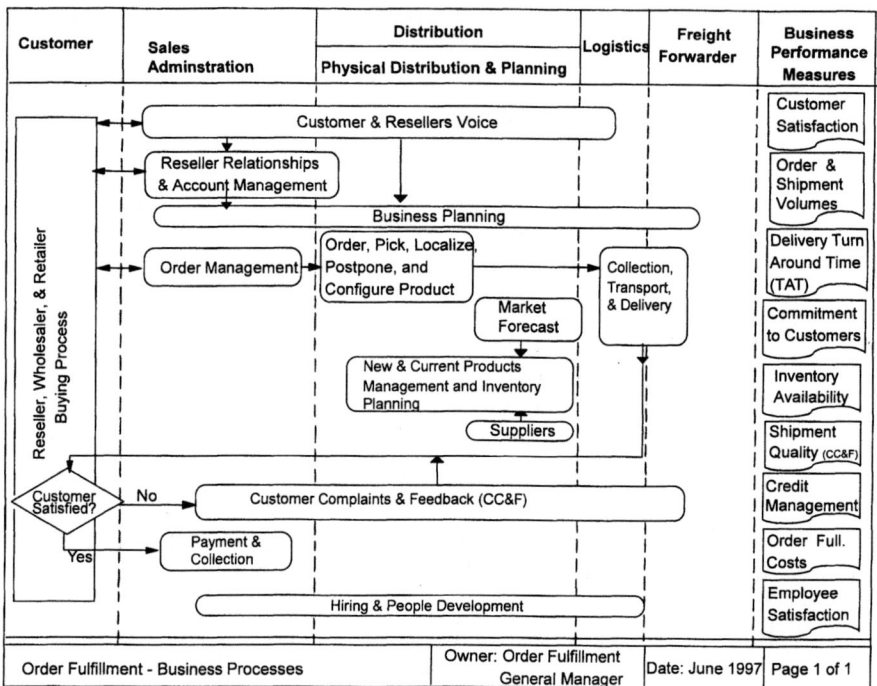

Customer	Sales Adminstration	Distribution		Logistics	Freight Forwarder	Business Performance Measures
		Physical Distribution & Planning				

Figure 5.9 Process map of an order fulfillment operation.

Customer-related. The most crucial are the customer-related measures, and these should include items such as collection of the customer's voice in the area of satisfaction and needs, customer complaints, and building of rapport with customers. Two specific customer measures based on customer surveys and inputs were included: delivery commitment and delivery turnaround time (TAT).

Operation-related. This can be measures that look at internal efficiency and productivity. Suppliers and vendors could also be included here.

Finance-related. This is to ensure that the operation runs at competitive costs.

People-related. Monitoring is done of employee health, education, and satisfaction.

Together these items measure the health of the organization. Or as we have called it, this is the business scorecard of the operation—at any time it can tell the how the key processes are working and can indicate how well the value delivery system is working.

Order Fulfillment Operations - Performance Measures / 1997	
Item	**Remarks**
CUSTOMER RELATED	
Commitment to Customers	
-Meet Ist. Acknowledge Date	Goal per current Daily Management Plan
-Meet Customer Delivery Date	★ *This is a Hoshin strategy for all countries*
Escalated Complaints & Claims to General Manager	Goal per current Daily Management Plan
Turn Around Time	Daily Management for all countries but a
-Order to Deliver (per Country)	★ *A Hoshin strategy for India and China*
Customer Satisfaction	Goal per current Daily Management Plan
	data comes from yearly independent survey
Shipment Quality (Delivery &	Goal per current Daily Management Plan
Packaging Failures)	
Internal/Operational	
Inventory Availability	
-System Fill Rate (SFR)	Goal in current Daily Management Plan
-Supplier on Time Delivery	★ *New, will be in FY97 Hoshin Plan*
Financial	
YTD Net Shipments	Goal in current Daily Management plan
Quarterly Expense Spending	Goal in current Daily Management plan
YTD Distribution Cost	Goal in current Daily Management plan
Credit Management - Account Receivables	Goal in current Daily Management plan
Total Order Fulfillment Cost	Goal in current Daily Management plan
People Related	
Performance Evaluation Timeliness	Goal in current Daily Management plan
Employee Communication Sessions	Goal in current Daily Management plan
Employee Satisfaction Surveys	Conducted every year

Order Fulfillment Operations - Daily Management Plan (FY97)				Date: 10 Jan, 97 Prepared by:	
	Goal	**Limit**	**Data Source**	**Nov**	**Dec**
Customer Related					
Commitment to Customers	92%		OP	93	94
Meet 1st. Ack. Date					
Turn Around Time	Per		Logistic		
Order to Deliver	Table				
Escalated Complaints & Claims to G. Manager	3	5	File	1	2
Quality of Shipments and Delivery	1000 ppm		Quality	745	825
Customer Satisfaction To be measured in July	#1	In 5 Countries	Mktg		
Internal/Operational					
System Fill Rate	93%	85%	ML	92.6	91.8
Financial					
YTD Net Shipments	Target	- 10%	BS	-4.9	-4.9
Quarterly Spending	Target	+ 5%	BS	0.98	0.8
A/R over 90 days	<1%		BS		
YTD Distribution Cost		- 10%		0.9	0.85
(% of net shipments)	1.0		BS	4.7	4.7
Order Fulfillment Cost (% of net shipments)	≤ 5.3		BS		
People Related					
Overdue Evaluation	0	0	Pers.	0	0
Comm. Sessions	3/qtr	2/qtr	File	3	2
Employee Satisfaction (Due May)	8.0	6.5	Pers.		

Figure 5.10 Performance measure table and daily management, or business fundamentals, plan for the order fulfillment operation discussed in Fig. 5.9. Note that several performance measures (see remarks in italic) do not get transferred into the daily management plan because there is no process to achieve them.

- In the Remarks column, we have indicated that some items are listed in the operation's daily management plan, while others are listed in the operation's annual Hoshin plan. Why in the annual Hoshin plan? The reason is that some new processes have just been added, and these processes are not understood and hence not well managed. This is what happens when you attempt to implement process management. Some processes will be in control and well managed; other processes will be new, not understood, and poorly managed; and still other processes may have moved from well managed to poorly managed, because of change or negligence. What you need to do for the new or poorly managed processes is to put them into your annual Hoshin plan to document and improve them. That is what we indicated in Fig. 5.10. You could also manage these processes via the operation's project list. We personally prefer, however, to stick them in the annual plan. Hence, if we look at Fig 5.10, the following three are new processes that need to be established before they can go in a daily management plan:

Meet customer delivery date, that is, date product will arrive at customer's doorstep

Order-to-delivery time for India and China

Supplier's on-time delivery performance (that is, is the supplier meeting commitments?)

- On the right of Fig. 5.10, we give an extract of the daily management plan for the order fulfillment operation. Actual performance for two months is shown. Note the following:

    The current processes being managed are shown in Fig. 5.10, in the daily management plan.

    The daily management plan is reviewed every month, and a formal review is done together with the operational annual Hoshin plan every quarter. The expectation is that all processes meet their target (within limits), all deviations and abnormalities from target are captured and understood, corrective action is taken, and their recurrence is prevented. For many of the performance measures, the review and analysis are done daily, especially at the next level. Hence our use of the term *daily management*.

    All measures have goals and limits. Refer to the discussion of limits in Chap. 3.

- In this order fulfillment operation, there is no next-level process map, although there could be. Since the operation has already come down to the third level of process maps, it was decided not to go any further. The next-lower levels in the operation, however, have daily management plans. These come from the process owners exploding each process in the map in Fig. 5.9 and managing the subprocesses at the next level. Finally, down one more level in the operation, the subprocesses are split and managed at a country level. Hence, we have moved from creating more layers of process maps to exploding the processes to the second and third levels.

Identifying and managing key processes

Each of the key processes from a process map must be managed. This will require that you go through these five steps:

- Process documentation
- Process performance measures
- Managing out-of-control conditions
- A training plan
- Continuous improvement of each process

From our experience we have identified several generic key

processes. We list them below and discuss each of them, either in this chapter or in App. C. Those shown in App. C are more esoteric and complex. Some of them are very important because they can help you predict and prevent problems, thus moving you to a zone of generating or creating quality.

The generic processes we have identified and wish to discuss are listed below. Some are discussed in this chapter, others in App. C or Chap. 2. Typically, the highest-level performance measure of each of these key processes will be in someone's daily management plan, while other submeasures are managed by the team or process owner.

- Competitive benchmarking. This was discussed in detail in Chap. 2.

- Product development process, specifically QFD. This was touched upon in Chap. 2 and is discussed in detail in App. C.

- Failure mode and effects analysis (FMEA). This tool allows prediction and prevention of problems to ensure product and service reliability, hence reducing potential customer complaints and inconvenience, and enhancing customer satisfaction. It is discussed in App. C.

- Project postmortems for both manufacturing and sales. This helps ensure that we learn and grow from an analysis of successes and failures. They are discussed in this chapter.

- Project postmortems using the T-type matrix. They are fairly esoteric and meant for analyzing and optimizing the process of design and testing of a product. They are discussed in detail in App. C.

- Setup and management of a companywide quality assurance system. This is crucial for any organization providing products and services and is discussed in this chapter.

- The sales process. Yes, sales is a process! A detailed discussion is given in this chapter.

Analysis of New Product Development Process via Project Postmortem

Problems in manufacturing can be detected, analyzed, and solved relatively easily because of several reasons: good process documentation and deviation analysis are usually available (refer to the section "Quality Assurance"), problems are stereotyped, and problems occur during a short manufacturing stage.

In new product development, however, the relationship between problem generation and problem detection is obscure. In addition, the new product development stage can take a long time, ranging from months to several years. Because of these issues, problems are often resolved on an immediate ad hoc basis with little concern for the root cause of the problem, which could have occurred years ago.

Typically, about 50 to 80 percent of problems discovered in the later stages of manufacturing and at the customer's location are created in the early stages of product development. This means that if we catch problems early, or better still eliminate problem generation, we will save the time, effort, and resources needed to fix the problem later. Customer satisfaction will also increase.

Here we discuss how the new product development process can be analyzed, with the purpose of reducing design changes and problems and hence providing continuous improvement. We offer two specific methodologies that will help:

- The project postmortem
- The T-type matrix for more detailed analysis. This is discussed in App. C.

The project postmortem

After a product has been designed, manufactured, and delivered to the customer, a review should be conducted to analyze the product in terms of market acceptance, the design process, costs, ease of manufacturing, and quality. Such an analysis, or postmortem, is useful to understand the product development process. The knowledge learned will be used to improve the entire product development process from collecting customer needs, designing and manufacturing the product, delivery, and use by the customer. This postmortem is done after product release. Let us go into some detail and review an actual example.

Postmortem objective and benefits. *Objective:* To analyze the product development cycle after product release, in order to understand strengths and weaknesses. The knowledge learned will be used to improve future product development cycles. The postmortem process will

- Strengthen cooperation of product development teams
- Facilitate sharing of experience and learning throughout an entity
- Allow understanding of the root causes of problem generation and design changes in new product development
- Reduce future product development time—measured by time to market, breakeven time, and so on
- Reduce future start-up costs and delays caused by redesigns and production changes, after product release to manufacturing or the customer
- Improve product quality and customer satisfaction for future products

■ Ensure a more efficient product development process

Postmortem process. Typically, a postmortem or project review team is selected about 2 months after the initial product release. The team includes representatives from the research and development, marketing, manufacturing, and quality departments. The team could be led by the project leader (of the product under review), a production manager, or a quality assurance manager. It will meet several times and discuss the issues shown below in the report format. All the required data will be available during the 3-month period after product release. Many of the R&D process data, however, will be available at product release; hence, these data can be collected earlier by R&D.

Postmortem Process Flow

	Process step	When done
1.0	Project released to manufacturing and shipped to customers	Product release = T
2.0	R&D reviews design process	T + 1 month
	Use of T-type matrix	
3.0	Production engineering or quality assurance appoints postmortem task force leader	T + 2 months
4.0	Postmortem meeting planned	T + 2 months
	Agenda	
	Premeeting questionnaire (request for data mailed out according to recommended format)	
5.0	Meetings take place	T + 3 months
	Data collected and discussed for each item in recommended report format	
	Deviations analyzed	
	Problems reviewed	
6.0	Issue report	T + 4 months
	Standard format	
	Analysis and proposals	
7.0	Review proposals for corrective action to improve processes and practices	T + 5 months
	Final meeting with senior management	
	Agreement on proposals	
	Short-term and long-term plans	

Note: Step 2 is driven by the R&D department, while steps 3 to 7 are driven by a task force leader.

Project postmortem—report format.   The following is an outline of a recommended format. Note that the specific items measured at different companies will vary according to product type, especially for items 7 and 8 shown below.

1. *Project identification*

   a. Product name and number
   b. Team members

2. *Objective.*   Standard "canned" phrase given in the postmortem objective shown earlier

3. *Marketing*

   a. Forecasts versus actual, and deviation analysis
   b. Orders and shipments
   c. Customer acceptance, customer feedback, and comments
   d. Planned profit versus actual, and deviation analysis
   e. Proposal and corrective action

4. *Research and development*

   a. Design time, planned versus actual, and deviation analysis
   b. R&D process analysis, that is, design changes and rework required (redesign). A design process analysis using a T-type matrix is recommended. The T-type analysis matrix is discussed next in App. C.
   c. Proposal and corrective action

5. *Manufacturing*

   a. Costs, planned versus actual, and deviation analysis
   b. Manufacturing difficulties due to design
   c. Number of design errors caught in manufacturing, that is, production changes that were made to correct for design errors. The T-type matrix in item 4 above will also provide some of these data.
   d. Supplier problems or issues
   e. Material and parts issues
   f. Production yields and scrap—planned versus actual and deviation analysis
   g. Proposal and corrective action

6. *Quality assurance*

   a. Test data
   b. Failure data
   c. Customer feedback, comments, and issues
   d. Proposal and corrective action

7. *Design error performance measures.* This is to compare design errors between this project and past projects. The comparison gives an indication of improvement over time.

    a. Rate of design changes (measures production changes after product release versus past trends)
    b. Rate of first run defects (measures errors during the first prototype run versus past trends)
    c. Proposals and corrective action

8. *Standardization issues.* The purpose here is to review progress of automation and purchase of approved parts versus past trends.

    a. Rate of automation (measures percentage of automated assembly versus past trends)
    b. Rate of use of approved or certified parts versus past trends
    c. Proposal and corrective action

9. *Overall summary*

    a. Summary of learning points
    b. What will be changed and corrective action to improve current processes

10. *Appendix*

    a. Table of contents
    b. Supporting data for all items discussed in main report, that is, items 3 to 9
    c. Include long-term trend data for item 7 (design errors) and item 8 (standardization goals)

Postmortem meeting guidelines. How should a postmortem meeting be conducted? Here are some guidelines, based on our experience, on how to ensure productive and effective postmortem meetings.

1. Prior to the first meeting, the team leader must send out a detailed questionnaire. The questionnaire should be very specific, based on the report format proposed earlier. The participants should be given enough time to collect the needed data.

2. At the postmortem meeting, remember to

    a. Identify successes and give credit.
    b. Listen to everybody.
    c. Prioritize problems.
    d. Identify weaknesses, analyze for root cause, and then propose corrective action.
    e. Keep good minutes.

> *f.* Stick to the facts, and use data.

And do *not*

> *g.* Blame anybody. Everybody is there to learn.
> *h.* Divide up into competing groups.
> *i.* Turn the meeting into a complaints session.
> *j.* Deviate from the agenda, allowing storytelling or "dog and pony" shows.
> *k.* Assign action items to a person who does not accept them.
> *l.* Try to solve all problems.

3. After the project postmortem, remember to

> *a.* Distribute the report.
> *b.* Share conclusions and get acceptance for proposals from senior management.
> *c.* Keep performance measures on postmortem process at your entity, such as
>   - Percentage of projects with postmortems
>   - Timeliness of postmortem activity
>   - A list of and number of contributions to the best-practice file or standard update reports
>   - A list of and number of problems that recur from one project to another
> *d.* Follow up on proposed corrective actions. That is, were they implemented? This is important.
> *e.* Ensure that the lessons learned are institutionalized.

For the last three items listed above, we suggest you have a central custodian for the process, for example, a R&D standards manager or a neutral quality assurance manager.

Sample project postmortem report. We show a portion of a report in Fig. 5.11*a* and *b*. Note the summary and recommendations. These will be used to improve the design process in the future.

Summary of project postmortem

Approximately 50 to 80 percent of problems and design changes discovered in the later stages of manufacturing and at the customer's location are created in the early stages of the product development process. The reasons for this must be analyzed and understood, and the problems reduced.

A project postmortem is crucial for understanding strengths and weaknesses in the new product development process. And within the

PROJECT POSTMORTEM   Page 1 of 15

Product: 5000X Components Tester   2 August 1990

1. Team Members:
   R&D Section Manager
   Marketing Manager
   Production Manager
   Engineering Section Manager

5000X Project Postmortem, Page 2 of 15

3. MARKETING REVIEW

   A) Forecast & Shipments

      The 5000X Components Tester has been well received by Customers.
      Sales, however, are at 71% of forecast during the last 6 months. The
      breakdown is as follows:
      * Sales to Market Segment A (off-line testing) = 120% of forecast
      * Sales to Market Segment B (in line testing) = 43% of forecast
      * Overall Sales are at 71% of forecast (Refer to appendix for more data)

   b) Profits

      The operating profit year to date is 8.3%, which is below forecast and is
      due to shipments below forecast (Refer to appendix for more data)

   c) Sales Deviation Analysis:

      Although we have exceeded forecasted Sales in Market Segment A, Sales
      to Market Segment B are below forecast due to Customer dissatisfaction
      with the feature set. The 5000X is not very appropriate for certain in-line
      testing. We interviewed customers in both market segments.
      All the needs of market segment A & B were captured in the top level QFD
      table (5000-A1), but in the second revision (5000-A2), some needs were
      dropped because of possible complexity and project delay. The decision
      to drop was made by the R&D engineer in the absence of the section
      manager. This severely impacted the needs/requests of Segment B/
      In line customers.

   Action: Establish a formal procedure for discarding any customer request.

2. Objective

Figure 5.11a A portion of a project postmortem report.

5000X Project Postmortem, Page 9 of 15

8. Standardization Issues
   * Rate of Automation:
   This is defined as $\dfrac{\text{number of fully automated component insertions}}{\text{number of total components}}$ X 100%

   The rate for the 5000X was 73%, which compares favorably with the division average of 65% and 1987 target of 70%.

   * Rate of use of Approved/Certified parts is defined as

   $\dfrac{\text{number of approved parts used}}{\text{number of total parts used}}$ X 100%

5000X Project Postmortem, Page 10 of 15

9. Overall Summary
   Recommendations

   1) During collection and collation of Customer needs & requests, (in the QFD top level table) none shall be discarded without prior authorization by the Marketing and R&D Managers.

      Action: a) Inform all Marketing and R&D staff
              b) Add checklist in preliminary stage and Final design review to discuss all customer requests proposed for deletion
                 a) by Marketing and R&D section managers
                 b) by Marketing manager

   2) We conducted a FMEA ergonomic study for the 5000X tester front panel. The resulting instrument panel received rave reviews by customers. In the future we will conduct a FMEA study on all new product front panels.

      Action: Quality Manager

   3) Paco was selected specifically as a supplier for the 5000X product. However, Paco was selected by R & D without consulting the Purchasing Department. Paco is a consumer electronics supplier and we have had numerous problems with them not being able to meet our specifications. In future, we must adhere to the policy of selecting suppliers through the Purchasing Department.

      Action : Purchasing Manager and R & D Manager

Figure 5.11b A portion of a project postmortem report.

project postmortem the T-type matrix will provide a very detailed analysis and understanding of design changes, problems, and errors. In fact, the project postmortem is a process that analyzes the effectiveness of all the processes in new product development.

Kaoru Ishikawa, the father of Japanese total quality, has this to say about the analysis of new product development: "Whenever I am asked to help introduce a total quality control program to a company, I choose for my case study one of the company's new product development projects that is full of problems."

The lessons learned from the postmortem can be used to improve the entire product development process from collecting customer needs, to forecasting, to designing, to manufacturing, and to use by the customer. This continuous improvement of the product development process is the key to building products which are robust and successful.

Quality Assurance System

Every product manufacturing division, operation, or entity should have a quality assurance system. Such a system should include

- Procedures for management of customer issues, feedback, complaints, and claims
- Standards for design, processes, purchasing, etc.
- Quality control procedures for the manufacturing process
- Overall coordination of the improvement process, after customer and failure data are collected.

We have discussed some of these items already, for example, the customer complaints and feedback system. We will not discuss items that have been mentioned, but we tie them here to the quality assurance system.

Look at Fig. 5.12, where we show the flow of quality assurance activities. The quality activities start immediately during design when the product and process must be designed; and they continue until after product delivery to the customer. Figure 5.12 has the following highlights:

1. Product specific standards are listed on the left of the figure. These are required during the design of each product and include product specifications, process flow and control, and job or work standards. The job standards refer to the actual work procedures for each step in manufacturing. These items will then influence the following:

   a. Specific standards for new, never-purchased-before material and parts

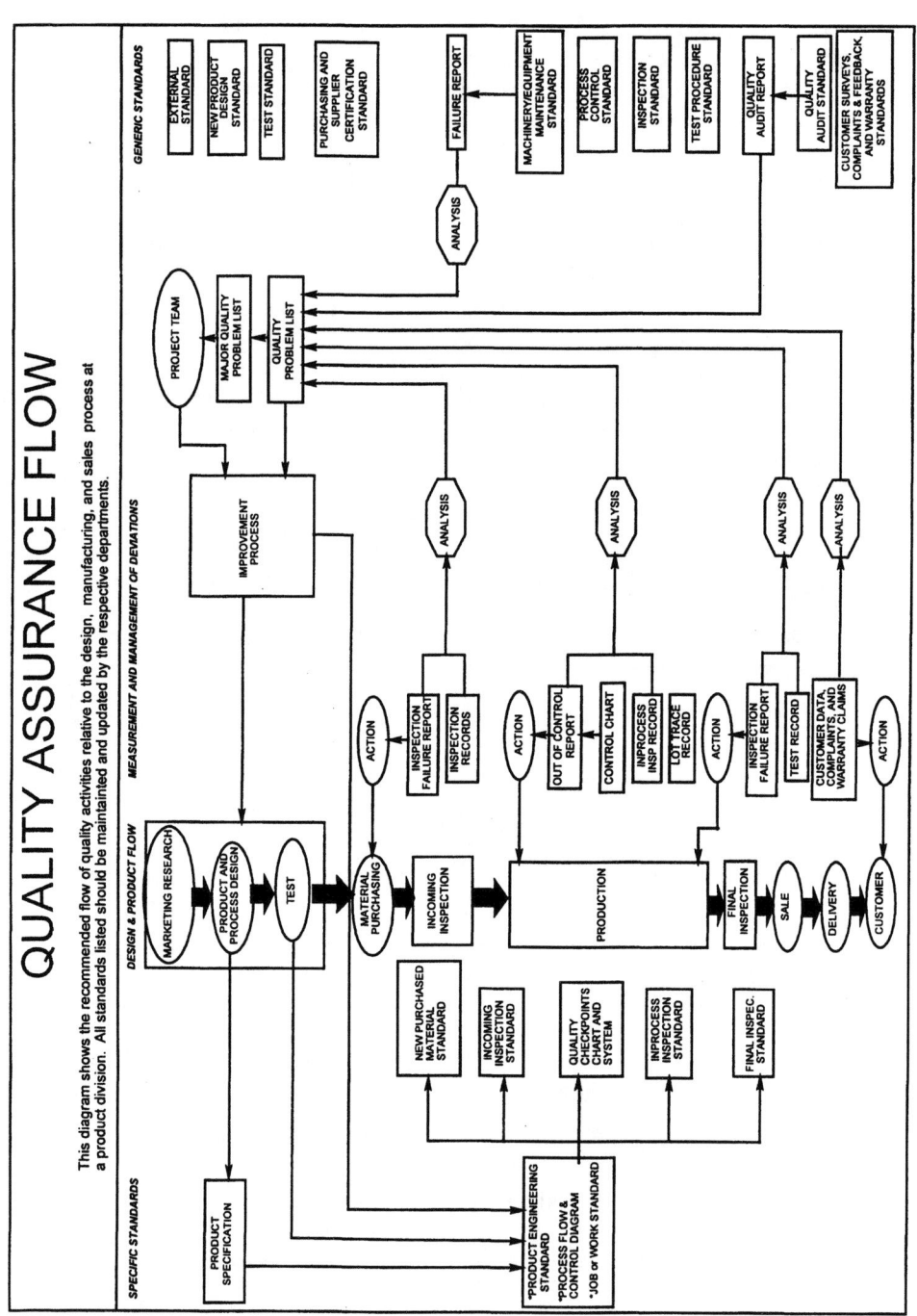

Figure 5.12 Quality Assurance System.

    b. Incoming, in-process, and final inspection standards

    c. Quality checkpoints chart and systems. This is discussed in great detail in the section "Quality Checkpoints Chart and System."

2. Generic standards are listed on the right and include

    a. External standards, such as measurement standards

    b. New product design standards

    c. Machine and equipment maintenance standards

    d. Quality audit standards which are used to audit both the current manufacturing process and products on a routine basis—to ensure these confirm to the agreed upon standards

    e. Customer complaints and feedback system standards discussed in Chap. 2

3. On the immediate right of the product flow (with heavy black arrows), we have a series of activities that measure and manage deviation from standards. Examples are

    a. Inspection failures

    b. Customer complaints and claims after product shipment, discussed in detail in Chap. 2

    c. Control chart analysis and deviation report. We discuss this in greater detail later in this chapter.

These deviations must be detected, analyzed, and eliminated. In addition, items for improvement are routinely identified.

Quality assurance improvement cycle

Look again at Fig. 5.12. The various deviations from standards require

- Immediate short-term corrective action
- Detailed analysis and understanding of the root cause
- Generation of list of quality problems, all of which need resolution
- Generation of a major quality problem list, which are given high priority and assigned to project teams for resolution
- Use of the resulting improvement cycle to influence marketing, product design, process design, and the current process

This closed loop in the quality assurance system is very important. This runs in parallel to the product postmortem that we discussed earlier, except that this is an ongoing activity. The review of quality problems and the appointment of project teams to improve them should be managed by the quality department, working with the quality steering committee of the company.

Quality checkpoints chart and system

The quality checkpoints chart is a very crucial and important item for any manufacturing environment. It is not commonly applied. When applied, it is not always managed with sufficient discipline. That discipline is what makes the difference between a company that makes products with high variable (or inconsistent) quality and a company that makes products with consistent quality.

Such a system is basic to all process planning and control. It provides a collation of all important and critical manufacturing data which should be documented, and it includes

1. A production process flowchart, including material status and its transformation to higher-level assemblies

2. Reference to job standards and other documents that are necessary to complete each step in manufacturing

3. Points in the process where measurement and control occur

4. Responsible parties for managing the process and correcting out-of-control situations. This step will be supported by a formal, documented out-of-control report.

The documentation is typically kept for each product manufactured. As manufacturing processes become simpler and more automated, the number of critical or control points will decrease. Nevertheless, the overall documentation and process must be maintained, managed, and improved by manufacturing personnel.

The process and quality checkpoints chart. In Table 5.1, we show a completed format for listing the process and quality checkpoints of a product during manufacturing. It is maintained on a Lotus 1-2-3 file or equivalent. Here is a brief explanation of each column in the figure:

- *Material.* This refers to the specific part or assembly that is being used.

- *Flowchart step number.* There are two parts to this. *Incoming:* This refers to incoming material or a subassembly flowing into the main assembly procedure. *Main:* This refers to the main assembly steps. This can be merged into one flow if preferable. Refer to Table 5.1 for an actual example. The step numbers refer to the actual step in a separate flowchart of activities.

- *Process specifications.* This is the actual step of the work standard that describes how each step is done by workers or machine. This should consist of actual work procedures, not theoretical ones.

- *Critical control parameters.* Here we list the critical parameters that must be managed or controlled at each process step.

TABLE 5.1 Actual Example of a Process and Quality Checkpoints Chart

PRODUCT: LAMP				TITLE: PROCESS ASSEMBLY AND QUALITY CHECKPOINTS (DAILY MANAGEMENT FOR PRODUCTION LINE)							
MATERIAL	FLOWCHART IN / MAIN	PROCESS SPEC. & JOB STANDARD	NO.	CRITICAL CONTROL PARAMETER	FREQUENCY OR SAMPLE SIZE	INSPECTION OR CONTROL METHOD (SPEC)	CHART USED	INSPECTION ITEM	SPECIFICATIONS LIMIT	RESPOND PERSONNEL	ACTION PERSONNEL
SILVER-EPOXY	1 / 2	DIE ATTACH 5956-5141-42	2.0	DIE SHEAR STRENGTH	ONE BOTTLE PER SHIPMENT SS = 90 UNITS	DIE SHEAR M/C	RECORD BOOK		MIN 80 GM	E.O.	IQA ENG.
			2.1	STORAGE IN REFRIGERATOR						U.O	PROD SUP
			2.2	KEPT AT ROOM TEMP FOR 4 HRS MIN BEFORE OPENING BOTTLE		TIME AND DATE	RECORD ON BOTTLE	SILVER EPOXY BOTTLE	MIN 4 HOURS	U.O	PROD SUP
			2.3	CHANGE EPOXY EVERY 72 HRS		CHANGE EVERY NIGHT SHIFT ON SUNDAY & WEDNESDAY				D.L.	PROD SUP
DICE	3										
LEAD-FRAME	4 / 5	DIE ATTACH 5956-5128-42	5.1	DA INSP CRITERIA	4 X 150 UNITS PER MACHINE/SHIFT	VISUAL @ 20 X MIN.	TREND CHART	DIE ATTACH UNITS	ACC = 1, REJ = 2	INSPECTOR	OPERATOR & TECH
					INSPECT AFTER MAJOR CONVERSION/REPAIR 1X/SHIFT	VISUAL @ 20 X MIN.	TREND CHART (WRITE AC/AR AT THIS INSPECTION POINT)	DIE ATTACH UNITS	ACC = 1, REJ = 2	INSPECTOR	OPERATOR & TECH.
	/ 6	DIE ATTACH CURE 5956-5141-42	6.1	OVEN TEMP	ONCE A WEEK	TEMP PROFILE	PROFILE CHART	IR OVEN	SOAK TIME = MINIMUM 6 MINS	E.O.	ENGR
					1X/SHIFT	TERMINAL	RECORDING BOOK	IR OVEN	ZONE 1:400+/-5°C ZONE 2:350+/-5°C ZONE 3:350+/-5°C ZONE 4:350+/-5°C ZONE 5:350+/-5°C ZONE 6:350+/-5°C ZONE 7:350+/-5°C (NOTE *=DEGREES)	D.L.	TECH/SUP.
			6.2	SPEED	1X/SHIFT	TERMINAL	RECORDING BOOK	IR OVEN	5 IN/MIN	D.L.	TECH/SUP
			6.3	DIE SHEAR STRENGTH	5 UNITS/SAMPLE 1X/SHIFT	DIE SHEAR M/C	X-R CHART	SHEAR STRENGTH	80 GRAM (MIN)	D/A INSP	PROD ENG.
GOLD WIRE	7		7.1	TENSILE TEST	3 SPOOLS PER SHIPMT	TENSILE STRENGTH METER	RECORDING BOOK	TENSILE STRENGTH	MIN 8 GM ACC = 0, REJ = 1	IQA	IQA ENG.
			7.2	VISUAL	SS AQL = 2.5%	VISUAL 20X	RECORDING BOOK	GOLD WIRE	AQL = 2.5%	IQA	IQA ENG.

INTERNAL PROCESS DOC # : OPT-G-001   REVISION: G   DATE OF ISSUE: 15 MAY 1989   PAGE: 1 OF 3

- *Frequency or sample size.* This explains how often the critical parameter is inspected.

- *Inspection or control method.* This explains the method of inspection of the critical parameter.

- *Chart used.* After the critical parameter is inspected or controlled, the data are entered and maintained in the document listed here.

- *Inspection item.* This is the actual item to be inspected. This is different from the critical parameter to be controlled. For example, the critical parameter could be dimension, while the inspection item could be gold bonding wire.

- *Specification limit.* This states the limit. When the limit is crossed, we have a failure and corrective action must be taken. These limits are based on experience and data, for example, the upper and lower limits in a control chart. During normal day-to-day operation, these limits would not be crossed. But when they are exceeded, we have an out-of-control situation that needs analysis and resolution.

- *Respond personnel.* This states the person who is responsible for checking the various critical parameters and raising the alarm when a specification limit is crossed.

- *Action personnel.* This states the person who must take corrective action when a specification limit is crossed. This activity includes an analysis of why the limits were crossed, resulting in an out-of-control situation, and the appropriate immediate and long-term corrective action.

Selecting quality checkpoints. Selecting the correct points for the quality checkpoints chart is crucial. How do we ensure that the correct points are selected? Here are some suggestions:

From experience. This is the most obvious method and relies on experienced production personnel. A documented list of critical production steps should be available for reference whenever a new product quality checkpoint chart is prepared.

Linkage to quality function deployment charts. If QFD has been used for product design, then a lower-level QFD chart can be prepared to link the production process to crucial product characteristics. This provides a link back to original customer needs. Refer to the discussion of the four phases of QFD.

Failure mode and effect analysis (FMEA). After a detailed product flow is drawn, all production steps are listed. To highlight potential

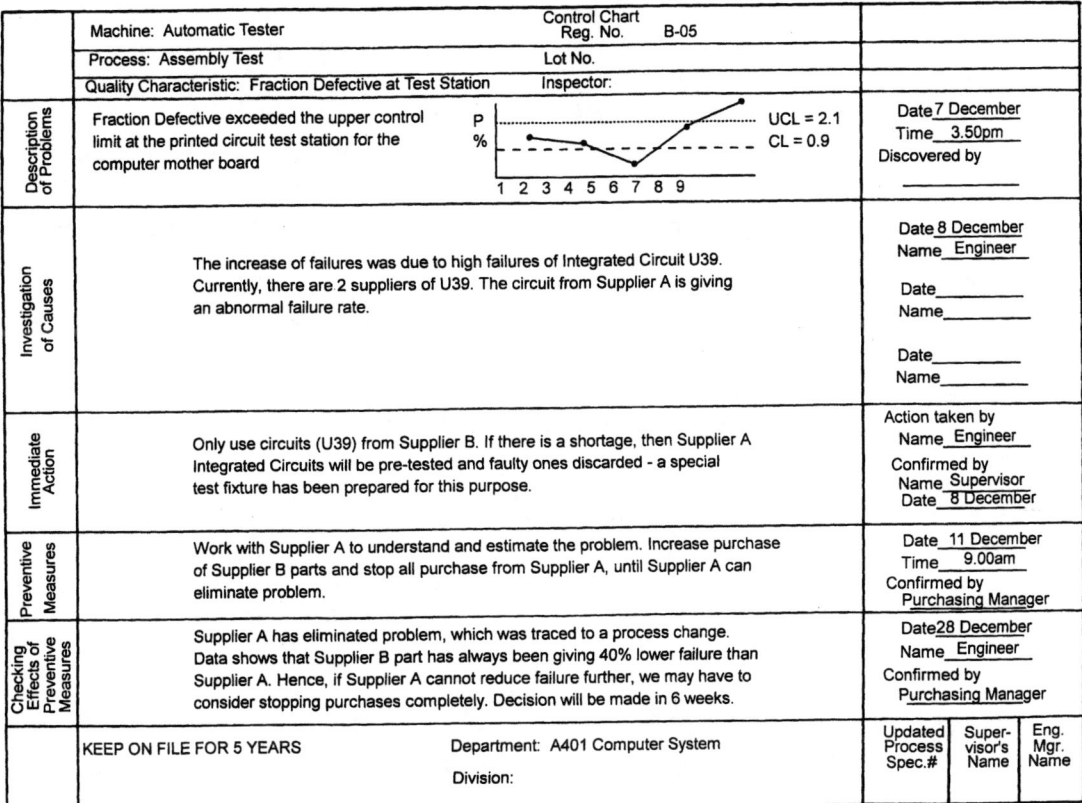

	Machine: Automatic Tester	Control Chart Reg. No. B-05
	Process: Assembly Test	Lot No.
	Quality Characteristic: Fraction Defective at Test Station Inspector:
Description of Problems	Fraction Defective exceeded the upper control limit at the printed circuit test station for the computer mother board	Date 7 December Time 3.50pm Discovered by
Investigation of Causes	The increase of failures was due to high failures of Integrated Circuit U39. Currently, there are 2 suppliers of U39. The circuit from Supplier A is giving an abnormal failure rate.	Date 8 December Name Engineer Date_____ Name_____ Date_____ Name_____
Immediate Action	Only use circuits (U39) from Supplier B. If there is a shortage, then Supplier A Integrated Circuits will be pre-tested and faulty ones discarded - a special test fixture has been prepared for this purpose.	Action taken by Name Engineer Confirmed by Name Supervisor Date 8 December
Preventive Measures	Work with Supplier A to understand and estimate the problem. Increase purchase of Supplier B parts and stop all purchase from Supplier A, until Supplier A can eliminate problem.	Date 11 December Time 9.00am Confirmed by Purchasing Manager
Checking Effects of Preventive Measures	Supplier A has eliminated problem, which was traced to a process change. Data shows that Supplier B part has always been giving 40% lower failure than Supplier A. Hence, if Supplier A cannot reduce failure further, we may have to consider stopping purchases completely. Decision will be made in 6 weeks.	Date 28 December Name Engineer Confirmed by Purchasing Manager
	KEEP ON FILE FOR 5 YEARS Department: A401 Computer System Division:	Updated Process Spec.# / Supervisor's Name / Eng. Mgr. Name

Figure 5.13 A completed out-of-control report.

process problems, an FMEA can be done. Refer to the discussion of FMEA in App. C.

Derived from customer issues. After a product is released to market, customers may give feedback on minor or major defects. This feedback is used to analyze the process and to determine areas that need better control. This information can be used to improve current process controls or to add new process control points.

The out-of-control report. When an out-of-control situation occurs, a detailed analysis and corrective action are required; refer to Fig. 5.13 for an example. Specifically, such a report must cover the following:

- *Description of the problem.* Describe what happened and where it happened; provide data.

- *Investigation of causes.*   Why was the specification not met, or why were the control limits crossed? This requires an understanding of the root cause.

- *Immediate corrective action.*   Here we list a quick fix or remedial action. This action will get things back to normal. This includes inspection for the defect or repair of the defect.

- *Preventive corrective action.*   Here we list the preventive action. This will include elimination of the root cause, hence preventing future recurrence of this problem. This is more important than the immediate action, but due to constraints the immediate action may be implemented first.

- *Checking effects of long-term measures.*   This step is to check on the effectiveness of the preventive measures and to ensure the root cause of the problem is eliminated.

In Fig. 5.13 we have given an example of a completed out-of-control report. The analysis and corrective action are fairly straightforward. Note that the problem was traced to a process change at one of the suppliers; although the problem has been solved, the supplier is in danger of being dropped.

The Sales Process

The sales process is often thought to be unstructured and not measurable, but it is really quite simple to structure the work into several activities and subprocesses. In Fig. 5.8, we have shown several activities occurring in the marketing and sales environment, ranging from understanding market needs and opportunities, all the way to delivery and support of the product. All the activities listed can be better understood and properly documented with performance measures, and training can be provided.

It would be foolish to suggest that a structured sales process is a panacea for all sales problems—there is no substitute for an intelligent and creative sales representative. The techniques we describe are those that many successful sales representatives or managers already use. The outcome of using these techniques can be a better-managed and more predictable sales process, resulting in higher productivity.

In this discussion, we are referring not to retail sales, but instead to items sold by a direct sales force—items such as instruments, computers, turbines, engines, and machinery. We do not discuss everything shown in Fig. 5.8; instead, we focus on some of the less well-understood activities. I am grateful to Shailesh Naik, sales and quality manager at Hewlett-Packard Australia, for his help in editing this section.

Quality Assurance Sacrificed for Earlier Time to Market

Recently, even reputable Japanese manufacturers have been having quality problems. *Business Week*[1] reported that Japanese television makers Matsushita, Sony, Pioneer, and Toshiba recalled dozens of models that smoked or caused fires. In fact, a Pioneer TV set may have caused Japan's first major high-rise apartment fire, according to a government disclosure. Epson recalled about 100,000 laptop computers. Even highly reputable Toyota had to recall its luxury Lexus sedan to fix brake light and cruise control problems. Here is a list of recent recalls in Japan, compiled by *Business Week*.

Company/Product	*Problem*
Toyota Lexus automobiles*	Cruise control, brake light
Mitsubishi Motors Pajero wagon	Accelerator
Isuzu trucks	Loose bolt
Fuji Heavy Rex minicars	Clutch contact
Yamaha Virago motorcycles	Fuel leak
Honda Horizon motorcycles	Transmission cog
Seiko Epson laptop computers	Circuit soldering
Toshiba TVs	High-voltage circuits
Pioneer Electric TVs	Circuit soldering
Matsushita TVs	Transformer insulation
Sony TVs	High-voltage circuits

Business Week questions whether Japan is losing its quality edge. The answer is, not yet. The sudden surge of problems can be traced to the following causes, compiled by *Business Week* and the author:

- Cheaper imported components from other Asian suppliers, who are not able to provide consistent quality.
- Lack of technical skill in new workers; or worse still, erosion of the work ethic.
- Lack of discipline in analyzing and correcting minor deviations.
- Introduction of high-technology parts and designs without proper understanding of how to control their quality or proper prediction of possible failure modes.
- The desire for shorter time to market, requiring a shorter design and manufacturing cycle. This may cause the elimination of certain tests or extensive tests, which were normally done previously.

*The only recall affecting the U.S. market.

The sales funnel

The concept of the sales funnel is well known to most marketing and salespeople. This was popularized, in part, in the book *Strategic Selling.*[2] A sales representative just does not wait for orders; instead, she or he must keep pouring prospects and suspects into a pipe or funnel and expect that some will come out the other end as orders.

Items in the funnel can be divided into those above the funnel, those in the funnel, the best few, and orders. This is shown in Fig. 5.14. Each potential order must be tracked from inquiry, through the funnel, until it becomes an order. The steps that the potential order goes through will form the process.

We can now list the steps in the sales process. These steps will occur above, in, and at the bottom of the sales funnel, shown in Fig. 5.14. More specific steps of the sales process are listed in Table 5.2.

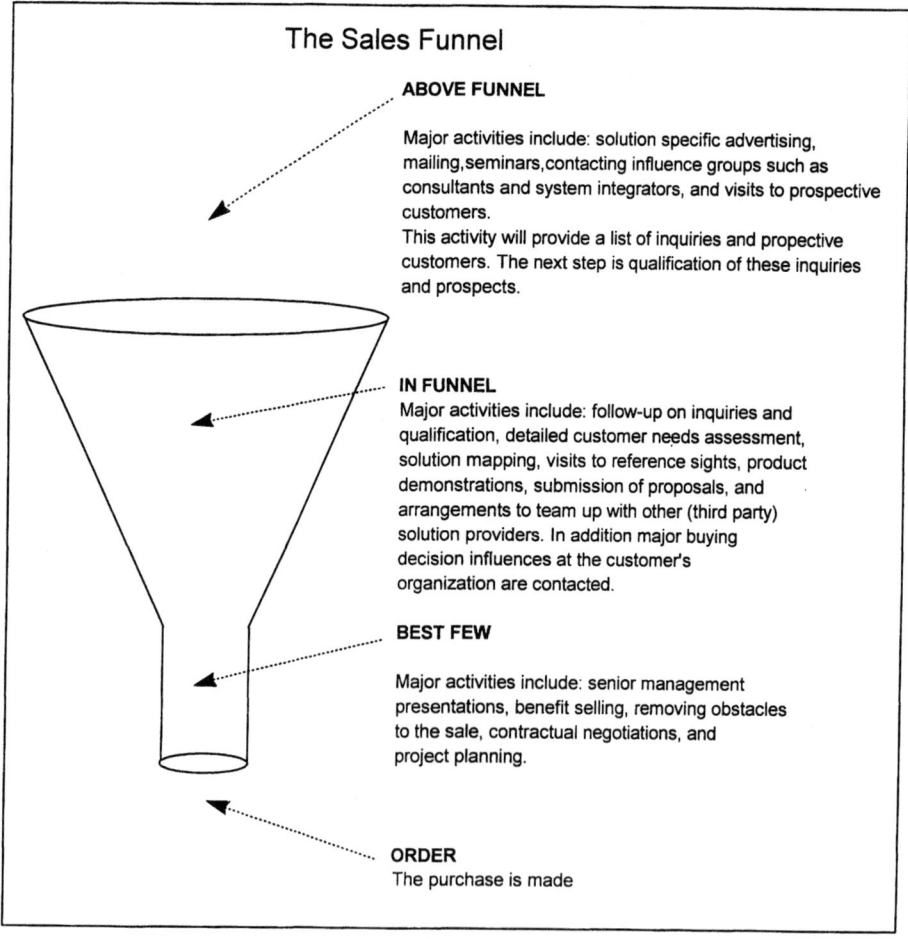

Figure 5.14 The sales funnel.

TABLE 5.2   Steps in the Sales Process

Process step	Detail of the step

Sales planning	Provide training in the sales process, products, and solutions
	    Business segmentation by products and markets
	    Territory quota planning
	Customer contact plan
	Product introduction and demonstration to develop customer inquiries
	Marketing programs to develop customer inquiries

Selling	Customer contact and visit
	    Understand needs
	    Review competition
	    Check budget
	    Check product availability
	Qualification of customer
	    Meet buying influences
	    Suggest solutions
	    Understand intricacies and politics of situation
	Review budget constraints
	Investigate
	    Study detailed needs
	    Provide solutions
	    Provide reference sites
	    Position competition
	Present detailed solution
	Close—get the order

Delivery	Deliver product
	    Keep commitments
	    Ensure trouble free delivery
	Install product

Sales won/lost postmortem	Analyze sales won or lost
	    Understand success and failure factors
	    Understand competitive offerings
	Improve current sales process

After-sales support	Provide after-sales support and service support Manage all accounts, ensure product and solution are meeting expectations; if not, understand why. Provide product support and service Measure customer satisfaction Collect and analyze all complaints and feedback

This entire process should be well documented and training provided; this is listed in the very first step of Table 5.2. Improvements to the process should be made based on analysis from the sales won/lost postmortem and other analysis.

Based on our discussion, we can now come up with performance measures for the sales process. But before we do that, let us look at the sales situation from a different perspective.

The sales situation

In Fig. 5.15, we show the relationship between sales forecasts and the actual results. Look at quadrant I in the figure. When a sale is forecast and won, we have success. What are the characteristics of such a situation?

- The needs of the customer were understood.
- Effective solutions were available for the customer.
- The sales activity and negotiations with the customer were well managed.
- If this success rate continues, sales productivity will be high.

In a similar fashion we can analyze the characteristics of all four quadrants in the sales situation. We show the characteristics of each quadrant in Fig. 5.16.

Now that we understand the characteristics, we can do a few things. First, clearly, as far as possible we want to be operating in quadrant I, where a sale is forecast and won. Alas, this cannot always be the case. So we have performance measures with appropriate targets for each quadrant. If we do that, we will be better able to control all four quadrants.

Forecast: \ Actual Results	Sale Won	Sale Lost
Forecasted Sale	Quadrant 1 Success	Quadrant 2 Failure
Non-forecasted Sale	Quadrant 3 A Lucky Win	Quadrant 4 Undeveloped Activity in seeking new customers

Figure 5.15 The four variables in the sales situation.

THE SALES SITUATION

ACTUAL RESULTS / FORECAST	WON	LOST
Forecasted Sale	**Success** * Understood customer needs * Effective solutions available * Good management of sales activity * Order productivity is improved	**Failure** * Beaten by competitor or postponed/stopped purchase. * Inappropriate understanding of customer needs. * Diminished relationship with customer.
Non-forecasted Sale	**Lucky Win** * Not aware of potential market * Lack of customer contact * Lack of customer information * Customer brought order forward	**Undeveloped Activity** * Need to develop new customers, markets and marketing communication *Need to understand customer requirements.

The above characteristics can be better managed by selecting appropriate performance measures as shown below

PROPOSED PERFORMANCE MEASURES

ACTUAL RESULTS / FORECAST	WON	LOST
Forecasted Sale	**Success** * In Funnel Rate $= \dfrac{\text{Qualified Funnel \$}}{\text{Sales Quota \$}}$ * Success Rate $= \dfrac{\text{Number of deals won}}{\text{Number of deals chased}}$	**Failure** * FTF Rate $= \dfrac{\text{Face to Face Time with Customer}}{\text{Total available time}}$ * Sales won/lost Postmortem
Non-forecasted Sale	**Lucky Win** * Lucky Win Rate $= \dfrac{\text{Non-Forecast \$}}{\text{Total Orders or Sale \$}}$ * Account Coverage Rate $= \dfrac{\text{Coverage of customers}}{\text{Total potential customers}}$	**Undeveloped** * Inquiry Rate $= \dfrac{\text{Total Inquiries (above funnel)\$}}{\text{Sales Quota \$}}$ * New Customer Call Rate $= \dfrac{\text{New customers contacted}}{\text{Total Customers contacted}}$

Figure 5.16   The sales situation.

Identifying appropriate performance measures

We now discuss some performance measures that can be used to manage the various subprocesses in sales. We have developed these measures from the sales process and the sales situation. The derivation of some of these measures is shown at the bottom of Fig. 5.16. More details about the performance measures are listed in Table 5.3, with typical targets and comments. The targets are typical of several sales managers we know. The performance measures and targets, however, will vary with the business and the management style.

The performance measures in Table 5.3 should be measured monthly by comparing actual results to targets. Typically the data are collected from sales representatives via a daily or weekly activity sheet. These data are entered into a database, such as Lotus 1-2-3. The performance measures can be plotted and compared to target figures. When there is a deviation, it may indicate a need to intervene. For example, if the inquiry rate starts to decline, it may indicate weak marketing programs or a lack of new products.

Sales won or lost postmortem

In the previous section, one performance measure discussed was a sales won/lost postmortem. The concept of a sales postmortem is similar to that of a project postmortem. What we wish to do here is to understand why product or service sales deals are won or lost.

Objective and benefits of a sales won/lost postmortem.   The objective is to better understand why sales are won or lost. The knowledge gained will be used to influence future sales, in order to increase sales productivity and improve the company's product development cycle. The sales won/lost postmortem will

- Provide a better understanding of the competition
- Enable you to better position your products or services against the competition
- Enable you to conduct a better advertising campaign or better product seminars
- Provide information on better product or service pricing
- Provide feedback to designers on product or service weaknesses
- Provide data on resources needed to pursue future sales deals
- Provide information on how to improve the training of sales representatives

TABLE 5.3   Performance Measures and Targets for the Sales Process (Continued)

	Performance measure	Typical target and limits (%)	Comments
1.	Inquiry rate 3-month average = inquiry $/sales quota $	700 (+200--200)	The inquiry or above-funnel volume must be kept high, via marketing programs, seminars, product demonstrations, etc. The inquiries that get qualified will move into the sales funnel. *Note:* Sales quota $ is the planned sales volume.
2.	In-funnel rate 3-month average = qualified funnel $/sales quota $	300–400 (+50--50)	This measures the rate of qualified customers in dollars versus the sales quota in dollars. The target will vary, depending on the types of product or deal. Small-value products could have higher targets, while large-value products could have lower targets. If this number drops, there is a possibility of not meeting the sales quota in the future.
3.	New customer call rate = new customers called/total customers called	30 (+10--50)	This is important to ensure development of new accounts or markets by a sales representative or organization. Also, it is especially useful when new products are introduced.
4.	Lucky win rate 3-month average = nonforecast $/total order or sale $	30 (+10--5)	These are unexpected orders, sometimes called *bluebirds*. If this rate gets too high, review the coverage of the business and market segments.
5.	Account coverage rate = coverage of customers/total potential customers	40–100	Current and potential customers must be covered and visited. All customers who buy small-value products could be contacted by a telephone sales force. Customers who buy large-value products should be visited regularly, but 100 percent coverage may be difficult, hence a lower target. This can be better covered by requiring a sales representative to have a customer visit schedule.
6.	Success rate = number of deals won/number of deals chased	25-35 (+10--5)	This measures a sales representative's or sales operation's productivity. The number could be smaller for low-value products but is expected to be higher for high-value products (big deals) because more time is needed for a big deal. If this gets too low, review the qualification process and account coverage process (performance measure 5).
7.	FTF rate = face-to-face time spent with customers/total available time	40 (+10--5)	This is the time a sales representative or manager spends with customers. If it is too low, there might be an administrative burden that requires attention. A possible alternative is the customer coverage rate, which includes customer face-to-face time and time spent working on customer needs. If this performance measure is used, the sales representative should ensure that time is spent with all potential customers instead of with only a few.
8.	Sales won/lost postmortem	Complete for all deals of value $100K or more	An understanding of why sales deals are won or lost is important. This will enable the sales force to improve productivity by institutionalizing the lessons learned.

The sales postmortem process.   Next, we show a flowchart of the sales won/lost postmortem process. Note that the data are collected on a routine basis, but meetings and analysis are conducted after 3 to 12 months of data are available. The actual frequency will depend on the amount of data available, which is a function of the volume of sales. The postmortem team will be led by the specific product manager or marketing manager.

The process starts after each sale is won or lost. The data are then consolidated and analyzed. The steps are illustrated below:

1.   Sales cycle is completed, and the sale is won or lost.

2.   Sales representative is to collect pertinent data immediately. Data can be tabulated manually or by computer and stored in central file.

3.   Appoint postmortem team led by specific product or marketing manager.

4.   Meetings planned

   a. Agenda
   b. Data requested in recommended format

5.   Meetings take place (after 3 to 12 months of data are available)

   a. Postmortem meetings held when sufficient data are available
   b. Data reviewed and analyzed

6.   Issue report

   a. Standard format
   b. Analysis and proposals

7.   Review of proposals for corrective action to improve current processes and practices

   a. Final meeting with senior management
   b. Agreement on proposals
   c. Short-term and long-term plans

Sales won/lost postmortem report format

1.   *Objective.*   Standard canned phrase, similar to the one given earlier in this section.

2.   *Win/loss Pareto chart*

   a. Pareto chart of reasons sales were won
   b. Pareto chart of reasons sales were lost

  c. Pareto charts grouped by systems or specific products

3. *Win / loss ratios*

  a. Win/loss ratios versus trend over the years
  b. Win/loss ratios against each competitor

4. *Competitive issues*

  a. Specific products, strategies, and solutions that we used that
  were successful against competitors
  b. Specific products, strategies, and solutions that competitors
  used that were successful against us

5. *Analysis*

  a. List our strengths and weaknesses.
  b. How can we use our strengths as leverage?
  c. How can we reduce or eliminate our weaknesses?
  d. List future product strategies for sales and factories.
  e. Is there a third party or consultant that can help us do better?
  f. How can we train our sales representatives better?

6. *Recommendations*

  a. Specific immediate action
  b. Specific long-term action

Postmortem meeting guidelines. We provided guidelines for project
postmortem meetings earlier in this chapter. Those guidelines are
equally applicable for the sales won/lost postmortem meeting.

Collecting information on sales won or lost. There are several ways to
collect this information. We recommend the use of a formal data sheet
for each sale won/lost. It is preferable for the customer to be visited
after the event. When a sale is won, this will be easy; but when a sale
is lost, it may be difficult. Hence the visit needs to be positioned very
carefully with the customer. The type of data to be collected will
include the items shown in the data sheet in Fig. 5.17.

The information needs to be collected routinely, by product. In a
large organization with numerous products, you may wish to collect
the information for products exceeding a certain dollar value. When
these data are collected over several deals, it will be possible to come
up with a summary report which extracts the information from the
data sheets and provides an analysis and recommendation. We show
a portion of a report in Fig. 5.18*a* and *b*. Note the recommendations.
These will be used to improve the sales process and future products.

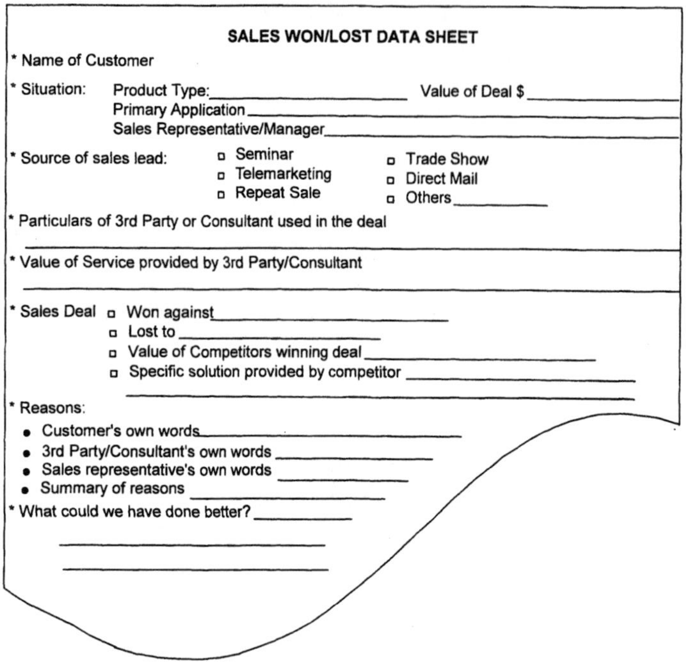

SALES WON/LOST DATA SHEET

* Name of Customer
* Situation: Product Type:______ Value of Deal $______
Primary Application______
Sales Representative/Manager______
* Source of sales lead: ▫ Seminar ▫ Telemarketing ▫ Repeat Sale ▫ Trade Show ▫ Direct Mail ▫ Others______
* Particulars of 3rd Party or Consultant used in the deal
* Value of Service provided by 3rd Party/Consultant
* Sales Deal ▫ Won against______
▫ Lost to______
▫ Value of Competitors winning deal______
▫ Specific solution provided by competitor______
* Reasons:
  - Customer's own words______
  - 3rd Party/Consultant's own words______
  - Sales representative's own words______
  - Summary of reasons______
* What could we have done better?______

Figure 5.17 Sales won or lost data collection sheet.

Summary of sales process

The sales process is often thought to be unstructured, but we have provided some ways to structure it. We have proposed several performance measures with typical targets for managing the sales funnel. In addition, we have proposed a formal sales won/lost postmortem. These activities will help provide a better-managed and more predictable sales process, resulting in a more competitive and productive sales force.

Questions and Answers on Process Management

Let us review some frequently asked questions on process management.

Question. Should every process in an organization be documented and managed?

Answer. No. It would be nice if you did that, but it is very difficult and could be counterproductive. You should focus on key processes—we discussed the concept of the process map of an entity or the next-level functions. Basically, whatever you can list on a one-page process map is what you should try to manage for an entity or function. This allows you to have a manageable and focused list.

Question. For any given process, how do you ensure it is operating at an optimum level?

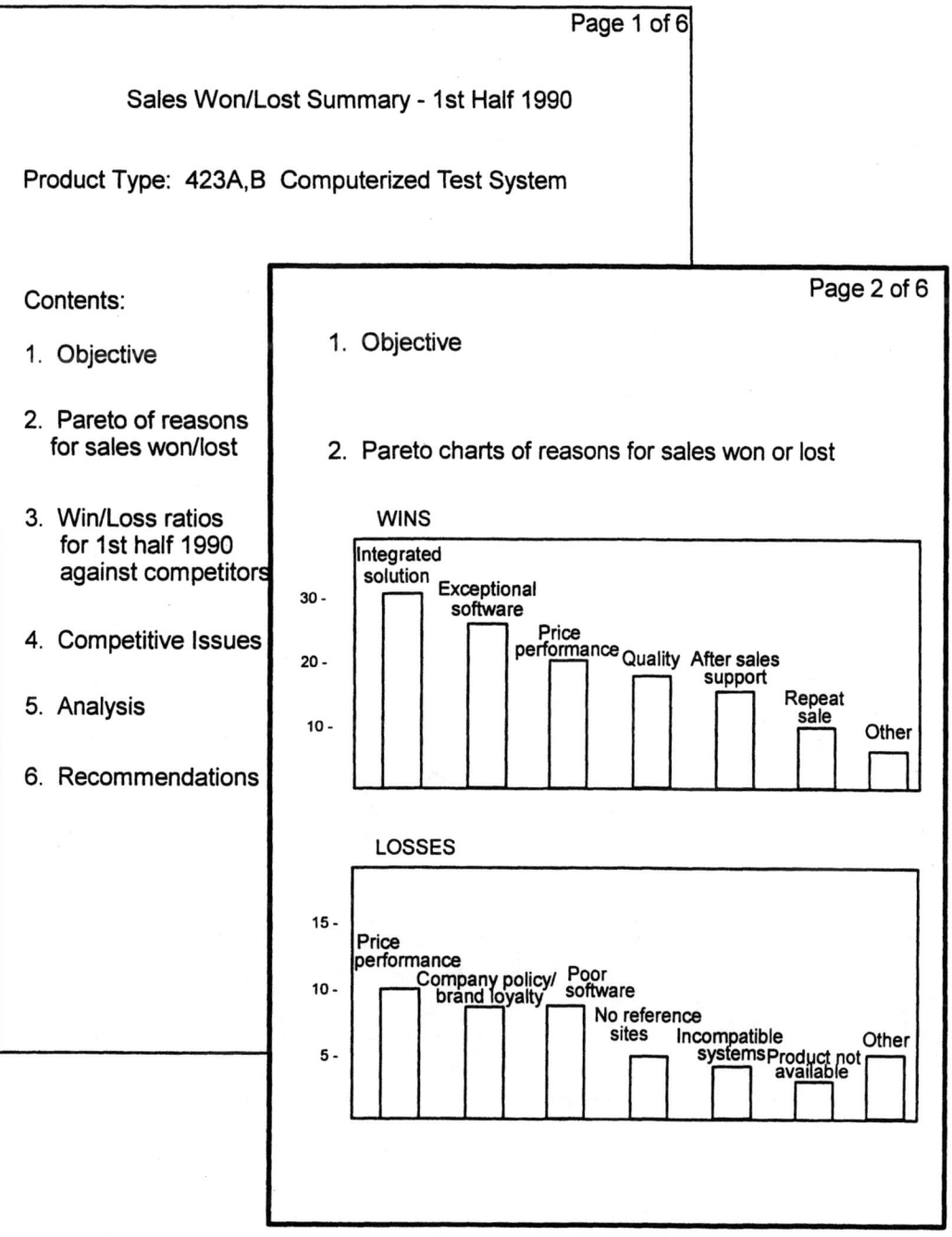

Sales Won/Lost Summary - 1st Half 1990

Product Type: 423A,B Computerized Test System

Contents:

1. Objective

2. Pareto of reasons
    for sales won/lost

3. Win/Loss ratios
    for 1st half 1990
    against competitors

4. Competitive Issues

5. Analysis

6. Recommendations

1. Objective

2. Pareto charts of reasons for sales won or lost

WINS

Integrated solution
Exceptional software
Price performance
Quality
After sales support
Repeat sale
Other

30 -
20 -
10 -

LOSSES

Price performance
Company policy/ brand loyalty
Poor software
No reference sites
Incompatible systems
Product not available
Other

15 -
10 -
5 -

Figure 5.18a A sample sales won or lost postmortem report.

4. Competitive Issues

Three of our competitors have the following specific superior features:

6. Recommendation's (continued)

b) For all deals over $400K, we recommend appointing a program manager to coordinate factory, consultants and 3rd party solutions. This will help overcome price perfomance and poor software concerns.

Action: Sales Manager

c) We need to prepare a complete list of available reference sites for all past solutions by product lines and markets.

Action: Marketing Manager

d) Start an aggressive Marketing communication program to convey our successful installations - via seminars and visits to potential accounts.

Action: Marketing Manager

e) The factory design team has been informed of the system incompatibility problem, which lost us several deals. It seems a solution will be available in 4 months.

Follow-up Action: Service Manager

f) A list of superior features from three of our competitors has been compiled. This was reported to us by our customers and includes superior/more friendly software. The information will be sent to the factory design team.

Action: Marketing Manager

Figure 5.18b A sample sales won or lost postmortem report.

Answer. There are several ways of looking at this. In the short term (1 to 2 years), you must ensure that the process operates within limits and targets. Any abnormalities and deviations must be immediately detected, analyzed, and eliminated. In the longer term (more than 2 years), there is no such thing as an optimum process. All processes must be improved, and whenever possible, subprocesses eliminated. This will come with continuous improvements, better designs, reengineering, and elimination of unnecessary activities (that were once thought optimum).

Question. You discussed the quality assurance system and recommended the use of control charts and out-of-control reports. But this implies managing a process that is generating defects. How do I go beyond that—toward zero defects?

Answer. You can, if your current level of defects is low. To move toward zero defects, consider the following data. Recently I visited a printer production line, with about 0.5 percent defects in the entire assembly line. The majority of the defects were caused by bad parts and workmanship. This leads me to recommend the following activities to move you toward zero defects:

- Ensure that all your purchases are good via an effective supplier management program.

- Eliminate workmanship errors by applying the concept of poka-yoke, discussed in Chap. 4. This also implies automating as many production steps as possible. These steps should have a high process capability index to minimize or eliminate defects.

- Automate inspection wherever possible. Where you cannot, do 100 percent inspection—but this has risks.

Question. It is extremely difficult to get sales and marketing people involved in process management. Any ideas?

Answer. Yes, it is difficult, very difficult, but do not give up. Here are some ideas to get started:

- Start by asking how they would measure their success, so the marketing or sales manager can start to list the key things that will measure his or her operation's success. It will probably be revenue or market share. This becomes part of the initial business scorecard, discussed earlier. But these items are lagging indicators of success. You need some leading indicators of success that will help achieve the revenue or market share.

- Leading indicators of success will be what come before revenues are achieved—items such as customer needs, customer satisfaction, product introductions, product delivery, and an educated and well-trained sales force. Put all these items down, and you have a business scorecard or business fundamentals (or daily management) plan.

- With the business scorecard you can then propose a simple process map that lists the key processes to be managed.

- Propose to manage the process map via the *business scorecard*. This will ensure ownership of the key processes—this will mean managing and eventually improving the key processes.

- This approach, proceeding from basic needs (getting results) to what is needed to get them, is a good way to generate enthusiasm and interest.

Question. I would like to get started on process management for my operation. Once I have a simple process map and business score-card, how long before I can see results?

Answer. About 4 to 6 months. You will need to do some simple documentation, start to measure results, take action for out-of-control situations, and get trend information. You will then understand your key processes and will be in a position to manage them.

Summary: Process Management

We have discussed the importance of managing key processes. Well-managed processes can be *leading* indicators which predict good results for a company. We have discussed the concept of key process maps, and if this map is done well, it represents the *value delivery system* of the organization, which can be measured, tracked, and managed. We have connected the entity's key processes to the entity's business plan (which will consist of an annual Hoshin plan and a daily management, or business fundamentals, plan). This daily management, or business fundamentals, plan becomes the *business score-card* for managing the fundamental processes of the business.

We have discussed some of the key processes in an organization that designs, manufactures, and sells products. These are typically what you would expect in a consumer electronics, industrial products, or systems company. The specific processes in your company or organization may vary—you may have different products or services. But the same concept of managing key processes will apply to your company or organization.

The project and sales postmortem processes are very useful because they allow us both to analyze the effectiveness of all our key processes in the design, manufacturing, and sales environment and to improve them. This will ensure that our organization learns from its mistakes. As we gradually decrease our mistakes, we become more efficient and lower our costs. In addition, design time and time to market a product can be reduced.

All these key processes must be managed well on a day-to-day basis. Process management means defining and monitoring a key process, ensuring it meets a target (within limits), discovering abnormalities, and preventing their recurrence. Success here means the ship (company) travels smoothly, while the captain (senior manager) focuses her or his effort on the destination (breakthrough objective). Managing your key processes and improving them will help increase efficiency, productivity, and quality. The results will be a more competitive and productive work force that provides increased customer satisfaction and higher profits.

References

1. Robert Neff, Neil Gross, and William J. Holstein, "How Japan is Getting Jumpy about Quality," *Business Week,* March 5, 1990, p. 40.
2. Robert B. Miller and Stephen E. Heiman, *Strategic Selling* (New York: Warner Books Inc., 1986).

6

Employee Development
and Participation,
and Leadership

Quality is everybody's job.
A. W. FEIGENBAUM

Overview

An effective total quality effort will require the participation of every
person in your company or organization. Never underestimate the
worth of each individual. Consider the following news article (extracted from *The International QC Forum,* November 1984):

**If you think you are working hard, read on...a new record of
suggestions—9310 suggestions per year**

Mr. Koji Nakayama, QC Section at Utsunomiya Plant of Matsushita
Electric TV Division, has established a record-high number of improvement suggestions reaching a total of 9310 in a year, all by himself. As to
the secret of his inexhaustible source of suggestions, Mr. Nakayama
says: "When an improvement idea hits me, I will go over it little by little
on a daily basis. I try to put my idea in a presentable form and suggest it
daily. I work on the idea on an average three hours every night after
returning home. Complaints from my wife and kids? No problem—
because I put myself at their disposal every weekend. So they don't bother me during weekdays. As to the source of hints, I'm a voracious reader
and read all those newspapers, magazines, journal articles and watch
TV programs that have something to do with my work directly or indi-
rectly.

"Good communication with other departments in the company is
important to get richer information. Another source of suggestions is to

develop (new) ideas for improvement after (each) improvement. I always keep a memo pad with me to jot down my ideas and observations."

This is just one of the ways in which individuals can contribute to the success of the company or organization.

In this chapter, we will discuss this and several other methods to harness the potential of every person. We will also discuss the importance of leadership and the expectations of a good leader. Our discussion will include the following:

- Quality circle, quality team, and project team activity
- Employee suggestion schemes
- Employee education
- Publicity promotion and recognition
- Leadership in the company or organization

Quality Circle, Quality Team, and Project Team Activity

Numerous comparisons have been made between quality circles and quality teams. In general, quality circles have been considered as ongoing voluntary employee groups (operating at the lower levels of the organization) that solve problems in their work environment. Quality teams are considered ad hoc groups set up to solve specific quality problems. We think the distinction is fine, but in today's competitive environment, it is difficult to accept voluntary participation.

Every employee needs to participate. Teams of well-trained employees on the production floor can solve process and product problems; teams of clerks in a bank can reduce paperwork or customer-related problems; management teams can solve service, product, or organization problems.

In many U.S. and European companies, the term *quality circle* has a stigma attached to it—in part this is due to initial false starts. Hence we suggest calling them anything you like—quality circles, quality teams, or project teams.

For the remainder of this section, we will discuss how you can foster quality circle and team activity in your organization.

Management of quality circle, or team, activity

A successful quality circle or team program is never an accident, but is the result of good management. Here are several key strategies that will help ensure a successful program:

1. *Organize a quality circle, or team, tactical committee.* This committee will best operate as a subset of the organization's manage-

ment staff. This committee will consist of senior managers with appropriate quality experts. At a department level it may simply consist of the department manager and staff.

2. *Set expectations.* The committee should meet monthly to review and guide the effort. On a yearly basis—preferably at the beginning of the planning year—it should meet and set expectations and targets for the year, for example,

   - Participation rate of 80 percent of all employees.
   - One project per year, per circle or team.
   - Review of projects or areas for improvement. At small organizations or at department level meetings, the specific projects should be reviewed, discussed, and approved.

3. *Identify and train facilitators.* All teams should get assistance from facilitators, except when there are experienced leaders. The facilitator is an experienced and well-trained individual who can assist the team in staying on schedule, understanding the PDCA cycle, developing practical solutions, seeking technical help, and educating members.

4. *Provide education.* We have listed a suggested education program in the question-and-answer section. In addition, the following will be helpful:

   - Maintain training data for all employees.
   - Identify ongoing training needs. With time and progress, new training needs will become obvious. The facilitator or leader must identify these needs and arrange for training.
   - As part of training, completed projects should be available and accessible as a learning tool. This includes projects from within and outside the organization.

5. *Monitor progress.* There are several ways of doing this.

   - Maintain a monthly status report of each circle or team. Figure 6.1 gives a suggested format. The schedule shown in the figure should show planned and actual progress of each project. The goal should be for all projects to be completed within 1 year.
   - Teams should select projects and commit to a schedule. Any deviations or slowdown will require intervention and analysis by management. The facilitators will track and discuss issues and problems with the steering or tactical committee.
   - Teams should be requested to keep minutes of all meetings and to track the steps of the PDCA cycle that they are following.
   - Reward and recognize teams that are meeting regularly, have good attendance, or are exceeding agreed upon schedules. This will require regular feedback of activity, probably by the facilitators. Rewards can be tokens of appreciation.

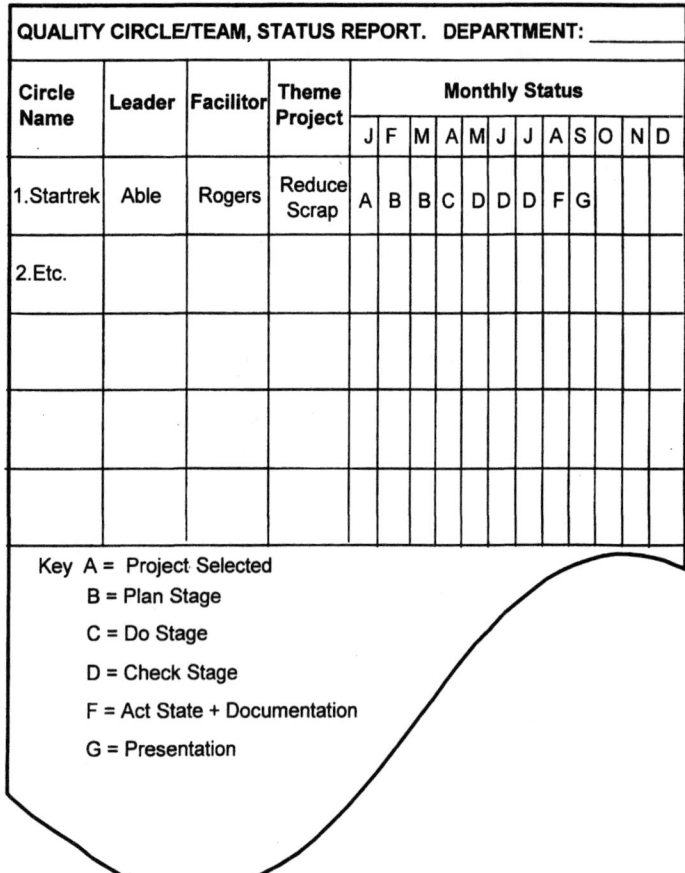

Figure 6.1 Quality circle or team status report.

6. *Promote, recognize, and reward.*   Ongoing promotion, recognition, and rewards are a must. Promotion will include the following activities:

- Display team members' photographs and project data on bulletin boards. Completed projects can be displayed and published in monthly newsletters.
- Regular quality circle or team conventions are a must. Departments should hold their own conventions in which circles or teams can present their projects. The best department projects can be presented at the organization's annual quality convention. A large organization can hold more than one convention per year.

EVALUATION & SCORING SYSTEM FOR QUALITY CIRCLES/TEAMS			
CIRCLE/TEAM: _____			
	EVALUATION CRITERIA	**MAX.**	**ACTUAL**
CIRCLE PROCEEDING (10 PTS)	1. Meetings are held regularly and members attendance has been good.	2	———
	2. Ideas contributed by members use of and improved upon.	2	———
	3. Difficulties experienced and overcome during team's proceedings.	3	———
	4. Help sought from other people, sections or departments inside and outside the organization when required.	3	———
PROJECT SELECTION (10 PTS)	5. Project selected based on study of background information constraint and past data.	5	———
	6. Project meets the needs of the people, sections or departments served by the team, i.e., the customers - or - 7. Selected from department objectives or Annual Hoshin plan.	5	———
PROBLEM DEFINITION (5 PTS)	8. Problems clearly identified and defined.	3	———
	9. Target explained.	2	———
ANALYSIS TECHNIQUES (20 PTS)	10. Techniques and methods such as fishbone chart, pareto diagram, graphs, check sheet, etc. effectively used.	10	———
	11. Systematic approach to identifying and verifying the most probable causes. (Team used PDCA approach correctly).	10	———
			page 1 of 2

Figure 6.2*a.*

- A few outstanding teams from the annual convention can go on to represent the organization at state, country, or international conventions. This not only provides recognition and esteem to the best teams, but also it exposes them to activities and new techniques outside the organization. This can be highly motivating.
- At annual quality conventions, the best projects should be graded and selected by a team of judges, comprising experienced managers. Recommended grading criteria are given in Fig. 6.2*a*

EVALUATION & SCORING SYSTEM FOR QUALITY CIRCLES/TEAMS

CIRCLE/TEAM: _____

	EVALUATION CRITERIA	MAX	ACTUAL
CORRECTIVE ACTION AND IMPLEMENTATION (15 PTS)	12. Alternative solutions considered.	2	———
	13. Solutions are properly evaluated.	3	———
	14. Recommended solution(s) is/are sound and practical.	5	———
	15. Implementation was effective.	5	———
RESULTS ACHIEVED (15 PTS)	16. Tangible results achieved.	5	———
	17. Intangible results achieved.	5	———
	18. Variation(s) between results & original target(s) is/are explained.	5	———
STANDARDIZATION (10 PTS)	19. Standardization carried out by changing procedures or through other arrangements.	5	———
	20. Follow up action taken to ensure that new procedures are maintained.	5	———
SELF-EXAMINATION AND FUTURE PLANS (5 PTS)	21. Team's next project stated and reasons given.	2	———
	22. Team is aware of its limitations and potential problems.	2	———
	23. Team gives proposals to overcome them.	1	———
PRESENTATION (10 PTS)	24. Team presents in an interesting manner.	2	———
	25. Presentation is well organized.	2	———
	26. Involvement by members in the presentation.	2	———
	27. Audio visual aids used effectively with important points highlighted and clearly explained.	2	———
	28. Presentation easily understood by listeners.	2	———
	TOTAL POINTS (100 MAXIMUM)		□

Note: The points awarded range from 1 to 10 depending on the importance of the criteria.   page 2 of 2

Figure 6.2b.

and b. We recommend use of these criteria. The judges should be experienced in quality circle or team techniques. Also, consider getting external, experienced judges.

An important afterword on quality circles and teams

When the concept of quality circles was first introduced, participation was voluntary. In today's highly competitive environment, everybody must chip in. Participation is no longer voluntary—it is essential for success. Hence, quality circles or teams must be encouraged at all levels in the organization; this includes formation of management or cross-functional teams.

They will work on different types of problems. The lower-level teams will solve problems or improve processes in their own group environment. The management teams will solve problems within their departments or across departments or divisions. In many cases, management projects will be items selected from the department or organization's annual Hoshin plan. The process used will be similar—the PDCA improvement cycle. To achieve high levels of participation, training of all employees and management commitment are essential.

Employee Suggestion Schemes

In the overview to this chapter, we quoted a very prolific person who contributes over 9000 suggestions per year. Certainly not every employee can do this, but the potential is enormous. Many U.S. and Japanese companies have such programs.

In Japan, many companies obtain an average of more than 20 suggestions per employee per year, and up to 90 percent of them are implemented. The results can be millions of dollars of savings and better-quality processes, products, and services. In 1987, about 350 large Japanese companies are estimated to have saved about $2 billion.[1]

In the United States, IBM has a successful employee suggestion program, in which the contributor gets a percentage of the savings. In Germany, Siemens Company has a very successful suggestion scheme.

The basis for a suggestion program is to allow all employees to have a say in improving things that they think are wrong, to allow bright ideas to be captured formally, and to recognize that the company's senior management does not have all the answers.

Guidelines for an employee suggestion scheme

Here are some guidelines for an employee suggestion scheme. The bulk of the guidelines come courtesy of Hewlett-Packard Division Singapore.

1. *Objectives of the employee suggestion scheme*

    a. To support the participative management philosophy of the company
    b. To allow employees to give logical and practical suggestions for improvement in their work environment including safety, processes, products, services, systems, quality, design, etc.

2. *Specific areas applicable for suggestion schemes*

   a. Production materials and flow
   b. Design and layout of production floor
   c. All systems and processes
   d. Design of equipment, tools, and fixtures
   e. Work environment
   f. Safety-related issues
   g. Quality and design of products and services
   h. Work procedures
   i. Information flow
   j. Customer services and customer relationships

3. *Specific areas not applicable for suggestion schemes*

   a. Personnel policies and guidelines
   b. Salary and wage administration
   c. Personal grievances
   d. Human conflicts
   e. Items within the direct job responsibility and assignment of the employee, that is, what the employee is supposed to do

4. *Rules and regulations*

   a. Suggestions should be submitted on a standard form. An example is provided in Fig. 6.3.
   b. All concerns must be accompanied by a suggestion before they can qualify for assessment. Proposer should approach immediate supervisor for assistance wherever necessary.
   c. The judges' decision is final.
   d. The company reserves the right to make changes to the suggestion scheme and its reward system whenever necessary.
   e. Points obtained in a suggestion scheme grading system (see next item) may or may not be accumulated. The decision to accumulate will depend on the types and amount of rewards given.

5. *A proposed grading system.* Refer to attached form in Fig. 6.3. All individual and group suggestions will be graded into thank you, bronze, silver, or gold awards, based on a points system. The points given will depend on the following criteria. The assessors are not required to complete the cost savings computation for suggestions which will be graded thank you.

   a. Idea
   b. Effort
   c. Customer satisfaction
   d. Net savings
   e. Safety
   f. Quality

BRIEF DESCRIPTION: _____   REGISTRATION NO.: _____

hp

EMPLOYEE SUGGESTION SCHEME

TQC

Name: _____

Employee No.: _____

Department: _____

Location Code: _____

Date of submission: _____

Supr-in-charge: _____

QC CIRCLE PROJECT: ☐ Yes ☐ No

PHOTO

(OPTIONAL)

Guidelines:

A. Briefly describe present condition, method or practice.

B. Details of your suggestion for improvement.

C. Please write neatly or type.

D. Use additional sheets if necessary.

E. For filing and tracking purposes, please give a brief description or your suggestion (not more than 10 words) at the top left corner.

PRESENT CONDITION:

_____

_____

_____

_____

_____

_____

_____

_____

YOUR SUGGESTION:

_____

_____

_____

_____

_____

_____

_____   Is your suggestion already implemented?

_____   ☐ Yes   ☐ No

_____   If yes, date of implementation: _____

MANAGER's COMMENTS:

_____

_____

_____

_____

_____

For implementation : Yes/No

Responsible person : _____

Date of completion : _____

Net Savings
(First Year only)   : S$ _____

Type of Prize

[   ]

Figure 6.3a.

to be completed by assessor
(For individual or group suggestion only)

			S$/year
(1)	No. of labour units or man-hour saved x labour rate (S$ /unit labour/mth)	X 12 =	
(2)	Man-hour reduction (hour/unit x no. units produced/mth x hourly rate (S$ /hr)	X 12 =	
(3)	Cost Savings/unit x no. of units produced/mth	X 12 =	
(4)	Other savings/mth: • Occupancy	X 12 =	
	• Auto-expenses	X 12 =	
	• Operating supplies	X 12 =	
	• Others	X 12 =	
(5)	Capital Investment x Depreciation rate/mth A Total Savings =		
(6)	Other incremental cost/mth: • Labour	X 12 =	
	• Materials	X 12 =	
	• Others	X 12 =	
		X 12 =	
	B Total Savings =		

A Total Savings/yr−   B Total Costs/yr =   C Total Net Savings/yr

Criteria	Measures	Grades			Score 1st Assessment	Score 2nd Assessment
	Possibility of immediate Implementation	Cannot or has been implemented 0 Point	Will be implemented can be implemented 10 Points			
Idea	Degree of originality	Negligible 0 Point	Some 8 Points	Significant 15 Points		
Effort	Amount of effort in generating the suggestion	Negligible 3 Points	Some 8 Points	Significant 15 Points		
Customer Satisfaction	Customer Satisfaction Level	Negligible 0 Point	Some 5 Points	Significant 15 Points		
Net Savings	First year net Savings as a result of implementation	Zero — 0 point 2 points for each nearest $1000				
Safety	Degree of safety improvement	Negligible 0 Point	Some 5 Points	Significant 15 Points		
Quality	Degree of quality improvement	Negligible 0 Point	Some 5 Points	Significant 15 Points		
Others	Additional points could be given for other criteria not mentioned above. In this case give reasons: (maximum of 15 Points)					

Points to be awarded if suggestion will be or already implemented.

TOTAL NO. OF POINTS

Reward System

Total score	D-14	14-49	50-79	> 80
Prize value	Nil	$10.00	$40.00	$80.00
Certificate	Thank You	Bronze	Silver	Gold

Date of final Assessment: _____

Figure 6.3*b*.

TABLE 6.1   Reward versus Total Score

Total score	0–14	15–49	50–79	80 and above
Prize value	0	$20	$100	$200
Award given	Thank you	Bronze	Silver	Gold

    g. Others

6. *Reward systems.*   Rewards systems can vary. The system shown here comes from the Hewlett-Packard Singapore suggestion scheme. We also mention some other reward systems.

    a. For individual and group suggestion only, Table 6.1 will apply. This gives the reward versus score achieved in the suggestion.

    b. For a group suggestion, the prize has to be shared amongst the group members.

    c. For every five suggestions (from thank you to gold award) submitted within each fiscal year, the originator will receive an additional reward of $20.

    d. Each year, the best six gold awards will get a special best suggestion of the year award—typically a plaque and token of appreciation, such as a designer watch.

    e. Each department must submit its claim voucher to the administration department at the end of each month and collect the cash awards.

Other reward systems can have such features as

    f. Gifts, tickets, dinner vouchers, etc.

    g. Accumulation of points during each year. At the end of each year, the total points accumulated can be used to collect awards.

    h. A percentage of the dollar savings up to a maximum dollar value, of say $25,000. The IBM Company uses such a system.

7. *Guidelines for all supervisors and managers*

    a. The immediate supervisor must decide whether the suggestion raised by the proposer is within the job responsibility or assignment. Suggestions that are within the direct control of the proposer will not qualify as suggestions.

    b. Response time to suggestion after submission varies. For a suggestion requiring the proposer's own department to assess it, it is 1 week. For a suggestion requiring another department to assess it, it is 2 weeks.

    c. If actual assessment time exceeds guideline, responsible supervisor must explain the reasons for the delay.

    d. The immediate supervisor can approve a suggestion deserving a $20 prize. The department manager can approve up to $100 prize. All $200 prizes must be approved by the functional manager.

e. Under the net savings assessment criterion, only the first-year savings will be considered.

f. The immediate supervisor is responsible to assist the proposer in offering a suggestion to resolve logical concerns which are beyond his or her capability.

g. The immediate supervisor should reject all frivolous suggestions or suggestions outside the company's control, for example, "build a bridge across a nearby river in order to reduce time taken to reach the company." This will save time and effort.

Promotion and measurement activity

Constant promotion and encouragement will be necessary to make the suggestion scheme successful. In a separate box, we have given a random sample of the type of suggestions you can expect if you start a suggestion scheme. Here we will mention how you can promote and measure success.

It will be useful to have department targets, in terms of the number of suggestions desired and constant encouragement by managers. Promotion should focus on

- Productivity gains, not cash gains, of employee
- Supervisor-worker meetings in order to solicit and encourage contributions
- Giving awards in public at department or company meetings

Success can be measured by tracking the suggestion scheme activity. The following performance measures can be used. They are given in order of implementation, that is, what you will measure initially and what you should measure as you get more experienced and sophisticated.

- Initial phase
  Number of suggestions per employee per year
  Percentage of employees participating
  Time to respond to contributor
  Time to implement suggestion
- Mature phase
  Yearly goals for suggestions per employee
  Quality of suggestions (one measure for this is percentage of suggestions implemented)
  Cash savings per year

Flowchart of Suggestion Scheme Tracking System

For the suggestion scheme form shown earlier, we provide a flowchart of activity. This is shown in Fig. 6.4. Note that suggestions with higher scores go to progressively higher levels of management. This ensures checks and balances, prevents abuse of the system, and allows senior management to be involved. Also included is a list of suggested performance measures.

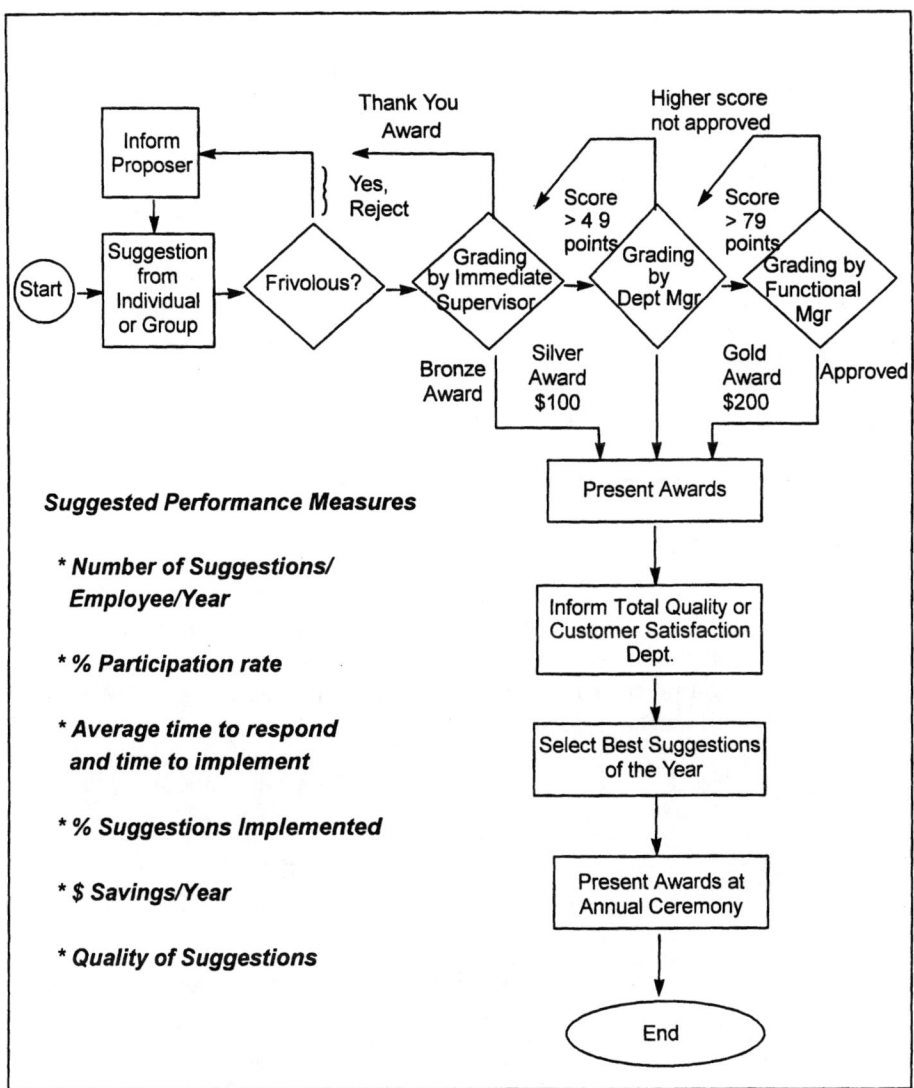

Figure 6.4 Flowchart of suggestion scheme activity.

What Type of Suggestions Can You Expect?

Suggestions will range from frivolous to excellent. With time and employee education, the quality and number of suggestions will improve. Here is a list of randomly selected suggestions that were taken from Hewlett-Packard Singapore and Hewlett-Packard Japan:

- Sell (in order to recycle) waste material instead of discarding it. Result: savings of $30,000 per year.

- Eliminate misinsertion of connectors, which causes 24 percent rejected connectors, by using a keyed connector. Result: dollar savings were marginal, but rejects reduced to 0 percent.

- Modify an assembly work holder to increase production speed. Result: savings of $12,000 per year.

- Improve copper plating process. Result: savings of $16,000 per year.

- Propose a new application for a Hewlett-Packard distance-measuring instrument. Result: increased sales of $550,000 per year.

- Reduce number of copies of accounts payable invoices. Result: savings of $5000 per year and reduced paperwork burden imposed on some managers.

Altogether, this is a very impressive list. It represents improvements above and beyond what an organization's management would even think of doing. It helps enormously, then, if all members of the organization keep their eyes and ears open and suggest ways to improve products and processes, resulting in increasing productivity, profits, and customer satisfaction.

Education in Quality Methodologies

Education in quality methodologies is crucial and necessary if a company is to succeed and achieve its quality vision, to increase its productivity, and to ensure increasing customer satisfaction. Listed here are some basic quality methodologies that should be part of an employee education effort.

Course and contents	Target audience
1. *Introduction to total quality.* Current trends and techniques of quality; the why, what, and how of productivity, quality, and customer satisfaction	All employees (special modification for senior managers, with a business and profit focus recommended).
2. *The PDCA improvement cycle*	All employees
3. *The proactive PDCA cycle*	Employees with experience in using the PDCA cycle
4. *Seven quality control tools.* Refer to App. B.	All employees
5. *Seven new quality control tools.* Refer to App. B.	All staff, but only after the seven quality control tools are well understood and used frequently
6. *Design of experiments and Taguchi methods*	All engineering staff

7. *Design for reliability techniques.* For example, training in thermal design, component derating, and FMEA. Other training as appropriate with the products you may manufacture

All engineering staff

8. *Quality function deployment (QFD) training.* Refer to App. C.

Marketing, R&D, and engineering staff. Management must understand concepts.

9. *Long-range and annual Hoshin planning.* Refer to Chap. 3.

All managers

Publicity, Promotion, and Recognition

Ongoing publicity, promotion, and recognition of the quality effort are extremely important. Here are some suggestions to get you on the right track.

- *Monthly quality newsletter.* This can give recognition to successful projects and good suggestions as well as convey new initiatives, ideas, and techniques.

- *Display boards.* Display successful projects, all accepted suggestions, plans for the year, and entity success or performance measures. Displays must be managed and updated regularly; otherwise employees recognize quickly that management is not serious.

- *Recognition.* Provide recognition for employees who have contributed to the quality effort. This can be done in the newsletter display boards and other activities, such as

  Departmental and company quality conventions, where successful projects can be presented and recognition given to project teams

  Sending the best project teams to country and international quality conventions

  Rewards for the best suggestions of the year

Leadership in the Company or Organization

There are many managers in a company, but what we need is leaders. What is the difference? We feel leaders can move an organization toward success and the right position in the marketplace. But managers simply direct and control the business, provide day-to-day direction, and achieve the basic results. Nevertheless, *leadership* is a fuzzy word, and we do not plan to enter the perennial debate of defining and discussing it. There are, however, some basic traits that a business leader must have; and if these are well executed, they can help drive the organization to success. Some of the traits are as follows:

- *Create a vision.* Leaders create a vision for the future, often by articulating a dream. Their vision will inspire their organization, provide a basis for decision making, and give a clear sense of direction. Leaders understand the need for a vision to harness, motivate, and guide the organization. Leaders also understand that a vision is more than just a goal in the future. A great vision incorporates core values, purpose, goals, and envisioned future. The core values are probably the most crucial because they represent the culture and essence of the organization—these will ensure that the organization remains strong over the long term. If they do not exist, the leader needs to develop and nurture them.

- *Set clear goals.* An aggressive vision will provide new challenges and excitement but will take many years to implement; hence, the strong leader must be consistent and unwavering and must continue to steer the organization through thick and thin toward the vision. This will require setting clear goals, with supporting strategies, that will help achieve the vision. In reality, however, more than the vision needs to be achieved, and this will require setting priorities.

- *Set priorities and then focus the organization to achieve them.* Leaders understand that there is more to do than work toward achieving the vision, for example, meeting customer and market needs, improving critical areas, and working on employee issues. So priorities must be set, after which the organization must focus its energy on achieving them. If this course of action is followed, results are achieved, employees are motivated, and the cycle can be repeated. Otherwise, the organization tries to do everything, and the result can be success in some areas and failure in others.

- *Stay in touch with customers and meet their needs.* Leaders must meet with their customers regularly and assess their needs. Leaders must strive not only to meet those needs but also to drive the organization to exceed them. Meeting customer needs includes providing basic functions such as impeccable product quality while exceeding needs includes providing the attractive, exciting, and unexpected. The leader recognizes that there is no such thing as being too customer-sensitive, but there is such a thing as too few customers.

- *Empower the organization.* An effective leader does not micromanage because there is always too much to do. The leader delegates and coaches and encourages a strong management team to emerge and grow, in order to implement the necessary strategies that will lead to success. This includes hiring strong people, training and educating employees, giving tough assignments, delegating strategies to lower-level managers and professionals, and setting up cross-functional teams to achieve organizational goals.

- *Encourage risks.*   Leaders take risks. They also build strong teams that are willing to take risks. When employees take risks without fear of failure, they tend to set aggressive goals and are motivated to achieve them. Leaders recognize that managers and employees grow when given tough assignments—there will be some failures, but this is acceptable. Otherwise employees take no risks because they are afraid of failure and its impact on their career.

For guidance in preparing a vision, setting clear goals, prioritizing, and exceeding customer needs, refer to Chaps. 2 and 3.

There is always a shortage of leaders in companies, organizations, and nations. We have no magic formula for leadership but have listed some basic traits of a leader. A leader who practices these traits will certainly be able to lead the organization down a successful path.

Some Questions and Answers on Employee Participation and Leadership

Here are some answers to some recurring questions on employee participation and leadership.

Question.   Can you tell us more about quality circles and teams?

Answer.   The answer to this question has several facets: The quality circle would come from the same work area for a lower-level team, for example, from production workers, technicians, engineers, material handlers, clerical workers, or waiters. This group would be led by its supervisor or another employee. A facilitator may be needed to help with the improvement process. Such teams could work on projects from the department's annual Hoshin plan or on specific customer-related issues.

Teams consisting of higher-level workers, professionals, or managers can have members from different work areas, for example, employees or managers that form teams from different departments, different divisions, or between a company and its suppliers. The team would be led by an experienced manager. Such cross-functional teams can be very powerful and typically should be working on cross-functional projects that show up on the annual Hoshin plan.

Question.   Why do some quality circles fail?

Answer.   All too often, a well-intended effort to start quality circles fizzles out. Listed below are some of the causes. All these causes apply to lower-level quality circles, but many are applicable to higher-level department or cross-functional teams of professionals or managers.

- Insufficient education or training. The right tools and techniques have to be provided before a start can be made.

- Overeducation. For example, there is no need to teach all seven quality control tools. There is a lack of trained facilitators or managers to help quality circles in the problem-solving process.

- Solving problems outside the control of their own work area or process. This is one of the most common causes of failure. When quality circles are formed, they should confine their initial activity to their own work area or process.

- Lack of time to meet. Management must allocate time for circles to meet—in slack times and during high-growth periods.

- High expectations by top management. Expecting quick results or large savings is quite common. When this does not happen, it leads to disappointment and often withdrawal of support to quality circle activity. Typically, a circle may complete one project per year. If two projects are completed per year, it is considered very good. In addition, it must be emphasized that quality circles are not a panacea for a company's problems. We estimate that only 20 percent of all quality problems are solved by quality circles; the remaining 80 percent are attributable to and controllable by management and can be solved by management, management teams, or cross-functional teams.

- Not accepting a quality circle's proposal. This can have negative effects—it is the best way to kill a quality circle effort! This can be avoided at the outset by confining a circle's activity to its own work area or process. Further, a well-trained leader and facilitator will ensure that the circle solves the right problems and provides good solutions. Management acceptance of the circle's proposal then becomes a fait accompli. We recommend working on problems originating from the annual Hoshin plan, customer issues, or process failures or wastage. Frivolous projects are not recommended—this is one of the key reasons why quality circles or teams have failed in the United States.

- Lack of management support. This is probably the major reason why quality circles fail. The reasons mentioned above are all controllable by management. Therefore, management must provide the resources and encouragement for a quality circle effort to succeed.

Question. Why do quality circles succeed?

Answer. There are numerous companies that have a high degree of participation—over 90 percent of all employees—in quality circles and teams. The basic ingredient is management commitment, support, and encouragement. We list the main reasons here:

- Give people, especially from lower-level circles, time off to meet. This is usually 2 hours per week.

- If a meeting takes place outside office hours, always entertain overtime claims. After all, employees are working to solve company problems.

- Hold regular in-house quality circle conventions—one or two per year to encourage participation, share successes, and offer recognition. All managers should attend these one-day presentations.

- Consider a reward scheme for quality circles for each project completed. This can be a small but significant token of recognition for their efforts.

- Each year, consider sending the best circle on a working holiday to visit and present in other parts of the organization, often in other countries.

- Education in quality circle techniques, the seven tools, and the PDCA cycle must be given.

- Set the expectation that all employees participate in quality circles. Their achievements can be discussed during their annual performance evaluation. This reinforces their belief that management is totally committed to making quality circles successful.

- Set high expectations in quality improvement for all departments. High-quality products and services must be the goal. Target for quality improvement every year. Get employees to understand that customers expect top quality and high satisfaction; therefore, everyone has to work to meet this requirement—through quality circles or in other ways.

- Managers should review a formal presentation of each circle's completed project in order to endorse findings and recommendations.

- Finally, management must exhibit lots of patience. Rome was not built in a day. Neither can quality circle success be. Managers have to support quality circle activities through good times and bad, through periods of intense activity and low activity, through crises and calm. And then success will be theirs!

Question. Can you comment on cross-functional and management teams?

Answer. Despite our enthusiasm for lower-level quality circles, we realize that they can only contribute to improving about 20 percent of the problems in an organization. The bulk of improvements will be management-driven, perhaps originating from the annual Hoshin plan. Many projects from the annual plan will be managed by quality (or project) teams and cross-functional teams. Hence, higher-level quality teams and cross-functional teams must be formed to run major company projects. These teams are essential for the success of a company.

Question.   Suggestion schemes work well in Asia, but does our culture not make it impossible to introduce such a scheme here?

Answer.   On a recent visit to Germany, we heard this comment from a senior German manager. On further investigation, we found that Siemens, a typical German company, has had a very successful employee suggestion scheme for over a hundred years. So much for the comment that German culture restricts suggestion schemes. Hence, such comments are typically a smokescreen to avoid doing something. Suggestion schemes will work in any culture or company, but they require constant promotion, encouragement, and an atmosphere in which there is no fear by lower-level employees. In addition, the quality department must facilitate this activity and help overcome roadblocks.

Question.   Our employees are not trainable. What can we do?

Answer.   Then you have a problem, and your company may fail in the marketplace. Your employees are your most valuable resource; if they lack basic education, give it to them and then train them in quality methodologies.

Question.   My priority is to get products out of the door. Won't I will be diluting my resources if I start to promote quality circles and suggestion schemes?

Answer.   If you run a small operation, say less than 100 employees, or if you are managing a start-up operation, your concern is justifiable. But if you run a larger or mature operation, then you need to reconsider. Employees today want to participate and contribute—the larger the organization, the greater the need. As we said earlier, there are only so many things that management can focus on, but employees can devote their ideas and energy to improving processes and products that they are close to. Also, unfortunately, management does not have all the answers—getting employees to contribute via the various systems suggested can only help, not hinder. The outcome will be management and employees working in unison toward common goals of higher productivity and increased customer satisfaction—and, of course, increased employee satisfaction!

Question.   Should the quality department conduct training?

Answer.   Preferably not. All training should be managed by a central training department, such as human resources. If you conduct quality training via the quality department, then quality is viewed as a separate program—often to be ignored. The quality department can help design the quality training program, but it must be managed by one central training department and preferably conducted by experienced employees and managers.

Question.   Why do you talk about leadership in a book about total quality?

Answer.   In this book we talk about breakthroughs, quality systems, processes, and organizational change. It takes courage and leadership to implement them and not treat them like another fad which fizzles out. On a broader scale, the company or organization needs leadership to progress and succeed. So, based on our experience, we have listed important leadership traits.

Summary

Employee development and participation, and leadership are a very important element of a total quality effort. Certainly, no total quality effort can be effective without employees contributing to improving products and processes. We have provided suggestions for education in quality methodologies, quality circle and team activity, and employee suggestion schemes. We have also discussed the traits of a strong leader, which can result in a more successful organization. The result of these activities can be higher morale and greater productivity, increased customer and employee satisfaction, and a strong, successful organization.

Reference

1. Data on suggestion schemes are quoted from the Japan Human Relations Association survey.

7

Getting Started and Ongoing Management

Quality is never an accident, it is always the result of an intelligent effort.

JOHN RUSKIN

Overview: Why Start a Total Quality Effort?

There are several reasons to start a companywide total quality effort. We list three of them:

1. *To increase profits.* As shown in Chap. 1, a total quality effort will result in less rework, increased productivity, and higher customer satisfaction and loyalty; the outcome of this can be higher profits. The company's management team will realize this and start the total quality effort.

2. *Driven by an enlightened chief executive.* An enlightened chief executive will realize that total quality will lead to a better competitive position and higher profits. The chief executive will then drive the entire company toward her or his vision.

3. *To save a company in a crisis situation.* In this case, company management realizes that there is a crisis—products do not sell, or there is a competitive threat—and adopts a total quality effort.

The most common reason seems to be the third reason. Here are some examples:

- Most Japanese companies started the efforts in quality after 1945, when the economy was ravaged. Japan had to export or perish, but because the word *Japanese* was synonymous with poor quality, they could not export. Through Edwards Deming's and Joseph Juran's efforts, which subsequently developed into the body of knowledge known as total quality, Japanese companies improved their quality. Today, Japanese exports provide such a high balance of trade that it is constantly embarrassed and requested to import more.

- After 1973, when oil prices increased dramatically, Japanese companies adopted a second wave of total quality. Yokogawa Hewlett-Packard (YHP), discussed in Chap. 1, was one of those companies. Today YHP retains strong market share in several businesses, despite being viewed as a U.S. company in the Japanese market.

- After seeing the benefits of total quality at YHP, Hewlett-Packard Company decided to—proactively—implement total quality throughout the organization.

- In the United States, Xerox launched its total quality effort because of declining market share and won the Malcolm Baldrige Award. Today, Xerox is a very aggressive company focusing successfully on innovative research and new businesses.

Today, reasons 1 and 2 are becoming commonplace—partly because management realizes that there is no choice, and partly because many large companies require their suppliers to improve quality, for example, the Deming Prize–winning companies in Japan and all automobile manufacturers in the United States.

In this chapter our discussion will focus on some of the ways to start and manage a total quality effort. Our discussion will include how to get started and how to move toward maturity.

Managing the Total Quality Effort

The entire total quality effort must be planned and managed by the company's or organization's management team. There must be a plan and a long-term commitment. The management team must dramatize the importance of quality by making it a strategic issue and by giving quality top consideration. Managers can demonstrate their commitment by the way they react to quality issues, the kind of people they promote, and the goals they set.

In previous years we proposed that a total quality effort be managed via a quality steering committee or quality council. No more. Based on our experience, the total quality effort must be driven by the management team. More important, the total quality effort must be integrated into the management process and must not be viewed or managed separately. Employees of a company or an organization learn very quickly how serious you are about quality. They will heed not only your words but also your actions. Hence, we stress the need for integrating the total quality effort into the management process. In the rest of the chapter we discuss how you can do this.

Setting Up a Total Quality or Customer Satisfaction Function

For a large organization, a total quality or customer satisfaction function is a necessity. For an organization of more than, say, 200 people, we propose two to four people. There are several activities to manage and facilitate, including the following:

- Facilitate development of a customer satisfaction model, and provide ongoing satisfaction measurements. Refer to the discussion in Chap. 2.
- Conduct a customer satisfaction survey, and facilitate follow-up corrective action.
- Manage a customer complaint and feedback system.
- Define and manage a quality education and training program for all managers and employees. A suggested list of programs was given in Chap. 6.
- Organize and facilitate an employee suggestion scheme.
- Organize quality circle or team conventions.
- Conduct publicity, promotion, and reward programs.
- Perform other activities as defined by management.

Phases of a Total Quality Effort

Based on our experience, we list the phases of a total quality effort with a discussion of each. If you already have some of these activities, you can move much faster.

1. *Introduction phase.* This involves the decision to launch the total quality effort and preparation of the detailed program. The details of the program will need to be worked out by the management team. The actual launch should be done with some fanfare and publicity. Specifics include

   - Setting up of a total quality or customer satisfaction function.
   - Creating the education and training program, including the PDCA cycle and planning process. You need to decide if you are ready for both long-range and annual Hoshin planning. You also need to review how this impacts or influences your existing planning processes.
   - Starting a customer complaint and feedback system.

- Conducting a customer satisfaction survey.
- Creating promotion, publicity, and reward systems

2. *Acceleration phase.* At this point the annual Hoshin planning process should be introduced. The annual Hoshin planning process should be used to drive and steer the company in the right decision. Specifics can include the following:

   - Introduce the annual Hoshin planning process, and deploy strategies down the organization.
   - Select cross-functional projects from the annual Hoshin plan, and review these regularly (say, quarterly).
   - Measure the progress of the annual plan by conducting open and honest Hoshin plan reviews.
   - Identify specific processes to manage or improve via the entity key process map. Some processes may be managed via the annual plan, and others could go immediately into your daily management, or business fundamentals, plan. Refer to Chap. 5.
   - Continue with promotion, publicity, and reward systems.
   - Begin to pursue and resolve customer issues from the customer complaints and feedback system and customer satisfaction surveys.
   - Introduce an employee suggestion scheme.
   - When the annual planning process is running smoothly, plan to introduce a long-range planning process.

3. *Cruising phase.* When the first and second phases are well managed, you will get to the cruising stage. You will reach this stage after several years. In this stage, you consolidate and fine-tune all the activities and programs launched. Specifics include coverage of the following:

   - Ensure employees and managers are customer-obsessed. The total product concept is well understood. Better still, you are using even more advanced concepts.
   - Ensure that your long-range and annual planning processes are in good shape and that you are progressing toward the entity vision.
   - Ensure you are addressing items to improve in your annual plan. These would be high-impact items that originate from customer needs. There should be good use and understanding of the PDCA cycle.
   - Ensure you have the key process map for several levels in your entity and that these key processes are managed via the daily management, or business fundamentals, plan. This is your business scorecard.
   - You are ready to conduct total quality reviews, discussed in the next chapter. These reviews will measure how well you are using quality systems to manage your business.

4. *Cruising at an even higher-level phase.* In the cruising stage, despite the best efforts, you will become complacent. So you need to push yourself and cruise at an even higher level. In this phase, you should be doing less routine management and quality improvements. You should be in a problem prevention-by-prediction phase. Refer to the discussion in Chap. 4 on the problem-solving hierarchy. *In addition, you have gone beyond total quality into total quality creation.* All your products and services should have attractive quality features, and you are creating products that fall in the outer zone of the total product concept. Refer to the discussion of total product concept in Chap. 2.

Total Quality and Time Management

One of the constant complaints of managers is that there is no time to implement total quality concepts. We believe that using total quality concepts and systems is a better way to manage, and so we show the relationship between the elements of total quality and a manager's basic activities.

The basic activities of a manager

For the answer to this question, let us look at a time management model from Itoh.[1] This is shown in Fig. 7.1. Let us discuss this model for a while. You read it from left to right. The specific activities are shown in the box, and as you glance to the right, you see the layer of

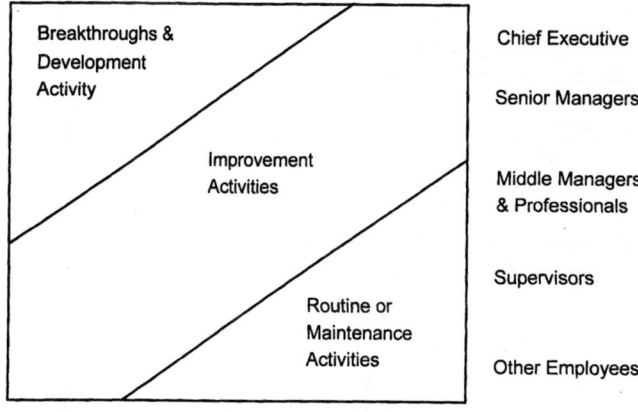

Figure 7.1 The Itoh model.

management or employees that is responsible for that activity. What are the main activities in an organization, and who does them?

Breakthrough or development activity.   This is the primary activity for a successful organization. This includes new development and introduction of new products and services; adoption of steps to steer the organization in the correct direction to the right destination; investments to grow, develop, and strengthen the organization; and so on.

The responsibility for this activity rests with the following:

- Clearly this activity is the work of senior managers. In fact, as shown by the segmentation of the activity, the bulk of this is done by senior managers.

- This chief executive of the organization probably does only this activity, and rightly so. If he or she does many other things, it may indicate that time is not optimally used.

- Many professionals, such as engineers, would also be involved in this activity but mostly in the creation, preparation, and design of new products and services.

Improvement activity.   Continuous improvement is another important activity. An organization cannot rely on a continuous stream of block-buster products and services. Many of the current processes, products, and services will require improvement—be it an automobile, a computer, a turbine engine, an instrument, or business-class service on an airline. These cannot be simply replaced with a brand new offering—so you improve them. Key internal processes will also require improvement, processes such as new product development, manufacturing, inventory management, supplier management, marketing, sales, delivery, or distribution. Where possible, benchmark them against the competition, and develop aggressive goals for improvement. The following are responsible for this activity:

- Senior management's role is to identify, prioritize, and review progress.

- The bulk of the work will be done by middle managers and professionals, such as engineers and accountants. They will form project teams, quality teams, and cross-functional task forces to facilitate this important activity.

- Supervisors and all the employees reporting to them will be involved. Some of the activities will be directed toward improving key internal processes as identified by senior management, but other activities will be performed at the micro level—by lower-level quality circles working on processes in their environment, for example, increasing the yield at a specific assembly point or reducing paperwork.

Routine or maintenance activity.   This refers to management of routine day-to-day processes. These processes must be maintained at their current level of performance. This would include the entire assembly line in an automobile, computer, or turbine engine factory; assembly in a diaper or cereal factory; or all the routine lending and borrowing activity in a bank.

These activities represent the fundamental processes of an organization—processes which are running routinely and require little management intervention, so long as they are in control. Responsibility for this activity is delegated as follows:

- This activity is done primarily by the workers, engineers, professionals, and their supervisors.

- Supervisors who manage the employees focus mostly on this activity and a little on the improvement activity, as shown in Fig. 7.1.

- There will be involvement by middle managers and professionals, such as emergency intervention and fine-tuning of the maintenance activity. But involvement should be minimal. If it is high, there is a possibility that the activity is not routine—it may indicate fire fighting to fix unstable processes.

Fitting the total quality elements in the Itoh model

We have discussed the five elements of total quality:

- Customer obsession
- Business planning
- Management of improvements and breakthroughs
- Process management
- Employee development and participation, and leadership

Within these elements we have discussed numerous quality tools and methodologies. Those tools can and will help manage the three critical activities shown in the Itoh model. In Fig. 7.2 this model is overlaid with the five elements of total quality. You can see that the various elements of total quality can help manage the three activities. For example:

1. *Breakthroughs or development activity.*   Long-range planning and Hoshin planning are used to plan for new products and services and

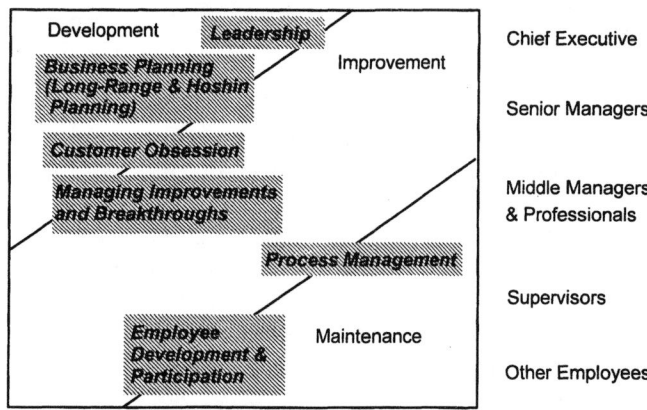

Portion of time spent on activity.

Figure 7.2 The total quality elements overlaid on the Itoh model.

to steer the organization in the right direction. In addition, many of the techniques mentioned earlier can help in identifying areas for possible breakthroughs in products, services, or processes. Some of the techniques are competitive benchmarking, capturing the customer's voice and needs, attractive quality, and the total product concept. In many cases, the proactive PDCA cycle will be useful for managing breakthrough projects.

2. *Improvement activity.*   Use Hoshin planning and long-range planning to help plan for what needs to be improved. Information on selecting items to be improved will come from the various techniques mentioned in Chap. 2, such as the customer complaint and feedback system, customer surveys, and customer satisfaction model. Processes to improve have been discussed in Chap. 5. The PDCA improvement cycle and the proactive PDCA cycle will be used to manage the improvement. The various employee participation activities, such as quality circles and suggestion schemes, fit well in this segment.

3. *Routine or maintenance activity.*   The key processes that must be routinely managed can be listed and monitored via the daily management, or business fundamentals, plan. These were discussed in Chap. 5 and include key processes in an entity and its functions. Each of these processes must be properly documented, everyone well trained, and the process well managed.

As you can see, total quality concepts and systems can be integrated into management's key activities as identified in the Itoh model. The three activities listed are crucial for management and employees, and total quality concepts and systems can help to manage these activities.

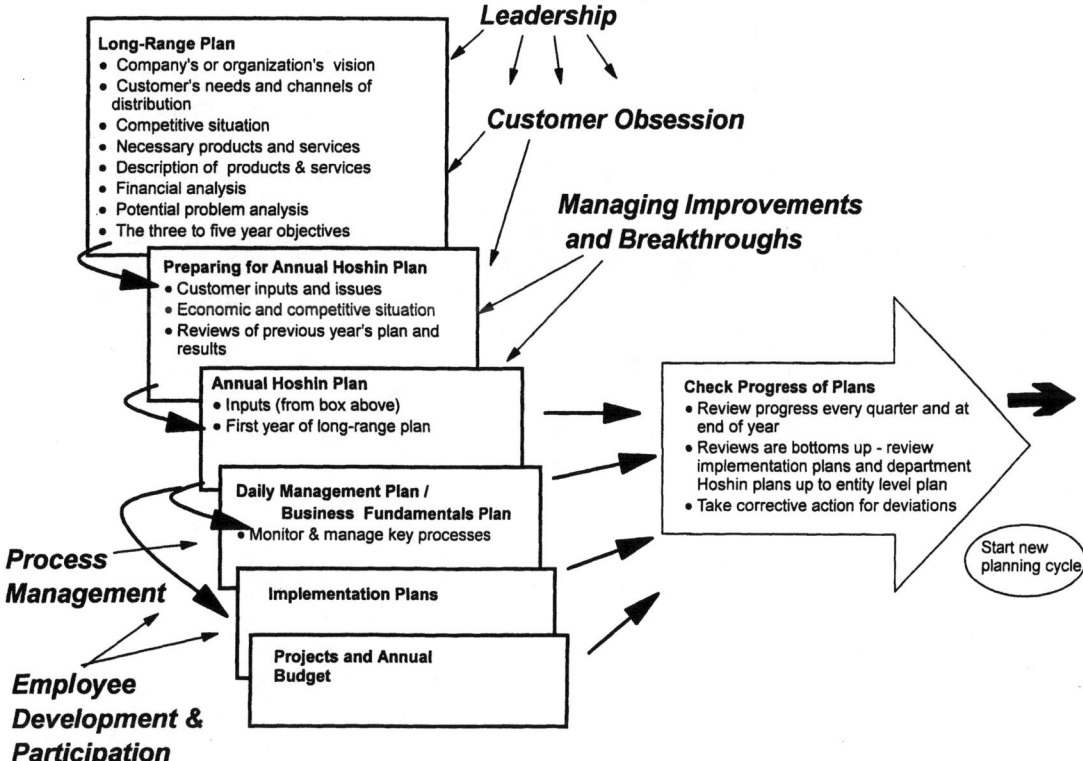

Figure 7.3 Integrating the elements of total quality into business planning. This figure illustrates business planning, which we discussed in Chap. 3. Note that the other elements of total quality, such as customer obsession, managing improvements and breakthroughs, process management, and leadership and employee participation can be integrated into planning. Strong leadership, of course, drives the entire activity.

Integrating total quality effort into management activity

The initial total quality effort will start with a management decision or initiative, but as quickly as possible, the total quality effort must be made a way of life. To do this well, we recommend that total quality be part of the planning process (see Fig. 7.3).

In Fig. 7.3, we identify the elements of total quality and show when to plan for them. For example, planning starts with the long-range plan, customer inputs, and competitive data. The annual Hoshin plan is prepared as well as the business fundamentals, or daily management, plan. At this point, employee participation and daily process management issues are reviewed. During the planning stage, items for improvement are identified; these are managed during the rest of the year. Finally, all these activities are reviewed for progress during the coming year. By integrating the tools and techniques of total quality into the planning process, we can make total quality a way of life at any organization.

Some Questions and Answers on Getting Started and Some
General Questions on Total Quality

Question. I run a small independent operation of about 100 people. Do you expect me to have a total quality department and a quality manager?

Answer. For a small organization of less than 100 people, it may not be possible to justify having a quality manager or a total quality department. We recommend you drive total quality in your organization via the management team. If you have about 100 people or more, you should be able to justify having a quality manager to facilitate total quality activity. And once you have more than 200 people, you can consider adding more resources.

Question. We started our total quality effort many years ago, but the effort continues to move slowly and many employees are unconvinced that it will help them. What's wrong?

Answer. The most common reasons why a total quality effort fails are lack of management commitment, involvement, and impatience. If the chief executive and her or his staff believe in it and drive the effort, there will be success. So if progress is slow, go back and examine your actions, behavior, and commitment.

What is commitment? It includes changing the policies of the company by adopting the five elements of total quality and emphasizing process over short-term results. This means making decisions that are right for the long term, although in the short term they will be tough and difficult. We have also suggested integrating the five elements into business planning.

Question. Why bother with starting a total quality program? We can be more competitive by cutting costs through laying off excess and unproductive employees.

Answer. Cost cutting by massive employee layoffs can be effective, but it gives only short-term results because it does not improve inefficient processes. A total quality effort, on the other hand, gives long-term results by providing efficient and productive processes, less waste, better products and services, and a high degree of customer focus. In addition, if you use the Hoshin planning process correctly, there is ongoing cost control. This means a lower likelihood that you will allow costs to get out of control.

Question. Our quality is already very good. Why do we need to start a total quality effort?

Answer. It is possible that your quality is very good—you may even be considered a world leader in the products or services you offer. But quality is a moving target. Today's leader can be tomorrow's follower. Just witness what happened in the automobile industry, the consumer electronics industry, or the camera industry. The list is endless. A leader is in a dangerous position because competitors want the leadership position. All the methodologies offered in this text help to keep you on your toes and moving ahead.

Summary

We discussed how to get started in a total quality effort and how to ensure effective ongoing management. Total quality methodology is not an alternative method of management, nor is it additional work or a separate program. Total quality methodology provides tools and systems to manage a business, in order to make it more efficient, productive, successful, and better prepared to face the future. We proposed integrating these tools into business planning. It is crucial to do this, because the quality systems and techniques will enable you to plan better and to be more successful and profitable.

For the total quality effort to start and succeed, management commitment and involvement are essential. The effort must be driven by management. With success may come complacency, and that must be avoided because total quality is a never-ending journey for continuous improvement and betterment of life.

Finally, progress must be monitored; and if there are roadblocks, they must be removed. In the next chapter, we discuss how to review a total quality effort, measure progress, and move forward to greater success.

Reference

1. We are not able to trace the source of this model; we have seen it many times without reference to its source. We have seen numerous variations of this, including the one used by the Kaizan Institute of America. The Itoh model predates the model used by the Kaizan Institute.

8

Conducting Total Quality Reviews

It is obvious that such reviews, conducted by upper level managers, can have a powerful impact throughout the company. The subject matter is so fundamental in nature that the reviews can reach into every major function.

JOSEPH M. JURAN

I learnt more about my division in 2 days of the total quality review than in my total 9 months' tenure as a General Manager.

A NEW HEWLETT-PACKARD GENERAL MANAGER

Overview

Why do a total quality review? To answer this question, let us look at an analogy in daily life. We often go for medical checkups. The purpose and requirements for a medical checkup are well understood. The doctor gives a diagnosis by measuring performance against what is possible. When we hear the results, we decide what to do with the recommendations. The result can be a better and healthier life.

The purpose of total quality reviews is similar to that of medical checkups. During the review a diagnosis is provided, and if the recommendations are followed, they can lead to a more efficient, productive, and successful organization, one that is better prepared for the future. In total quality, however, the requirements or standards are not well known—but in this text, we have defined the minimum requirements.

In this chapter we will provide detailed checklists and guidelines for conducting a total quality review.[1]

Objective of a Total Quality Review

From the discussion above, we can summarize the objective of a total quality review.

The objective of the total quality review is to discover and encourage the use of quality tools and systems to manage a business, in order to make it more efficient, productive, successful, and better prepared to face the future.

A Total Quality Review Procedure

The total quality review process consists of

- An agenda for the review
- A detailed checklist for the review, with a scoring system
- A trained review team to conduct the review
- The total quality review report with recommendations for improvement

We discuss each of these items in detail and provide guidelines and recommendations.

An agenda for the review

We provide two agendas for conducting a review:

1. A generic agenda for conducting a review for almost any entity, organization, company, or business unit. This will take about 2 days.

2. An agenda for reviewing a function within an entity, say, the marketing department. In this case the review can be shortened to about 1 day with the same items being reviewed.

Agenda 1. The following is a sample agenda for reviewing a business unit.

Item reviewed	Who or function	Length (hours)
Opening remarks by review team (review objectives, role of review team)	Team leader	0.25
Overview (entity vision, markets and customers, products and services, business environment and current performance)	General manager	0.25
Customer obsession	General managers and staff	3.00
Business planning	General manager and staff (this will require involvement of the next-level functions)	3.00
Process management and managing improvements and breakthroughs	General manager and staff (this will require involvement of the next-level functions)	5.00
Employee development and participation, and leadership	General manager and staff (where called for, next-level functions must be reviewed)	2.00
Break (review team prepares comments)		1.00
Verbal report of review team (observations, recommendations for improvement, explanation of scoring procedures)	General manager and staff	1.00
Total time (approximately)		16.00

Agenda 2.   The following is a sample agenda for a function.

Item reviewed	Who or function	Length (hours)
Opening remarks by review team	Review team leader	0.25
Customer obsession	Function manager and staff	1.5
Business planning	Function manager and staff	2.00
Process management and managing improvements and breakthroughs	Section managers	2.50
Employee development and participation, and leadership	Function manager and staff	1.00
Break (review team prepares comments)		1.00
Review team gives report		1.00
Total time (approximately)		8.00

Comments on the agenda and approach to reviews.   There are several points that need clarification.

- The terms *entity, organization, company,* and *business unit* are used to mean the same thing; that is, they refer to a business unit with many functions. A *function* refers to an operation that is focused on one activity, such as marketing or manufacturing.

- We recommend reviewing the managing improvements and breakthroughs and process management sections together. Start the discussion with process management, and note the plans for improvement. Then review several improvement projects, including those from process management.

- Many questions will overlap—this is difficult to avoid. The approach, however, will be different. For example, during the question on improvements in process management, we want to know if and what improvements were made. But in the managing improvements and breakthroughs section, we go into detail on several projects and review the rigor of the improvement methodology.

- When an entity is reviewed again, progress is measured against the deficiencies noted in the previous review. Hence, it is appropriate to start the review with a discussion of improvements since the last review.

- Formal reviews can be conducted every 2 to 3 years. Informal entity reviews or self-reviews should be conducted annually by your quality manager or preferably by other business managers.

- In-depth reviews of the entity's major functions can also be conducted annually, using the shorter agenda, agenda 2.

Scoring System

The scoring matrix, illustrated in Fig. 8.1, is designed to highlight the entity's strengths and weaknesses and to facilitate feedback. It is intended to help monitor progress, which will come from conducting regular reviews.

Guidelines in using the scoring system. When you look at each question in the checklist, select the cluster of comments that is most appropriate. To aid in selection, we have segmented the comments into approach, effort, and results. Then choose the score associated with the cluster of comments—this is the score for the question. Here are some additional guidelines:

Choosing a score value of 1. A score value of 1 will be very obvious to you. It represents the existence of basic or very minimal activity.

Choosing a score value of 2. This value is given if the entity is doing the basic activity in several areas.

Choosing a score value of 3 or 4. A score value between 3 and 4 indicates that the entity is competitive. The entity has gone beyond the basics, is successful, and is deploying quality tools and systems in many areas. Per the scoring guidelines, this will indicate that the

	Overall Performance: Basic ⟶		Competitive ⟶	Leading ⟶	
Approach	Using very little data and quality management tools.	Using some data and quality management tools.	Knowledgeable use of data and qualitity management tools.	Good use of data and quality management tools.	Excellent use of data and quality management tools.
Effort	Evidence of effort in a few areas. Opportunities exist throughout entity.	Evidence of effort in several areas. Numerous opportunities exist.	Evidence of effort in many areas. Further deployment possible.	Evidence of effort and deployment in most areas.	Evidence of effort and deployment in all areas.
Results	Little success, lots of opportunity.	A few successes, lots of opportunity.	Many successes, but more possible.	Successful in most areas. Some innovative approaches.	Good to excellent results in all areas. Many innovative approaches.
Score:	1	2	3	4	5

Figure 8.1 The scoring matrix.

entity is competitive with the industry or leading in some cases (for a score of 4).

Choosing a score value of 5. A score value between 4 and 5 represents a leading or advanced entity. It should be used only when it is clear that the entity is ahead of the competition. Ask for or get data to verify this. If in doubt, for a value of 5, give a lower value of 4 or 4.5, which is very good. We expect that only a world-class or extremely experienced reviewer—one who has reviewed or visited several world-class companies and most of the entities within an organization or large company—will be able to give a score value of 5.

Score value for questions with multiple items. Several questions in the checklist require you to look at multiple items. Faced with a choice of having a long list of questions or a short list, we choose the latter. This requires many questions to look at a cluster of items. This is what we suggest: *If all items in a multiple-item question are done extremely well and done in all functions of the entity, then give a score of 4.* If only one item (out of, say, three items) is done well, then give a score of one-third of that, or 1.3. On the other hand, if one function in the entity does all items in a multiple-item question well, but two other functions do not, then give a score of, say, between 1 and 2 instead of 4. Such a situation could arise if, say, an enthusiastic function manger is an innovator and does a great job in process management, but other functions do not manage their processes at all. *Note:* The scoring system prompts for such situations, wherever possible.

Choosing a score value of half point. Sometimes, you may feel that the score falls in between two clusters, for example, between 2 and 3. In this case, choose a score value of 2.5.

Summarizing the scores. The summary of the score for each element can be displayed on a radar chart, with five axes. Each axis represents one of the elements of the total quality review. In our proposed system, we retain the scale of 1 to 5 for each review element.

The summary radar chart is shown in Fig. 8.2. This is a convenient mechanism for displaying long-term targets (for example, 4.0 for all elements) and actual scores for successive reviews. An alternate system to display results, with comments on each element of the review, is shown in Fig. 8.3. This uses the scoring system that we have proposed and gives a short verbal comment to indicate how each element is ranked, that is, basic, competitive, or leading.

Which should you use? Both have their merits, and we suggest using both. The radar chart allows you to display specific areas of weakness and measure progress over the years. The scoring comments provide a description of the current status of each element.

Once scores are determined for each category and calculated, they will be displayed as a radar chart like the one shown below. The radar chart will give managers a snapshot of how their entity scored in each area; it can also show progress over the years.

1. Business Planning _____ (maximum 5)

2. Customer Obsession _____ (maximum 5)

3. Managing Improvements and Breakthroughs _____ (maximum 5)

4. Process Management _____ (maximum 5)

5. Employee Development and Participation, and Leadership _____ (maximum 5)

Total Score: _____ (maximum 25)

OVERALL SCORE: _____ (maximum 5)

To compute: $\dfrac{\text{Total Score}}{5}$

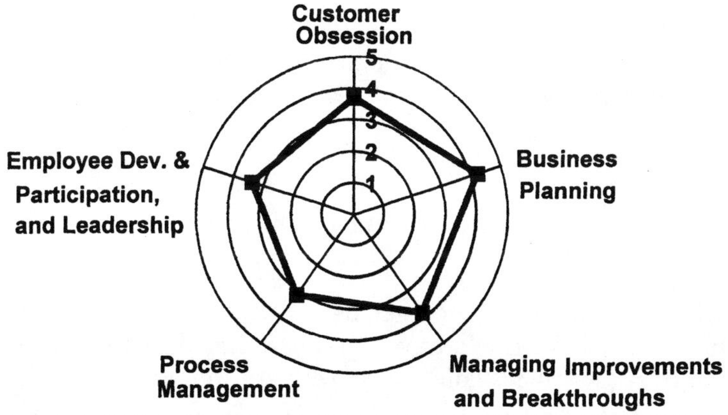

Figure 8.2 Scoring summary.

Scoring system comments. In the end, the scoring system is used to reveal relative performance between entities or to measure progress. It is not an absolute indicator of success in total quality. For that, we recommend calling in an expert who is experienced in total quality in several industries. In this way, you can also get a relative ranking across industries or with other companies or organizations.

Conducting reviews without scores. Some entities, however, may not want to go through a scoring process. In that case, we suggest you use

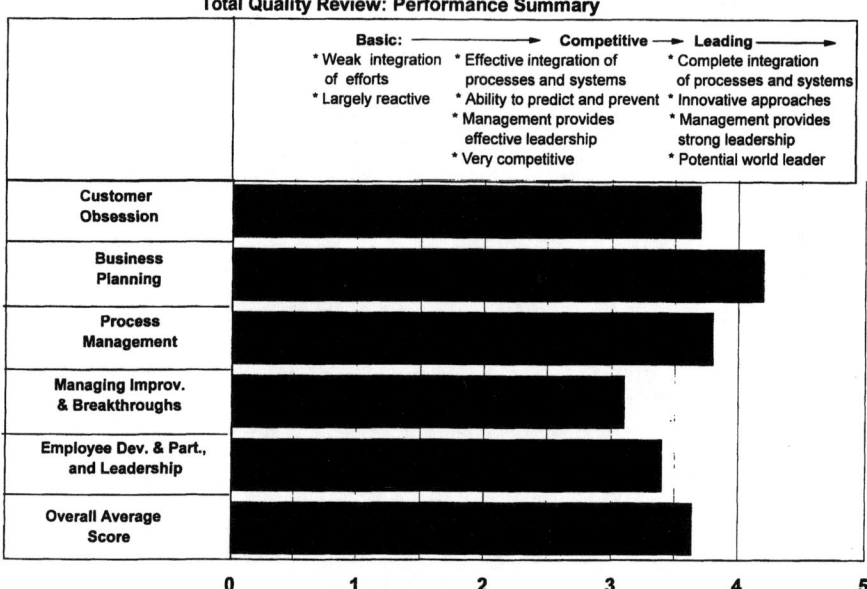

Total Quality Review: Performance Summary

Figure 8.3 This chart gives a performance summary for a total quality review. Both a score and a short verbal description are provided for each of the five elements.

a chart similar to Fig. 8.3, without scores but with a focus on the relative positioning of each element—basic, competitive, or leading.

Review Checklist

We will now give a review checklist. The format is as follows for each review element:

- There is a checklist of questions. These have been kept to 10 questions per element. We have, however, structured many of these as multiple questions. This helps to ensure simplicity and consistency.

- For each question, a short discussion is given, and where appropriate, notes are provided to give more help to the reviewer.

- A separate page is used to allow scoring of each element. This includes the scoring matrix discussed earlier.

Customer obsession checklist. In this section, we review the entity's obsession with customers and management of customer issues and how these impact products and services.

1. Customers are identified and stratified.
   Discussion: Here we look at the entity's external and internal

customers, and we determine if they are defined and stratified by needs, products, and services. The importance of this item is so obvious, yet some will do poorly here.

2. Product and service quality data collection processes are well established. Data are used to drive improvement.

   Discussion: Here we look at the process of collecting failure data on products and services from customers after the sale, and we review how the analysis of these data influences products and services and future plans. Look for a formal system to collect this information, to track it, and to drive corrective action.

   Notes: Items to look for include

   - Product and service failure information.
   - A customer complaints and feedback system.
   - Quick resolution of customer complaints.
   - Is resolution limited to damage control or recovery (by providing compensation, etc.)?

3. Customer interface activities are well established.

   Discussion: Here we determine proactive methods of visiting and meeting with customers to get their inputs, concerns, and suggestions. Items to look for include formal meetings and visits. Our focus should be on the entity's ability and desire to keep in touch with the customer. Check that the data are captured systematically. Actual follow-up activity is important and is checked here and later, in item 6.

   Note: Where possible, we have deliberately separated data collection from follow-up activities. We have reviewed many entities that are rich in customer satisfaction data and concerns but poor at follow-up. This separation also allows for easier scoring and identification of strengths and weaknesses.

4. Customer satisfaction levels have been determined.

   Discussion: Here we look at the process of determining satisfaction levels. We look at customer satisfaction surveys and other means of satisfaction feedback.

   Notes: Specifically we look at

   - Inputs via response cards mailed in after purchase of product or service
   - Data from visits, meetings, etc.
   - Analysis of the available data and understanding of how the satisfaction levels were determined and prioritized

5. Competitive data are available and analyzed; the analysis is used to influence both current and new products and services.

   Discussion: Here we look at competitive benchmarking and other means to get competitive data.

   Notes:

   - Specifically, we look for data for both products and services.

■ The entity should be getting such data, understanding its market position, and making decisions on how to respond.

6. The entity responds to customer satisfaction issues by taking follow-up corrective action.

Discussion: Our purpose here is to look at follow-up corrective action after data are collected and analyzed from surveys, customer interface activity, and competitive benchmarking.

Notes: Our focus should be on

■ What is done with the data

■ How priorities are set

■ Plans to improve current products and services and future products

7. The entity has developed a customer satisfaction model.

Discussion: We are looking at how this entity goes beyond a customer satisfaction survey to measure the various elements of a customer's satisfaction with a product or service. The expectation is that a full spectrum of attributes or elements is tracked, measured, and managed. Most entities will have parts of the model we discussed in Chap. 2, but we are looking at the maturity of the entity in managing all aspects of customer satisfaction.

8. New customer needs are solicited and understood.

Discussion: Here we are looking at the process of gathering customer needs. This can be via many methods such as focus groups and visits. Often this process could be a subset of other methodologies such as surveys. But typically customers need to be prodded to understand what's missing in today's offerings and what their expectations are.

9. The entity responds to new customer needs that have been identified.

Discussion: If and when new customer needs have been identified, we probe to see how they respond to these needs. Is there a process to capture these needs and translate them to products or services, or is this done randomly?

10. The entity is proactive and is stretching itself to provide new products and services that delight customers.

Discussion: Our purpose here is to determine if the entity strives to exceed customer expectations. Does it only look at improvements and competitive benchmarking, or does it go beyond that to provide attractive products and services?

Notes: Refer to the total product concept discussed in Chap. 2. Does this entity consciously strive to operate in the outer zones? The entity may not use the total product concept; nonetheless, we check whether the entity confines itself to thinking in a box or stretches its imagination.

Customer obsession scoring. This page is set aside for scoring the items reviewed. Use the scoring matrix and questions provided to determine the scores.

	Score
1. Customers are identified and stratified.	———
2. Product and service quality data collection processes are well established. Data are used to drive improvement.	———
3. Customer interface activities are well established.	———
4. Customer satisfaction levels have been determined.	———
5. Competitive data are available and analyzed; the analysis is used to influence both current and new products and services.	———
6. The entity responds to customer satisfaction issues by taking follow-up corrective action.	———
7. The entity has developed a customer satisfaction model.	———
8. New customer needs are solicited and understood.	———
9. The entity responds to new customer needs that have been identified.	———
10. The entity is proactive and is stretching itself to provide new products and services that delight customers.	———

Total score: (Maximum 50 points)   ———

Rescale: $\dfrac{\text{Total score}}{10}$ (Maximum 5 points)   ———

	Overall Performance: Basic ———————→		Competitive ——————→	Leading ——————→	
Approach	Using very little data and quality management tools.	Using some data and quality management tools.	Knowledgeable use of data and qualitity management tools.	Good use of data and quality management tools.	Excellent use of data and quality management tools.
Effort	Evidence of effort in a few areas. Opportunities exist throughout entity.	Evidence of effort in several areas. Numerous opportunities exist.	Evidence of effort in many areas. Further deployment possible.	Evidence of effort and deployment in most areas.	Evidence of effort and deployment in all areas.
Results	Little success, lots of opportunity.	A few successes, lots of opportunity.	Many successes, but more possible.	Successful in most areas. Some innovative approaches.	Good to excellent results in all areas. Many innovative approaches.
Score:	1	2	3	4	5

The scoring matrix.

Business planning checklist. In this section, we review business planning at the entity being reviewed and its influence on business results.

1. There is a long-range plan that addresses the strategic direction of the entity.

   Discussion: Review the entity's vision. Does it set the overall strategic direction and influence the business planning process? Ideally we expect to see the features of a vision that were mentioned during the discussion in Chap. 3. We look for the existence of core values in the organization, the purpose, the goal, and (hopefully) a specific description of the envisioned future. If any of these is missing, a discussion is appropriate. Also check how the vision is routinely communicated and if it is understood by all employees.

   The other items in the long-range plan are also important. Although we have given a suggested content for the long-range plan, the entity may have a different format. Look for the suggested content, and if it is missing, question why it was not addressed or considered important. The key point to look for is the entity's strategic direction. Where is it going? What are the competitive pressures and challenges? How will it overcome these challenges and pull ahead?

   Notes: This plan should cover items such as

   - The entity's vision
   - Customer needs and channels of distribution
   - Competitive situation
   - Necessary products and services
   - Also, all other items mentioned in the chapter on business planning, that is, development of partners, financial analysis, potential threats, and the 3- to 5-year strategies.

2. There is a well-established annual planning process.

   Discussion: Review the annual planning process. The key point to look for is breakthrough strategies over the next year. Are they convincing, and will they respond to immediate challenges or threats?

   Notes: We are looking for an annual planning process that uses the Hoshin planning methodology or an equivalent. The plan should have the following features:

   - It is driven from strategies developed from the long-range plan (unless this is a lower-level operation not requiring a long-range plan).
   - It is influenced by the previous year's issues and lessons learned.
   - If the entity does not do long-range planning, then the plan is driven by customer inputs, which are linked to the customer obsession checklist and higher-level plans.
   - Quality, cost, education, and delivery issues have been addressed.

■ Check for the features mentioned above; and if they are missing, understand why.

3. The annual plans are robust and focused.

   Discussion: When plans are reviewed, focus on one or two objectives, considered the most crucial, then check for robustness, thoroughness, and focus. Be especially wary if management has too many objectives or vague goals, and be prepared to discuss this issue and its potential problems.

   Notes: The following are important features to look for:

   ■ At the top level there are a few breakthrough objectives, with robust supporting strategies.

   ■ Goals and performance measures are realistic and measurable.

   ■ The top-level plans are properly deployed down the organization.

4. Cross-functional issues have been addressed in the annual Hoshin plan.

   Discussion: Strategies that are cross-functional and require teamwork in this entity and with other entities, partners, suppliers, and customers must be addressed; ownership must be clear. Check on how these are addressed, monitored, and managed.

5. Implementation plans exist for the annual Hoshin plans; plans are reviewed and ownership, time, and resources are defined.

   Discussion: This is a crucial step. We look at the lower levels in the organization and review their implementation plans or alternate processes. As discussed during the business planning section in Chap. 3, every Hoshin strategy in the organization eventually ends up in a detailed implementation plan, with an owner and a time line. This is where the rubber meets the road—hence this step is crucial for success.

   Notes: We look for the following items:

   ■ A plan that shows how the strategies in the annual plans will be implemented

   ■ Who implements the detailed strategy and when?

6. A robust daily management, or business fundamentals, plan exists and includes regular tracking of performance.

   Discussion: The top level and every function and department must have such a plan. Remember to ensure that the entity managers understand and distinguish breakthrough (Hoshin) from daily management issues. This plan should reflect and manage day-to-day activities or key processes from the entity process map. This will be reviewed in greater detail in the process management section.

7. All plans are reviewed regularly for progress, and corrective action is taken, including changes to the plan, when necessary.

Discussion: Check for the review process, progress to plans, lessons learned, and follow-up corrective action. Reviews should be done frequently, say, every 3 months.

8. It is clear that the vision and planned strategic direction drive the entity.

Discussion: We clarify here whether the entity is driven by its vision and planned strategic direction, that is, its business plans. The strategic direction is an extension of the entity's vision. The entity's team understands clearly the reason for its existence; it continuously listens to customers, initiates changes in direction as the environment changes, and understands critical success factors. There is strong alignment to higher-level plans (unless, of course, this is the highest level) and with partner plans. The business plans convey the current and future position in the marketplace. There has been sustained progress in market share, growth, and customer satisfaction.

9. The business planning process includes managing the linkages to other areas of the value delivery chain.

Discussion: No business unit can operate as an island. It will be part of a value delivery chain and have activities upstream (from suppliers, etc.), downstream (customers, etc.), and sideways (joint ventures, contractors, etc.). Hence we determine how the business unit is managing these partners.

Notes: We are looking for strong linkages here. Items to consider include

- Management of performance and relationships with partners.
- Satisfaction levels of partners.
- Typically, at most business units, there are opportunities upstream and sideways, while downstream activities are managed best.

10. Business results have been achieved as a result of good business planning.

Discussion: We have reviewed the entity's business planning process. Now we would like to understand if the objectives and strategies have helped generate success. This should be evident as we go through the entire review. At the end of the session, we need to collect our thoughts and confirm if the entity is moving aggressively in the direction stipulated by its vision, and is successful.

Notes: Key items to consider:

- The entity's vision has been a galvanizing force moving it in the right direction.
- Costs are under control; the entity is profitable and competitive.
- Success is a result of astute business planning and responsiveness to the market and customers, not luck.

■ The entity has successfully implemented most of its strategies, and this has created market success.

Business planning scoring.   This page is set aside for scoring the items reviewed. Use the scoring matrix and questions provided to determine the scores.

	Score
1. There is a long-range plan that addresses the strategic direction of the entity.	———
2. There is a well-established annual planning process.	———
3. The annual plans are robust and focused.	———
4. Cross-functional issues have been addressed in the annual Hoshin plan.	———
5. Implementation plans exist for the annual Hoshin plans; plans are reviewed and ownership, time, and resources are defined.	———
6. A robust daily management, or business fundamentals, plan exists and includes regular tracking of performance.	
7. All plans are reviewed regularly for progress, and corrective action is taken, including changes to the plan when necessary.	———
8. It is clear that the vision and planned strategic direction drive the entity.	———
9. The business planning process includes managing the linkages to other areas of the value delivery chain.	———
10. Business results have been achieved as a result of good business planning.	———

Total score: (Maximum 50 points)   ———

Rescale: $\dfrac{\text{Total score}}{10}$ (Maximum 5 points)   ———

	Overall Performance: Basic ⟶ Competitive ⟶ Leading ⟶				
Approach	Using very little data and quality management tools.	Using some data and quality management tools.	Knowledgeable use of data and qualitity management tools.	Good use of data and quality management tools.	Excellent use of data and quality management tools.
Effort	Evidence of effort in a few areas. Opportunities exist throughout entity.	Evidence of effort in several areas. Numerous opportunities exist.	Evidence of effort in many areas. Further deployment possible.	Evidence of effort and deployment in most areas.	Evidence of effort and deployment in all areas.
Results	Little success, lots of opportunity.	A few successes, lots of opportunity.	Many successes, but more possible.	Successful in most areas. Some innovative approaches.	Good to excellent results in all areas. Many innovative approaches.
Score:	1	2	3	4	5

The scoring matrix.

Managing improvements and breakthroughs. In this section, we review several improvement and breakthrough projects—about two each. Our purpose is to understand how managers, team leaders, and employees manage improvements and breakthroughs.

1. Improvement and breakthrough projects are linked to the annual plan or crucial issues.

   Discussion: Our purpose here is to understand how improvement and breakthrough projects are selected. Ideally, improvement projects should be identified during business planning, from customer issues or internal problems. A major weakness is to focus only on breakthroughs—new and exciting stuff at the expense of current weaknesses, which require improvement. Another related and common failing is to manage improvement projects separately from business planning. This must be avoided because of several reasons: Improvement projects will be treated as less important and given lower priority, and there will be no resources left over for improvements if all resources are consumed for business planning strategies. Remember that many improvement projects can be megaprojects—refer to the solder wave example (improvement of printed circuit production process) in Chap. 4.

   Notes: The following questions will help start the discussion:
   - What improvement projects were selected during business planning, and were they prioritized?
   - What breakthrough projects were identified during business planning?
   - Was there a conflict in selecting improvements versus breakthrough projects, and how was it resolved?
   - Note that some improvement projects can be ad hoc and not generated from business planning. These projects can occur because of crucial business issues, for example, customer complaints from a poor product or service.

2. Improvement projects are clearly defined with objectives, goals, and schedules.

   Discussion: We look for clear project definitions. This should include clear and aggressive goals and project schedules. What is a good improvement goal? In Chap. 4, we recommend a rule of thumb of 50 percent reduction—for defects or time to do something.

   Notes: To summarize, we look for
   - Project definition
   - Aggressive project goal
   - Project schedule

3. There is good data collection, analysis, and understanding of root cause.

   Discussion: We look for good collection of data and appropriate use of quality tools such as Pareto diagrams, graphs, and cause-and-effect diagrams. We also check if there is good understanding of root causes.

   Notes: We should check whether most likely causes have been identified, and verified with data.

4. Alternative solutions are reviewed, and results have been achieved. Deviations from goal are understood.

   Discussion: Alternative solutions are reviewed, evaluated, and implemented. Results have been achieved and measured; deviations from goals are well explained and acted upon.

   Notes: Specifically we check for
   - Explanations for deviations from goals
   - Sustained performance after project completion

5. Review is standardized, and future plans are made. Changes or the new process is standardized, and employees are trained.

   Discussion: New or modified processes are documented and employees are trained; future plans exist. Look for a method to update standards across the organization, for example, a method similar to the SUR format, discussed in Chap. 4.

6. There is a systematic approach to problem solving, with no recurrence of problems.

   Discussion: During review of the improvement projects, check for adherence to the steps of the PDCA cycle and documentation of the entire project.

   Note: Most important, we look for lessons learned and reasonable assurance that this problem will not recur at the entire entity—probe extensively for this!

7. How are breakthrough projects selected and initiated?

   Discussion: Our purpose is to understand the thinking behind breakthrough project selection. This topic may have been covered during the business planning or customer obsession sessions. If so, we can move quickly through this item. This item could be discussed in item 1, above. Note, however, that items 7, 8, 9, and 10 refer to breakthrough projects, not continuous improvement projects.

   Notes: We check for the rationale used behind project selection, for example,
   - Are projects customer-driven?
   - Are projects driven by competitive pressures?
   - Are projects driven by future, potential needs?

- Will these projects move the entity ahead of the competition? How far ahead?

8 There is a systematic approach to managing breakthrough projects.

Discussion: Here we try to understand how well the project was managed, and we focus on the approach used. The expectation here is that a process similar to the proactive PDCA cycle is used. *Note, however, that our intention is not to stifle creativity by imposing a straitjacket on the team. Instead, we look for some basic things that should be done.*

Where time permits, review a less successful product or service. Valuable insights can be gained in understanding why the project was not successful or did not meet expectations.

Notes: We look for

- Objective, clear goals for the new product or service
- Detailed schedules
- Whether the product or service was tested to specifications
- If possible, review of a less successful project and the lessons learned

9. For breakthrough projects, is there a process to prevent problems by prediction or testing?

Discussion: It may be challenging to have improvement projects that improve product or service reliability and acceptance by customers. But this is considered rework. Here we specifically check to see if an effort was made to prevent problems by prediction during the design of breakthrough projects. We discussed this during the proactive PDCA cycle, in Chap. 4.

Notes: The areas to look at are

- Testing the product's or service's reliability prior to product release
- Testing the product's or service's market acceptance via a pilot program or in a test market

10. Improvement and breakthrough projects have met expectations and achieved results.

Discussion: Have breakthrough projects had an impact on customers? This question will lead to a fruitful discussion with the project teams.

Notes: The following items help to answer this question better:

- Internal productivity gains
- Sales or market share data
- Customer feedback on both improvements and breakthrough projects

Managing improvements and breakthroughs scoring. This page is set aside for scoring the items reviewed. Use the scoring matrix and questions provided to determine the scores.

	Score
1. Improvement and breakthrough projects are linked to the annual plan or crucial issues.	————
2. Improvement projects are clearly defined with clear objectives, goals, and schedules.	————
3. There is good data collection, analysis, and understanding of root causes.	————
4. Alternative solutions are reviewed, and results have been achieved. Deviations from goal are understood.	————
5. Review is standardized and future plans exist. Changes or the new process is standardized, and employees are trained.	————
6. There is a systematic approach to problem solving, with no recurrence of problems.	————
7. How are breakthrough projects selected and initiated?	————
8. There is a systematic approach to managing breakthrough projects.	————
9. For breakthrough projects, is there a process to prevent problems through prediction or testing?	————
10. Improvement and breakthrough projects have met expectations and achieved results.	————

Total score: (Maximum 50 points) ————

Rescale: $\dfrac{\text{Total score}}{10}$ (Maximum 5 points) ————

	Overall Performance: Basic ————————→ Competitive ———————→ Leading ——————→				
Approach	Using very little data and quality management tools.	Using some data and quality management tools.	Knowledgeable use of data and qualitity management tools.	Good use of data and quality management tools.	Excellent use of data and quality management tools.
Effort	Evidence of effort in a few areas. Opportunities exist throughout entity.	Evidence of effort in several areas. Numerous opportunities exist.	Evidence of effort in many areas. Further deployment possible.	Evidence of effort and deployment in most areas.	Evidence of effort and deployment in all areas.
Results	Little success, lots of opportunity.	A few successes, lots of opportunity.	Many successes, but more possible.	Successful in most areas. Some innovative approaches.	Good to excellent results in all areas. Many innovative approaches.
Score:	1	2	3	4	5

The scoring matrix.

Process management checklist. In this section, we review process management. We probe for understanding of the process concept and then go beyond that to understand how management deals with key processes throughout the entity.

1. The process concept is understood. Key processes are identified, documented, and understood.

   Discussion: Our purpose is to start a discussion to understand if the entity managers and staff understand the process concept. The key point is: *A well-managed process can be a leading indicator of good results in any company or organization.* Often, processes are managed routinely with little purpose, the original objective having been forgotten. We should look out for such a situation. This flaw could be due to insufficient training or education in process management.

   Next, we request for and review the list of key processes. We ask and understand why they are key processes and how they are linked to customer needs and satisfaction. Keep an open mind when reviewing the list. Then review the processes and check all the parameters mentioned below. Plan to check only a short list of the processes—those that seem most important or impact customers the most.

   Notes: We look for
   - Understanding of the process concept
   - Identification or list of key processes
   - Linkage of the key process to customer needs or satisfaction
   - Process flowcharts and documentation
   - Performance measures to measure and track the process
   - Routine and ongoing employee training

2. How does the entity manage and coordinate all its processes?

   Discussion: The process concept may be understood, and processes might be managed, but we check to see how the entity views the big picture of process management and how it links to the value delivery chain or system. We discussed this in detail in Chap. 5. We look for the features listed below or review the rationale for a different approach. The key is to look for an approach that optimizes and prioritizes processes in the entity. *One of the worst things we have seen is an entity proudly managing several hundred processes—needless to say, our comments were negative.*

   Notes: Items to look for include these:
   - How is the big picture of processes viewed and managed?
   - Is there an entity process map—or an alternative—that shows the flow of activities, or processes in the organization?
   - Is there linkage of key processes to suppliers and customers? More on this later, in question 9.

- How does the entity optimize and prioritize its processes? One way, of course, is to use our approach of starting with an entity top-level process.

  If the above features exist, then review the entity process map and understand how the key processes and performance measures were selected and defined. If these features do not exist, discuss the concept and the benefits of managing via an overall process map. Note, however, if the entity already manages via an entity process map, the discussion would have started in the very first question listed above. This question and most that follow can be redundant, but this should not be an issue—it means we are reviewing a mature entity.

3. Key processes are monitored, and performance against targets is checked routinely.

    Discussion: We check if the process is monitored and performance measured routinely. For all processes reviewed, check how performance measures and targets have been established. For example, for sales process management (Table 5.3), review how the measures were established. For a process that is triggered by external activity, such as sales postmortems, check for existence responsiveness.

    Notes: When you review performance measures, it is important to note the following:
    - How were the performance measures and targets established?
    - Are some of the performance measures related to customer needs or satisfaction?
    - Do customers agree that the customer measures are meaningful for them?

4. Deviations from established targets and action limits are recorded and analyzed, and corrective action is taken; or look for rigor of application.

    Discussion: Look for analysis, corrective action, and rigor of application. The truly analytical managers will have an out-of-control report for deviations—this is essential in manufacturing processes, but less so in other environments.

    Notes: Look for
    - Cases when actual performance crosses targets and action limits
    - What analysis and corrective action occur when limits are crossed
    - When some performance is activity-based, for example, conduct of a sales postmortem, rigor of application.

    Here are some powerful questions to ask and steps to take when process deviations are detected:
    - When did you last see this problem? And if it occurred before, probe as follows:

- What was the fix to the problem?
- Was it a good fix?
- Why has the problem recurred?
- Review the previous corrective action or out-of-control report. Does it exist? Is it complete?
- Look for a good PDCA cycle for every situation.

5. Is the concept of standards well understood?

Discussion: Our purpose here is to see if critical processes are documented. This item may never show up during the earlier questions listed above. Look for the current list of standards and adequate documentation. Often the critical processes are agreed upon but never standardized. This can lead to an intense discussion on the pros and cons of documenting and providing education and training in such items.

Notes: Standards refer to routine work activity that should be documented. Examples are

- Business planning, design, manufacturing, sales, and marketing standards.
- For design, manufacturing, and sales environments: an entity quality manual that documents the overall quality assurance system. Other types of organizations should have something very similar, that is, a quality or customer service manual that documents how they manage the quality of their processes, products, and services.
- For the items identified, how are the standards established, revised, and abolished?

6. How does the entity deploy key process management down the organization?

Discussion: If the entity manages processes via an entity process map, then check if it also manages the next levels or functional areas in the organization via process maps. An entity may have only one level or several process maps. We recommend not more than a total of two or three levels. More than that gets too bureaucratic and may be overkill. How and why this is done will reflect on the maturity of the organization. A manufacturing division should have two or even three levels, others probably fewer. Even if there is only one entity process map, the next levels can manage the key processes by exploding them and managing them in greater detail. Both alternatives are good.

Notes: This can be done in several ways.

- Look for next-level process maps and how they are managed.
- Look for exploding of a key process (managing it in greater detail) at the next level by the process owner.

7. The entity key processes are managed systematically.

Discussion: In the questions above, we checked how the key processes are tracked and managed. The important question to ask is: If these are key processes, how do you link them to your business or business plan? A sophisticated approach is to have them tracked via a daily management, or business fundamentals, plan. This plan then becomes the business scorecard, which is what we ideally want to see. Alternatively, we can review how else the entity links key process management to the business and evaluate the pros and cons compared to our suggested approach.

8. The process has been routinely improved and documented, including the update of standards. Future improvements are evaluated.

    Discussion: It is usually not possible to improve every process, every year. Hence look for past, current, or future plans. Over time, processes will be deleted or added. In addition, key processes should be competitively benchmarked and a response initiated for deficiencies. Look for this.

9. The key processes are linked tightly to the value delivery chain.

    Discussion: Our purpose here is to understand how well the entity manages its linkages and processes with its partners, up and down the value delivery chain. This entity will be part of a value delivery chain and have activities upstream (from suppliers, etc.), downstream (customers, etc.), and sideways (joint ventures, contractors, etc.). Hence we determine how the entity links up with its partners.

    Notes: We are looking for strong linkages here. Items to consider include

    - Management of process performance with partners.
    - At a high level in the entity, say, that of the general manager, there may be just one or two high-level performance measures in the general manager's daily management, or business fundamentals, plan. At a lower level, however, someone who deals directly with the partner or customer will have more performance measures.
    - In all cases, we also look at how the entity manages and improves the relationship with the partners. This item was also discussed in the business planning checklist.

10. Customer satisfaction has been achieved as a result of process management.

    Discussion: We wrap up the session by discussing the benefits of process management at the organization under review.

    Notes: We need to understand the following:

    - Did managing and improving key processes result in benefits at the entity?
    - What gains did we see?
    - Did customers benefit? How do we know if they did?

Process management scoring. This page is set aside for scoring the items reviewed. Use the scoring matrix and questions provided to determine the scores.

		Score
1.	The process concept is understood. Key processes are identified, documented, and understood	_____
2.	How does the entity manage and coordinate all its processes?	_____
3.	Key processes are monitored, and performance against targets is checked routinely.	_____
4.	Deviations from established targets and action limits are recorded and analyzed, and corrective action is taken; or look for rigor of application.	_____
5.	Is the concept of standards well understood?	_____
6.	How does the entity deploy key process management down the organization?	_____
7.	The entity key processes are managed systematically.	_____
8.	The process has been routinely improved and documented, including the update of standards. Future improvements are evaluated.	_____
9.	The key processes are linked tightly to the value delivery chain.	_____
10.	Customer satisfaction has been achieved as a result of process management.	_____

Total score: (Maximum 50 points)   _____

Rescale: $\dfrac{\text{Total score}}{10}$:   (Maximum 5 points)   _____

	Overall Performance: Basic ⟶		Competitive ⟶	Leading ⟶	
Approach	Using very little data and quality management tools.	Using some data and quality management tools.	Knowledgeable use of data and quality management tools.	Good use of data and quality management tools.	Excellent use of data and quality management tools.
Effort	Evidence of effort in a few areas. Opportunities exist throughout entity.	Evidence of effort in several areas. Numerous opportunities exist.	Evidence of effort in many areas. Further deployment possible.	Evidence of effort and deployment in most areas.	Evidence of effort and deployment in all areas.
Results	Little success, lots of opportunity.	A few successes, lots of opportunity.	Many successes, but more possible.	Successful in most areas. Some innovative approaches.	Good to excellent results in all areas. Many innovative approaches.
Score:	1	2	3	4	5

The scoring matrix.

Checklist for employee development and participation, and leadership. In this section, we review how employee skills are developed and how employees participate in the organization. We also review the leadership style of management.

1. Project or quality team activity is well managed.

   Discussion: Here we review activities by project teams and look for the degree of involvement by employees. We try to understand what team activities and projects employees are involved in. We check for project linkage to business plans, key processes, and customer issues; also, overall coordination and results of this activity.

   Notes: Look for the following:

   - Project teams.
   - Cross-functional teams—within the organization and especially those that involve outside partners, suppliers, and customers.
   - Quality circles, quality teams, and project teams.
   - Management teams.
   - Check what projects these teams are working on and linkages to business plans and customer issues.

2. Systems and processes are in place to capture and use employee suggestions and contributions.

   Discussion: Here we assess how employee contributions are encouraged and implemented.

   Notes: Areas to look at include

   - Employee suggestion schemes
   - Other employee contributions, such as patents and customer satisfaction
   - Management of above activities

3. Management reviews and recognizes project and quality team activity and employee contributions.

   Discussion: We review how management recognizes and rewards employee participation and activity. We check for a forum for team presentations, activities, and awards to employees for their contributions. In all cases, senior management should be involved.

4. Education and training programs are defined, implemented, and sustained.

   Discussion: Our purpose here is to review how employees are trained and educated for current and future job requirements, and prepared for growth. Review education program for completeness, coverage of all employees, and proper implementation, and ensure program is sustained.

Notes: Look for

- Training and education to improve current and future skills
- Education for employee growth
- A well-established and managed training and education program

5. Management shows leadership by responding quickly to change.

Discussion: We review how management responds to the changing business environment and prepares the organization and employees for change.

Notes: We look for the following:

- What changes are occurring in the business environment?
- What are the critical issues impacting the organization?
- How is management responding to these critical issues—for the organization and for employees?

6. Employees are empowered to make decisions and proposals.

Discussion: We review how management shows leadership by delegating decision making down the organization.

Notes: Areas to look for:

- Front-line staff, who deal with customers on a day-to-day basis, are empowered to make decisions to resolve customer grievances.
- Employees and managers are able to take calculated risks without fear of repercussions.
- Proposals and decisions are made by teams.
- Strategies are delegated to lower-level managers, teams, and individuals.

7. Senior management develops a strong team and plans for the future via succession planning.

Discussion: We review how senior management builds a strong team and does succession planning for senior staff.

Notes: Areas of focus include

- Tough assignments for management staff and professionals
- New assignments that develop and enhance skills and knowledge
- Education to develop managers to think broadly and prepare for the future

8. Management shows leadership in ensuring that the organization always provides quality products and services.

Discussion: Management must ensure that products and services of the highest quality are generated in the organization. This message—quality first—must also be conveyed to the entire organization so that it is incorporated into the culture of the organization. Much of this information will be apparent as we go through the entire review.

Notes: Look for

- Product and service reliability compared to that of competition and industry norms
- The entity's focus and commitment to quality issues that impact the business and customers, as conveyed in business planning
- Customer and industry perception of the organization
- Employee perception of the organization's quality effort

9. Senior management shows leadership via a strong vision.

Discussion: We should have reviewed the vision in the business planning section. We mentioned the four components of a great vision—they are core ideology (core values and purpose) and envisioned future (goal and specific description). Our focus here is to check how the vision is propagated, maintained, and strengthened.

Notes: We look for

- Frequent communication of the vision by senior management, preferably the top executive of the entity.
- Regular communication of the strategies that support the vision.
- Nurturing and maintaining of the core values in the organization. This is a crucial process, because as the organization grows, the core values need to be maintained.
- Special attention is paid to how the above is done with new employees.

10. The entity's top executive and senior managers prepare for the future by planning and initiating forward-looking strategies.

Discussion: Our purpose here is to see if the management team is forward-looking and is preparing the organization and its employees for the future. The focus of the discussion should be on the organization and its employees, not on customers, market share, etc.

Notes: Specifically we check for these items:

- Does management prepare for the future?
- Are there forward-looking strategies?
- How does management prioritize these strategies?
- How will the future direction impact the organization and its employees?
- What is management doing to prepare the organization and its employees for the future?

Scoring of employee development and participation, and leadership. This page is set aside for scoring the items reviewed. Use the scoring matrix and questions provided to determine the scores.

		Score
1.	Project or quality team activity is well managed.	————
2.	Systems and processes are in place to capture and use employee suggestions and contributions.	————
3.	Management reviews and recognizes project and quality team activity and employee contributions.	————
4.	Education and training programs are defined, implemented, and sustained.	————
5.	Management shows leadership by responding quickly to change.	————
6.	Employees are empowered to make decisions and proposals.	————
7.	Senior management develops a strong team and plans for the future via succession planning.	————
8.	Management shows leadership in ensuring that the organization always provides quality products and services.	————
9.	Senior management shows leadership via a strong vision.	————
10.	The entity's top executive and senior managers prepare for the future by preparing and initiating forward-looking strategies.	————

Total score: (Maximum 50 points)   ————

$$\frac{\text{Rescale: Total score}}{10}$$ (Maximum 5 points)   ————

	Overall Performance: Basic ——————→ Competitive ——————→ Leading ——————→				
Approach	Using very little data and quality management tools.	Using some data and quality management tools.	Knowledgeable use of data and qualitity management tools.	Good use of data and quality management tools.	Excellent use of data and quality management tools.
Effort	Evidence of effort in a few areas. Opportunities exist throughout entity.	Evidence of effort in several areas. Numerous opportunities exist.	Evidence of effort in many areas. Further deployment possible.	Evidence of effort and deployment in most areas.	Evidence of effort and deployment in all areas.
Results	Little success, lots of opportunity.	A few successes, lots of opportunity.	Many successes, but more possible.	Successful in most areas. Some innovative approaches.	Good to excellent results in all areas. Many innovative approaches.
Score:	1	2	3	4	5

The scoring matrix.

The Review Team

The review team can consist of two managers. These managers should be familiar with quality systems and quality methodologies, such as those mentioned in Chap. 6. And certainly, at a minimum, they should be familiar with everything discussed in this text. They should be mature managers, and it would be ideal if one were an entity general manager or an experienced business manager. It is important that you develop this type of expertise within your organization. The organization's quality manager must organize this activity.

Review techniques. Review team members can start by practicing on several functions per agenda 2 and then can move on to review an entity. During an actual review, the review team should allow the presentations and discussions to flow smoothly. It is important that the review team avoid the temptation of interrupting and proposing a better way—it would be far better to probe at the weakness and allow the person or manager reviewed to understand and admit that weakness. According to W. J. Harmayar[2] in *Audit Interview Techniques:*

> It is not a (reviewer's) function to make instant evaluation and comment during the interview. If you suddenly ascertain that these people haven't yet discovered the wheel, make haste slowly. There might be something better than the wheel for the job at hand! If you have, in fact, unearthed the crime of the century, it will keep. Don't rush to advertise it.

Hence, during a review, the review team first looks for the item on the checklist. If present, the reviewers note it and move on. If not, the reviewers probe for it and ask questions and move several layers down. Let us review some examples:

- You are reviewing the section on customer obsession, question 4. You have seen the results of a customer satisfaction survey. The next thing you look for is a corrective action plan, that is, what action will be taken to resolve the customer issues. You can do this immediately or wait until question 6—but it is best to do it immediately by asking for action taken and noting the results. You would be looking at actual implementation of the corrective action plan and the feedback to the customer. After this, you can move to the next question, 5.

- You are reviewing a project in the section managing improvements and breakthroughs; you typically wait until the presentation is completed before asking any probing questions. The following discussion explains why.

  Refer to question 2 in the section on managing improvements and breakthroughs. During the presentation, you might have been satisfied with the problem definition and schedule of activities. But you had some concerns with the goal—maybe it was not aggressive. You wish to probe more deeply, and you want to recommend the rule of 50 percent reduction in defects, discussed earlier—but you will do this after you see the results and understand how the reviewees set their goal. Maybe they were not aggressive and discover that they have far exceeded the goal. Now you can convince them that they need to set aggressive goals and propose the 50 percent rule. Alternatively, they may have set too aggressive a goal of, say, 80 percent defect reduction. If they fail to achieve this, you again have a discussion point. You would take the same tack if the reviewees had no PDCA schedule of activities—resulting in an open-ended project, or no goal, or no verification of causes from a cause-and-effect diagram.

To reiterate, the reviewer must avoid the temptation of jumping in with advice at the first opportunity. Rather, the reviewer must let the discussions flow and must probe at weaknesses. If this tack is taken properly, the reviewees will understand their weaknesses. At that point, the reviewer can make recommendations for improvement. *Remember, someone who understands his or her weakness will be more willing to learn, improve, and accept the review team's recommendations.* One more point: The reviewers should not hesitate to ask questions during presentations to obtain clarification on any item.

Polite but probing questions.   The reviewer needs to be polite and firm. Discussions should stay on course and focus on first the process, and then the content. The time shown in the schedule is tight, but will be sufficient.

The reviewers need to ensure that there are no dog and pony shows. The discussions need to be focused and should stay within the review checklist and agenda. Below is a list of questions that can help reviewers control the flow of discussions and extract information quickly but politely.

- We have a very tight agenda, so we need to move on....

- Were you able to determine the root cause of the problem?

- How will you share what you have learned?

- How do you convey your vision (or plans)?

- You are doing a great job of responding to customer problems and concerns. How do you link these to future plans and products?

- What would it take to have 100 percent of your customers satisfied, or for your customer to become a reference site for potential customers?

Preparing and issuing the total quality review recommendations

Immediately after the review, we suggest that the review team give a verbal report. In the agendas that we provided earlier, we left about an hour for the review team to prepare a verbal report. The verbal report will consist of

- A discussion of the strengths and weaknesses of each review element and recommendations for improvement
- About five key recommendations—preferably one from each review element

We suggest giving a summary of key recommendations because, typically during a review, you may uncover numerous weaknesses. But you need to prioritize recommendations into just four or five items; this ensures that the entity being reviewed has a manageable list of recommendations.

A written report on the total quality review should be prepared and issued within a few weeks. The written report should be formatted as follows:

- Summary of the key recommendations. Keep the list short, with a maximum of five items.
- A detailed list of strengths and weaknesses, for each of the five review elements.
- Examples of recommended best practices, similar to those provided in this text.

A sample total quality review report is shown in Fig. 8.4. In it, we follow the proposed format discussed above.

Follow-up corrective action.   After the report is received, the reviewed entity prepares and implements a corrective-action plan. The plan should preferably be incorporated into the entity's annual plan. Successful implementation will result in a stronger and more competitive organization.

Common problems discovered during the review process, and some guidelines

What are some of the common problems or weaknesses that you will find during a review? We give a short list below. In addition, we give some guidelines on focusing your efforts if you are conducting a review.

I. Reviewing customer obsession

Date: December 5, 1997
Subject: Total Quality Review

To: The General Manager
Apex Division

We conducted a Total Quality Review at Apex on November 26 and 27.
Listed below are the key recommendations. In addition, we have attached
a detailed report, giving your strengths, weaknesses and opportunities
for improvement. A scoring summary is also attached.

2. Business Planning: **Page 2 of 10**

Strengths observed:
Your entity vision is well conceived and measurable, and you are moving
toward achieving the vision's goal. The long-range plan was well articulated
with strong strategies that will help to achieve your vision. The annual plans
were specific and reviewed regularly and progress to plans ranged from good
to moderate for most functions.

Opportunities for improvement:
 There was weak linkage between the long-range and annual plans. Most of
the annual plan strategies, for the current year, were not mentioned in the
long-range plan and seemed intuitive. We are not convinced that they can
help achieve your vision and goal in the short term, although they solve some
short-term problems. This needs to be resolved.
We observed that several functions did not have implementation plans
for their strategies. It seems the process is simply to "just do it." Some of the
departments without these plans were struggling to complete their strategies.
This is an area that can be strengthened with detailed implementation plans
 for managers and professionals who own these high-level strategies.

Recommendations:
Your annual plans should look at short-term and long-term strategies. The
short term strategies should address your current burning needs (these are
in your annual plan), improvement plans (which you are managing as
projects), and important strategies that you must implement in the current
year (these you do not have). The bottom line is, you cannot expect
improvements to be done separately and strategic issues to be done in later
years. Having a consolidated annual plan will also resolve your priority and
budget issues—a topic which came up several times.
We also recommend that owners of the strategies from your functional
managers have an implementation plan, that lists schedules and steps to
be taken to achieve the strategies.
Examples on how to do both these items are attached.

Figure 8.4a A sample total quality review report.

3. Managing Improvements and Breakthroughs   Page 3 of 10

Strengths observed:

The entity is very aggressive in pursuing and solving customer and process problems. All the projects reviewed and discussed were driven by the management team. The improvement process was rigorous and well managed. The PDCA improvement process was well used.

Opportunities for improvement:

There was weak linkage between the annual plan and the projects selected for improvement. Most managers keep a list of items to improve and these improvement projects are pursued separately from the planning cycle. This causes resource and priority conflicts.

The annual plans focus on new or breakthrough activities or projects. When we reviewed some of these large projects — specifically, an IT (information technology) system for product customization, new model handphones, and a telesupport service — we did not see the same rigor that we saw during the improvement cycle.

The result was that the IT system had numerous bugs and the solution took several months of troubleshooting before it worked. The handphone project was exciting and the product was well received in the market, with rave reviews and good sales. However, the failure rates during usage of these products was high and caused severe customer dissatisfaction. The telesupport project was innovative, but there were some difficulties and confusion after its launch. Most of these issues could have been avoided.

Recommendations:

You need to shed the mentality that fire-fighting or rework is the accepted practice and the way of life. You even tend to reward employees for fixing problems that should not have occurred in the first place. Our specific recommendations are:

1. When you prepare your annual (and your long-range plans), you should include major improvement needs (driven by customers and internal inefficiencies) in the plan. This will ensure that you better coordinate improvement projects, breakthroughs, and other activities in one plan, with a review of resources and budgets. This will avoid the resource and priority conflicts you are seeing today.

2. Introduce more rigor in managing your breakthrough, or new, projects. We suggest you use the proactive PDCA cycle. This will insure that you go into a predict and prevent mode by testing the products and services both internally and in test markets.

Figure 8.4b   A sample total quality review report.

A. Review the entity's approach to measure and improve customer satisfaction, loyalty, and competitiveness.
B. Look for understanding of customer needs and concerns and follow-up in the areas of

1. Customer satisfaction surveys
2. Customer complaints and feedback management
3. Customer meetings and visits
4. Competitive data and benchmarking
5. Predicting and meeting customer's hidden needs

C. Common problems or weaknesses observed

1. Weak system to collect and collate customer and competitor data
2. Insufficient follow-up or corrective action after getting the data
3. Not looking over the horizon to future needs

D. During the review, focus on

1. The analysis of data from the various systems such as customer surveys, customer complaints, competitive benchmarking, and future needs
2. Follow-up activity and corrective action as a result of the analysis
3. Two or three items (one from customer complaints, one from the survey, one from future needs, etc.) that you follow all the way through resolution

II. Reviewing business planning

A. The general manager and her or his staff should present their plans and discuss progress toward their plans.
B. Look for

1. A strong vision
2. A standardized planning process, including

   a. The long-range plan
   b. The annual plan
   c. A process to generate key issues driven by customers and future needs
   d. Separation of annual plans into Hoshin, or breakthrough-type, objectives and daily management plans, or routine day-to-day activity
   e. Implementation plans

3. Robust supporting strategies
4. Proper cascading of plans down the organization

5. Compact and concise plans

6. Regular reviews of plans

C. Common problems and weaknesses observed:

1. Poor targets and metrics

2. Too many objectives—an attempt to do too much

3. Weak or no detailed implementation plans for each strategy

4. Nonrigorous reviews or worse still, no reviews of progress of plans

5. Corrective action or management intervention when it is apparent that plans will not be achieved

D. During the review, look first at the process, then the content. Be prepared to discuss strategies for robustness and their contribution to achieving the objectives of the organization. Most importantly, look at how the plans support the vision and link to the value delivery chain.

III. Review of managing improvements and breakthroughs

A. Select four to five projects from process management and the management of improvements and breakthrough sections.

B. Look for a systematic process in problem solving and breakthrough project management:

1. Go through the presentation.

2. Note sequence—does it follow PDCA cycle or proactive PDCA standard?

3. Discuss any deviation from the standards.

4. Discuss the aggressiveness of targets.

C. Common problems and weaknesses observed:

1. Approach to problem solving not systematic

2. Fire fighting or genius/intuitive approach

3. No schedule or goal for project

4. New improved process not standardized, and possibility of problem recurrence

5. Not predicting and testing for potential problems with breakthrough projects

D. During the review focus on

1. The problem-solving process

2. Understanding of the PDCA and proactive PDCA cycle methodologies

IV. Review of process management

A. Review management of key processes

B. Look for process thinking:

1. Understanding of process concept
2. The concept of standards and process documentation
3. Tracking, monitoring, and management of processes
4. Analysis and control of deviations

C. Common problems and weaknesses observed:

1. Lack of standards
2. Loosely managed and poorly documented processes
3. No process performance measures and targets
4. No or weak analysis of deviations in processes
5. Weak corrective action when deviation occurs
6. Insufficient linkage to partners in the value delivery chain

D. During the review focus on

1. Benefits of each process
2. Good management of process
3. Improvement plans (past, present, or future) for each key process
4. Linkages in the value delivery chain, both upstream and downstream

V. Employee development and participation, and leadership

A. Review employee development and participation, and leadership.
B. Look for specific employee development and participation activity, and quality and strength of leadership

1. Employee development
   a. Quality circle and team activity
   b. Employee recognition and suggestion schemes
   c. Education and training programs
   d. Employee empowerment

2. Management leadership in
   a. Developing a strong team
   b. Supporting a quality-first attitude
   c. Planning for the future

C. Common problems and weaknesses observed

1. Lack of management leadership in setting and managing future direction
2. Weak management of quality circle and team activity.
3. No specific individual employee development program

D. During the review focus on management's leadership in employee education, participation, and organization development.

Presidential Audits or Executive Reviews

The total quality review process we have proposed is fairly robust and can be very effective in probing for weaknesses and improving business results in an organization.

In addition to the total quality review, we recommend a presidential audit or executive review. Presidential audits are common in many large Japanese companies and several U.S. companies. The presidential audit can be a very effective way for a president, chief executive, or business unit manager to review an organization, business unit, or entity.

The objective of a presidential audit or executive review is very similar to that of a total quality review. It allows the president or chief executive to convey her or his commitment, support, and leadership in using quality concepts, tools, and systems to manage the entity in order to steer it toward business success. This is a powerful method for the chief executive that allows him or her to understand the entity's business environment better. It will, however, tend to be a shorter and more focused review. This is obvious from the agenda that follows.

Review agenda for the presidential audit

The review can be structured in many different ways. Since time and resources are limited, we suggest the following:

What is reviewed	How long
Business planning	2 hours
Customer obsession	2 hours
Managing improvements and breakthroughs	2 hours

The audit should focus first on the process and then the results—it should not be a dog and pony show. The audit can be completed in about 4 to 6 hours. Such a focused agenda will allow the president or chief executive to address other topics, in a busy day.

We now provide you a review checklist for a presidential audit. It consists of a list of key questions, followed by a discussion that sets expectations.

Business planning checklist for presidential audit. This section allows the president or chief executive to determine if his or her objectives or direction has been picked up by the entity being reviewed and is being translated to implementation plans at some lower level. Additionally, progress of these plans is checked, and business results are reviewed.

1. There is a long-range plan that addresses the strategic direction of the entity.

   Discussion: Review the entity's vision. Does it set the overall strategic direction and influence the business planning process? The key point to look for is the entity's strategic direction. Where is it going? What are the competitive pressures and challenges? How will it overcome these challenges and pull ahead?

   Notes: Items to look for include the following:
   - The entity's vision
   - Customer needs and channels of distribution
   - Competitive situation
   - Necessary products and services
   - Development of partners
   - Financial analysis and projections
   - Potential threats
   - Three- to five-year strategies

2. Review the annual plan.

   Discussion: Review the planning process. The key point to look for is breakthrough strategies over the next year. Are they convincing, and will they respond to immediate challenges or threats?

   Notes: We are looking for an annual planning process that uses Hoshin planning methodology or some equivalent. The plan should have the following features:
   - It is driven from strategies developed from the long-range plan (unless this is a lower-level operation not requiring a long-range plan).
   - It is influenced by the previous year's issues and lessons learned.
   - If the entity does not do long-range planning, then the plan is driven by customer inputs, which are linked to the customer obsession checklist and higher-level plans.
   - Quality, cost, education, and delivery issues have been addressed. Goals and performance measures are realistic and measurable.
   - At the top level there are a few breakthrough objectives, with robust supporting strategies.
   - The top-level plans are properly deployed down the organization.
   - Check if the above-mentioned features exist. If not, understand why.

3. A robust daily management, or business fundamentals, plan exists and includes regular tracking of performance.

   Discussion: The top level and every function and department must have such a plan. Remember to ensure that the entity managers understand and distinguish breakthrough (Hoshin) from daily management issues. This plan should reflect and manage day-to-day activities or key processes from the entity process map.

Notes: We look for monitoring and management of the items on the daily management plan.

4. All plans are reviewed regularly for progress and corrective action taken, including changes to the plan when necessary.

Discussion: Check for the review process, progress to plans, lessons learned, and follow-up corrective action. Reviews should be done frequently, say, every 3 months.

5. The business planning process includes managing the linkages to other areas of the value delivery chain.

Discussion: No business unit can operate as an island. It will be part of a value delivery chain and will have activities upstream (from suppliers, etc.), downstream (customers, etc.), and sideways (joint ventures, contractors, etc.). Hence we determine how the business unit is managing these partners.

Notes: We are looking for strong linkages here. Items to consider include

- Management of performance and relationships with partners
- Satisfaction levels of partners

6. Business results have been achieved as a result of good business planning.

Discussion: We have reviewed the entity's business planning process. Now we would like to understand if the objectives and strategies have helped generate success. This should be evident as we go through the entire review. At the end of the session, we need to collect our thoughts and confirm if the entity is moving aggressively in the direction stipulated by its vision, and if it is successful.

Notes: Key items to consider:

- The entity vision has been a galvanizing force moving it in the right direction.
- Costs are under control; the entity is profitable and competitive.
- Success is the result of astute business planning and responsiveness to the market and customers, not luck.
- The entity has successfully implemented most of its strategies, and this has created market success.

Customer obsession checklist for presidential audit. The purpose here is to ensure that the entity being reviewed pays a great deal of attention to its customers' needs and is staying close to customers. The key items include customer interface activities, follow-up on customer surveys, customer inputs, resolution of customer complaints, and introduction of new products and services that will meet customers' current and future needs.

1. Product and service quality data collection processes are well established. Data are used to drive improvement.

    Discussion: Here we look at the process of collecting failure data on products and services from customers after the sale and at the review of how the analysis of these data influences products and services and future plans. Look for a formal system to collect this information, track it, and drive corrective action.

    Notes: Items to look for include
    - Product and service failure information
    - Plan for resolution of product and service failures
    - A customer complaints and feedback system
    - Quick resolution of customer complaints

2. Customer satisfaction levels have been determined.

    Discussion: Here we look at the process of determining satisfaction levels. We primarily look at customer satisfaction surveys, but other means of satisfaction feedback can be reviewed, for example, inputs via response cards mailed in after purchase. Check the analysis of the available data, and understand how the satisfaction levels were determined and prioritized. Having reviewed this information, we should look for the entity's response or follow-up corrective action.

    Notes: We specifically look for
    - Collection and analysis of customer satisfaction data
    - Prioritization of customer satisfaction issues
    - Plan to respond to customer dissatisfaction issues

3. Competitive data are available and analyzed; the analysis is used to influence both current and new products and services.

    Discussion: Here we look at competitive benchmarking and other means to get competitive data. This should be available for both products and services. The entity should be getting such data, understanding its market position, and making decisions on how to respond.

    Notes: Specifically, we look for
    - Competitive issues that impact the business
    - Response to competitive issues

4. New customer needs are solicited and understood.

    Discussion: Here we are looking at the process of gathering customer needs. This can be done via many methods such as focus groups and visits. Once needs are understood, we check on the entity's plan to respond to these needs.

Notes: We look for

- Identification of new customer needs
- Proposal for new, attractive products and services

Checklist for managing improvements and breakthroughs. Approximately two projects, which are linked to the annual plan of the entity, should be reviewed. We suggest one improvement and one breakthrough project. Where possible, one of the projects should be linked to, or originate from, the president's or chief executive's objective—for example, a request to cut costs or to improve product and service quality.

1. Improvement and breakthrough projects are linked to the annual plan or crucial issues.

   Discussion: Our purpose here is to understand how improvement and breakthrough projects are selected. Ideally, improvement projects should be identified during business planning, from customer issues or internal problems. A major weakness is to focus only on breakthroughs—new and exciting stuff at the expense of current weaknesses, which require improvement. Another related and common weakness is to managed improvement projects separately from business planning. This must be avoided because of several reasons: Improvement projects will be treated as less important and given lower priority, and there will be no resources left over for improvements if all resources are consumed for business planning strategies. Remember that many improvement projects can be megaprojects—refer to the solderwave example (improvement of printed circuit production process) in Chap. 4.

   Notes: The following will help start the discussion:

   - What improvement projects were identified during business planning, and were they prioritized?
   - What breakthrough projects were identified during business planning?
   - Was there a conflict in selecting improvements versus breakthrough projects, and how was it resolved?
   - Note that some improvement projects can be ad hoc and not generated from business planning. These projects can occur because of critical business issues, for example, customer complaints from a poor product or service.

2. Improvement projects are clearly defined with clear objectives, goals, and schedules.

   Discussion: We look for clear project definitions. This should include clear and aggressive goals and project schedules. What is a good improvement goal? In Chap. 4, we recommend a rule of thumb of 50 percent reduction—for defects or time to do something.

Notes: We look for

- Project definition
- Aggressive project goal
- Project schedule

3. There is good data collection, analysis, understanding of root causes, and results are achieved.

    Discussion: We look for good collection of data and appropriate use of quality tools such as Pareto diagrams, graphs, and cause-and-effect diagrams. Most likely causes have been identified and verified with data. Alternative solutions are reviewed, evaluated, and implemented. Results have been achieved and measured; deviations from goals are well explained and acted upon.

    Notes: We look for

    - Good data collection and analysis
    - Understanding of root causes
    - Alternative solutions proposed and results achieved
    - During review of the improvement projects, check for adherence to the steps of the PDCA cycle. *Most important, look for lessons learned and reasonable assurance that this problem will not recur at the entire entity—you must probe extensively for this!*

4. There is a systematic approach to managing breakthrough projects.

    Discussion: Our purpose is to understand the thinking behind breakthrough project selection. This topic may have been covered during the business planning or customer obsession sessions. If so, we can move quickly through this item.

    Notes: We check for the rationale behind project selection and how the project is managed. For example:

    - Are projects customer-driven?
    - Are projects driven by competitive pressures?
    - Are projects driven by future, potential needs?
    - Will these projects move the entity ahead of the competition? How far ahead?
    - Are there objective, clear goals and schedules for the new product or service?
    - Is the product's or service's reliability tested prior to product release?
    - Is the product's or service's market acceptance tested via a pilot program or in a test market?

5. Improvement and breakthrough projects have met expectations and achieved results.

    Discussion: Have breakthrough projects had an impact on customers? This question will lead to a fruitful discussion with the project teams.

Notes: We can also check for this by reviewing

- Internal productivity gains
- Sales or market share data

Most importantly, look for customer feedback on both improvements and breakthrough projects.

Issuing recommendations after a presidential audit or executive review

Recommendations, concerns, or issues that are raised at a presidential audit or executive review should be conveyed during or immediately after the review. The president or chief executive must convey her or his commitment, support, and leadership in using quality concepts, tools, and systems to manage the entity in order to steer it toward business success. Hence, immediate verbal feedback is essential. The review techniques are similar to those in our earlier discussions.

Summary

The objective of the total quality review is to discover and encourage the use of quality tools and systems to manage a business, in order to make it more efficient, productive, successful, and better prepared to face the future.

The total quality review is a very powerful tool to maintain or increase the momentum of a total quality effort. It can help uncover weaknesses in the business environment, ensure that managers and employees understand the weaknesses, and provide opportunities for improvement.

We also recommend the use of presidential audits or executive reviews as a means to ensure the chief executive's support, commitment, and leadership in the total quality effort. In the final analysis, the total quality effort is directed by the chief executive, and such reviews allow him or her to understand and direct the effort.

The total quality review and presidential audit are among the best methods that you can select to ensure a strong, successful, competitive, and more successful organization.

References

1. The total quality review procedure discussed here is a modification of the Quality Maturity System, developed by the author at Hewlett-Packard Company, Palo Alto, California.
2. W. J. Harmayar, *Audit Interview Techniques.*

9

Some Thoughts on the Essence of Total Quality

To return to the root is repose; it is called
going back to one's destiny. Going back to
one's destiny is to find the Eternal Law.
LAO-TSE
Tao Teh Chin (The Way)

What is the essence of total quality? By now, you probably guessed that it includes a focus on the process or "the way," rather than only on achieving results. Kaoru Ishikawa called it *thought revolution in management*; and that is what total quality is—a better way to manage, a better way to run a business.

We offer a few concepts that get to the heart, the very essence, of total quality.

Quality First

Quality comes before anything else. There are several items of importance here:

Continuous improvement

This is a fundamental concept of total quality—we strive for continuous improvement and aim for perfection. The concept of continuous improvement is predicated on the knowledge that breakthroughs will be few and far between—hence incremental improvements must be encouraged in all processes, products, and services.

Often, there is a concern about investing in continuous improvement—because one feels that one is good enough, or already competitive, or that there is balance between cost and quality. This complacency is dangerous in any organization, because complacency is the very enemy of quality. This is succinctly expressed as the whispering of Satan by Takoshi Hokake:

> ..."balance between cost and quality" is such a phrase that affects us like opium. Tearing off this veil of this beautiful phrase "balance of cost and quality"—the "quality first" policy should be adopted. This is the very energy source. But I have no weapon to shut out this whispering of Satan.

Hence, we stress the need for driving continuous improvement. But when problems or defects run at a very low level, it is time to look at the *second dimension of quality*—the area of attractive quality.

Attractive quality

We discussed this in detail in Chaps. 1 and 2. In the spirit of continuous improvement, when problems and defect levels are low, we must start providing more attractive quality items. An example of this is found in the automobile industry. Initially, Japanese manufacturers kept ahead of U.S. and European manufacturers by providing high-quality and (almost) defect-free cars. By 1990, the U.S. and European manufacturers were catching up. But the Japanese have leaped ahead, by going beyond providing defect-free cars, by focusing on *attractive* quality items—items such as ergonomics, extraordinary "touch and feel" features, superior service, and so on. In Japanese, this attractive quality feature is called *miryokuteki hinshitsu*. This attractive quality is what we define as the second dimension of quality. *If we are routinely operating in this second dimension of quality, we will be operating in a new zone called total quality creation. Clearly, this is where we want to be. Always.*

Zero-defect products and services

We have stressed that at low defect levels, priorities should turn to providing more attractive quality features. But as product volumes or service transactions increase, a low defect rate will multiply into a large number of customer complaints and issues. In today's competitive environment, this is not acceptable. Hence, expectations for zero defects in products and services are increasing, and several businesses are already there. Examples exist: television sets by Mitsubishi that last 10 years with no failures; automobiles from Toyota that run for the first 5 years with no defects (but require routine maintenance); laser jet printers from Hewlett-Packard that run flawlessly for more than 5 years; and impeccable service on every business-class flight at Singapore Airlines. This will be the trend from leading companies, and this is what gets complete customer satisfaction, which is the key to securing customer loyalty that generates long-term superior financial performance. To repeat, *the leading companies will and must aim for zero-defect products and services.*

Be mindful of quality

There are two ways to practice quality: You do it because you have been directed to achieve a goal, or you do it because you are mindful of it. *When you are mindful of it, quality is second-nature.* For example, when you wash the dishes, you should think only of washing the dishes—you focus, you concentrate, and you do it well. Similarly, whatever you do—clean house, bring up a child, design a product, manage a project, sell a product, or run a business—you must be mindful of doing it well and instilling quality in the way you do it. Take no shortcuts and aim for perfection—or zero defects in products or services. This is being mindful of quality.

Supporting methodologies and techniques

Some of the supporting methodologies and techniques include the PDCA cycle, the proactive PDCA cycle (and within it, quality function deployment and FMEA), computer simulation of products and processes, training and education (including computerized training), the total product concept, customer complaints and feedback system, customer surveys, competitive benchmarking, postmortems, employee suggestion schemes, and quality circles and teams.

Customers First

Customers always come first. They pay our wages and bills and help us to be successful. "The next process is our customer" is an extension of the customers-first concept. In any organization everyone is part of a process, and everyone has an internal customer. That customer must also be satisfied, not only the external customer. This creates a chain of suppliers satisfying each customer until we reach the end customer. Remember, there is no such thing as being too customer-sensitive, but there are such things as unhappy customers or insufficient customers. Therefore, when we design and manage our products and services, our goal is to aim for complete customer satisfaction, which gives customer loyalty and leads to repurchase.

Supporting methodologies and techniques

Some of the supporting methodologies and techniques include the customer complaints and feedback system, customer surveys, competitive benchmarking, postmortems, proactive PDCA cycle, quality function deployment, the total product concept, customer satisfaction model, and zero-defect products and services.

The Importance of the Way, or Process

Getting results is very important, but they should come as an outcome of well-managed processes. Well-managed processes can ensure repeatable and predictable results, and this requires that management and employees focus on using standards. An example is the use of standard methodologies for design rules, planning, improvements, breakthroughs, and other processes. *Most importantly, well-managed processes can be leading indicators that predict good results for a company or an organization.*

When experienced managers or employees leave, they are irreplaceable. But the loss is less severe in a process-oriented organization. Hence, we capture their experience and best practices into standards at our organization. In addition, well-managed processes can be a leading indicator of good results in a company. The outcome is decreased dependence on the genius of individuals and increased dependence on the system, the way, or established processes to get results.

Balance between standards and creativity

In implementing standards, there must be a balance between using standards and having time to be creative. Often, the argument against using standards—such as formal planning or the PDCA improvement cycle—is that they stifle creativity. *In total quality philosophy we emphasize the use of proven processes or standards. We save our creative energies for new unexplored areas. This is how we get maximum output and efficiency in an organization.*

Supporting methodologies and techniques

Some of the supporting methodologies and techniques for managing processes are business planning, the PDCA cycle, the proactive PDCA cycle, competitive benchmarking, customer complaints and feedback system, computer simulation of processes, and process management—including management of the entity's key processes.

The Organization That Learns and Grows

It is crucial that the organization learn and grow. This implies a willingness to take risks, to learn from mistakes and successes, to train and educate continuously, and to drive change in the organization.

Organizational learning is extremely important. This includes learning from mistakes and from the lessons of history. Whenever an error or problem is encountered, we do not just fix it and move on. Instead, we must do the following:

- Understand why the problem was created.

- Understand why the problem was not caught at creation.

- Ensure that the problem will never occur again, in our department.

- Ensure that the problem will never recur in our entire organization, in all our future products and services. This is very important.

Another unique way to ensure learning is via cross-functional task forces. These task forces consist of managers and employees who form a team that analyzes and resolves issues that cross functional boundaries. This is one of the best ways to develop employees and managers in an organization. Finally, best practices that occur in one part of the organization must be quickly propagated and emulated throughout the rest of the organization.

Supporting methodologies and techniques

Some of the supporting methodologies and techniques include cross-functional task forces, the PDCA cycle, proactive PDCA cycle, postmortems and the T-type matrix, Hoshin plan reviews, out-of-control reports, corrective action report and standards update request formats, training and education, and especially total quality reviews.

Total Quality: Organization and People Running at Peak Efficiency

The outcome of being strong in the essence of total quality is to have an organization and people operating at peak efficiency. A good analogy is to look at the automobile industry. General Motors' market share continues to decline, while Toyota's continues to increase. General Motors continues to invest heavily in automation and robotics, but it seems that money disappears into a black hole. Toyota, with less capital investment per employee, is able to have higher productivity and operate more effectively: Its average time to design a car is 3 years versus 5 at U.S. manufacturers; the Toyota Lexus luxury car is assembled in one-sixth the time of a similar luxury Mercedes model. Toyota has succeeded partly due to its total quality effort, which includes its continuous improvement philosophy and its world-famous just-in-time manufacturing system. Toyota and some of its subsidiaries have won the Deming prize for their quality and customer focus. Toyota's operations and people run at the highest efficiency in the automobile industry.

Total Quality, Creativity, and Success

Does total quality methodology ensure business success? To answer this question, we need to review the attributes of a successful company. The following is a list of important attributes.

- Keep close to the customer, and earn your customers everyday.
- Generate good products and services that are a solid contribution to the marketplace.
- Recognize that people are the most critical asset.
- Provide strong leadership that creates a focused organization and sets the right priorities.
- Be financially strong and the low-cost supplier.
- Leverage activities through partners—do not do everything yourself.

Total quality methodology helps to provide or influence many of the above attributes. It certainly ensures keeping close to the customer, influences good products and services, recognizes the value of people, and helps the leadership to focus its activities. These four attributes provide a good basic system. The other attributes of financial strength and leverage must also be provided by management.

With these basics in place there is stability; this provides ample opportunity for creativity in new areas, in beating the competition, and in planning for the future. The combination of all these provides a strong, dynamic, competitive, and successful company.

Conclusion

The necessity and benefits of total quality have been extolled throughout this book. By the time you, the reader, get to this point, a good understanding of total quality concepts, techniques, and methodologies has been conveyed. You are definitely ready to start implementing total quality. If you have already started, you will be able to improve and accelerate your current effort.

A vital point to remember is that the goal is continuous improvement. Total quality is only a philosophy, a tool, a means. Remember: *A total quality effort is a journey, not the destination.* Once started, you will be on an unending journey toward better quality. With a quality focus, you will achieve increased sales, increased productivity, and increased profits, and you will be better prepared to face the future. Quality is the key to survival, success, prosperity, and happiness.

10

An Afterword:
The Chief Executive's Role
in the Total Quality Effort

Men, at some time, are masters of their fates.
The fault (dear Brutus) is not in our Stars,
but in ourselves, that we are underlings.
                    WILLIAM SHAKESPEARE
              Cassius (from Julius Caesar)

What is management's role in a total quality effort? It is crucial and of the utmost importance because management is responsible for the success or failure of the company. Our discussion has centered on how total quality provides quality tools and systems to manage a business, in order to make it more efficient, productive, successful, and better prepared to face the future. The tools exist—now management has to use them. Nevertheless, many an attempt by management to start a total quality effort has failed, so management goes ahead and adopts another fad. Why?

We believe that some managers seem naturally inclined to get caught up in complicated ideas and new fads—the result can be failure. Consequently, organizations move from total quality, to reengineering, to something else. Instead, management needs to show leadership and commitment; it must articulate the fact that total quality—and within it, continuous improvement—is essential for success. An organization cannot simply rely on a few blockbuster products or services for long-term success. What must management do? More important, what must the chief executive do? We discuss this next.

We dedicate the following discussion to chief executives and senior managers. To understand your role, there are several key questions that you, the chief executive or senior manager, must answer.

What are the benefits of total quality?

Why must the chief executive and senior managers drive total quality in an organization?

How do you manage the dimensions of total quality in an organization?

Conversely, what must you do to fail to implement a total quality effort?

What are the benefits of total quality?

We discussed the benefits of quality in detail in Chap. 1. We discussed the generic model and how it contributes to profits and customer satisfaction and loyalty, we discussed the PIMS study, and finally we showed specific results from one company. The data are indisputable—in summary the benefits of total quality are

- Higher employee morale
- More efficient processes
- Higher productivity
- Less fire fighting, resulting in more time for innovation and creativity
- Lower costs
- Zero-defect products and services
- Increased customer satisfaction and loyalty
- Increased market share
- Higher profits

The successes of companies that practice a total quality effort are very clear. Some of these companies are Motorola, Hewlett-Packard, Sony, Toyota, Matsushita (Panasonic), and Singapore Airlines.

Why must the chief executive and senior managers drive total quality in an organization?

Throughout this text, we have extolled the benefits of total quality. The chief executive and the senior management team need to drive this effort. There are many reasons for this:

- It is the chief executive's and senior manager's responsibility to ensure complete and total customer satisfaction and loyalty.

- It is the chief executive's and senior manager's responsibility to set clear goals and priorities.

- It is the chief executive's and senior manager's responsibility to reduce waste, prevent product and service failures, and reduce costs in an organization.

- It is the chief executive's and senior manager's responsibility to reduce organizational complexity and ensure that the organization runs like clockwork, with the highest morale.

- It is the chief executive's and senior manager's responsibility to ensure managers, staff, and the organization operate at peak efficiency.

- To delegate this responsibility is to abdicate—the result in the end is an uncompetitive organization. In all the requirements listed above, the total quality effort provides many of the tools and systems to make them happen.

How do you manage the dimensions of total quality in an organization?

The chief executive and senior managers must integrate total quality into the organization. So, you must do the following:

1. Get personally involved in customer and vendor visits, to understand their challenges, issues, and concerns.

2. Understand the successes, challenges, and difficulties that customers have with your current products and services.

3. Get personally involved in the business planning process—both long-range and annual. Do not delegate this activity. Good business planning will ensure that the organization
   - Responds to customer issues and needs
   - Improves in areas of customer concerns, and poor products and services
   - Is less complex and more efficient
   - Generates breakthrough products and services
   - Prepares for the future

4. Measure the success of annual plans by
   - Conducting reviews of progress—check if you are delivering what you promised
   - Reviewing if the customer is seeing improvements and appreciating your new products and services

5. Initiate total quality reviews to help diagnose and understand the organization's strengths and weaknesses and to move it to greater

success. This is the best way to propagate total quality thinking throughout the organization. What is total quality thinking? We addressed this during our discussion of the essence of total quality. To repeat, the essence of total quality is as follows:

- Quality first—attractive quality and zero-defect products and services
- Customers first—aim to achieve complete customer satisfaction, which helps generate customer loyalty and repurchase of products and services
- The importance of the way (or process)—ensuring sustainable and predictable results, products, and services
- The organization that learns and grows

6. Conduct presidential audits or executive reviews. The above-mentioned activities take time—lots of time! A way to give you more time and to maximize your productivity in rolling out total quality in your organization is to conduct these audits or reviews. The presidential audit or executive review is a very powerful tool, and its objective is very similar to that of a total quality review. It allows you to convey your commitment, support, and leadership in using quality concepts, tools, and systems to manage the entity in order to steer it toward business success.

What must you do if you wish to fail in implementing a total quality effort?

Having discussed the benefits and success factors of a total quality effort, we believe it is now appropriate to discuss some factors that can contribute to failure. If you do any of them, you will fail.

- *Make a total quality effort into a program.*   It cannot be a program or another fad, because employees will give priority to managing the business and when they have time, they will work on total quality. You must integrate total quality into all business activities. The successful companies which believe in total quality have done this and have moved on to bigger and better things.

- *Delegate the total quality effort to a quality manager.*   The only way to succeed is to get the entire management team involved, not to give the responsibility to the quality manager. The quality manager is, at best, your facilitator.

- *Delegate total quality to lower levels in the organization or, better still, to quality teams or quality circles.*   A total quality effort is a holistic approach to managing the business and must be driven by management (and not lower-level teams). Furthermore, making the organization more customer-conscious and changing its mind set to make it more process-oriented are your job.

- *Believe that ISO 9000 certification is a total quality effort, which will benefit the customers or business.*   Its focus is on doing things per the documented specifications and has no influence in generating high-quality products and services. If you must do it, do it for political reasons!

- *Try to do too much too quickly.*   Here a chief executive may decide to implement everything known about total quality at one go. This will cause overload and failure. We recommend a focused, step-by-step approach.

- *Let's improve every process.*   Here management decides to improve every process in the organization—sometimes several hundred. The result can be a focus on the trivial, which is wasted effort. Instead, the focus should be on key process management—driven top-down by management.

- *If total quality does not work, try something else.*   Here management gives up a total quality effort and moves to reengineering or something else. Reengineering came after total quality and is one of many techniques proposed by consultants. Its key ideas are nothing new—improving processes and an employee focus (team activity, empowerment, rewarding employees). While reengineering has many good ideas, its weakest link is that it is divorced from an organization's strategy. This can result in an organization's doing the wrong things. In total quality, however, we take a holistic approach; we also ensure that all the activity is driven by the organization's strategic direction. Finally, moving from one fad to another is a recipe for employee confusion and lower morale, management indecisiveness and lack of commitment, and another failed fad.

Enough said! We have given you the background, benefits, and success factors for creating an organization that uses total quality techniques to focus on continuous improvement, in order to make it more efficient, productive, successful, and better prepared to face the future. So, go and create your vision, stay in touch with customers, set clear goals and priorities and implement them, prepare the organization for the future, and be successful. Remember that a total quality effort is a lifelong journey of learning and business success.

A

Improvement Project to Reduce Dissatisfied Customers Using the PDCA Cycle

Plan Stage

Project theme

During the recent customer satisfaction surveys, our overall satisfaction score was quite good, at 7.4 on a scale of 1 (worst) to 10 (best). We scored poorly in the category of sales interactions. This is an area of concern, as this means that our customers are dissatisfied with how our company and sales representatives interface with them.

Our project theme was to "reduce customer dissatisfaction in the area of sales interactions." The target will be set after analysis of the data.

Project schedule.   The following project schedule was agreed upon:

Activity	Timing
Select project theme and prepare schedule of activities.	January
Grasp the present status and set target.	January
Analyze the cause and determine corrective action.	February to May
Implement the plan.	June to October
Check the results, via another survey.	March
Standardize the results.	April, May

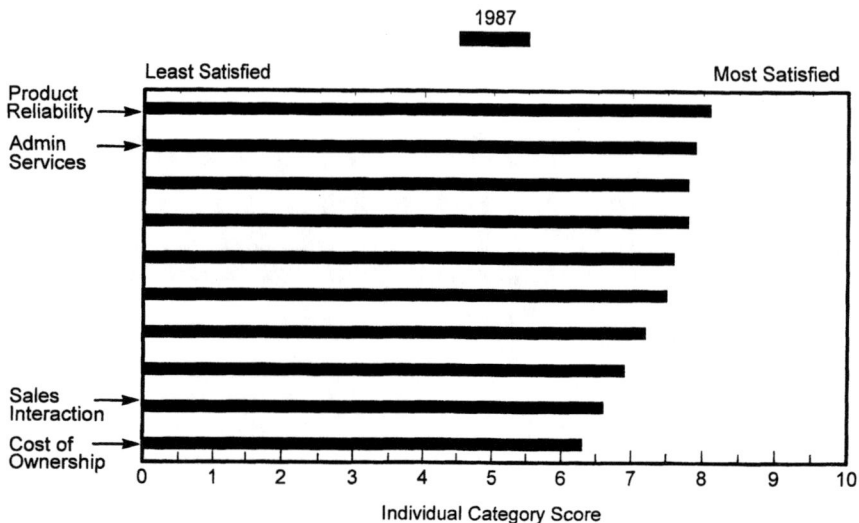

Customer Satisfaction Survey

Satisfaction Ratings by Categories

Figure A.1 Customer satisfaction by categories.

Grasp current status

In the customer satisfaction survey there are 10 categories. The results of the survey are shown in Fig. A.1. The figure shows that sales interaction and cost of ownership were the two lowest-scoring categories. We focused on sales interaction, as we felt we had an opportunity to improve; cost of ownership was also very important, but was passed over as there was already a separate project to reduce costs in the company. *Sales interaction* refers to how well our sales representatives relate and interface with our customers.

We were able to stratify the sales interaction data by satisfaction levels. This is shown in Fig. A.2. From Fig. A.2, it is evident that 33 percent of our customers were dissatisfied with sales interactions; that is, they were unhappy with the way our company or sales representative interfaced with them. We set our project theme and target as follows: To reduce dissatisfied customers in the sales interaction category by 50 percent.

Our analysis continued, and we reviewed the survey forms that each customer had completed. We noted customer comments for the sales interaction category. Comments ranged from "I am very happy with my sales representative" to "After the sale is made, your sales representative disappears." The sales interaction category had seven subcategories, and they were each rated in the survey on a scale of 1

Customer Satisfaction Survey - Sales Interaction Category
Stratification by satisfaction level

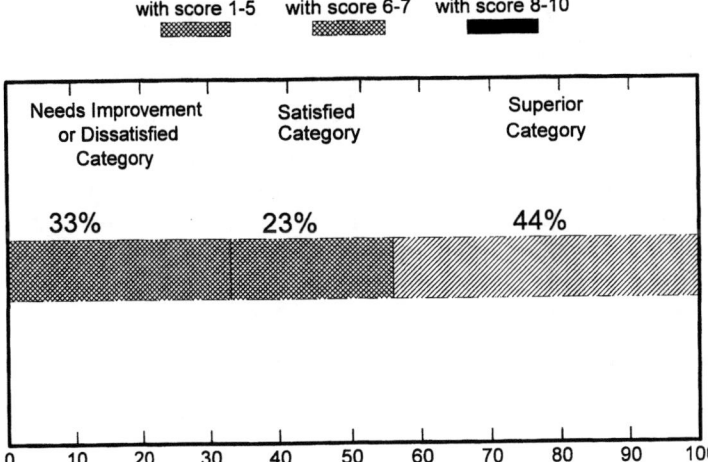

Figure A.2   Customer satisfaction for sales interaction.

Sales Interaction - dissatisfied customers
% Dissatisfied customers by Sales Interaction sub-categories

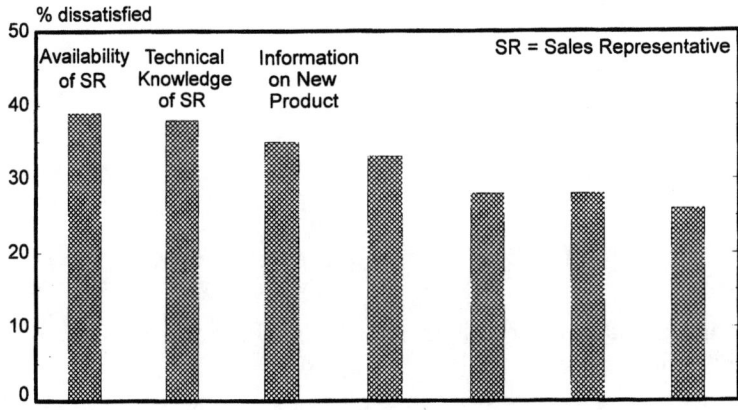

Figure A.3   Dissatisfied customers for sales interaction.

to 10. Data were available for each of the subcategories in terms of percentage of dissatisfied customers, that is, those rating us 1 to 5 (the need-improvement or dissatisfied scores). These data are shown in the Pareto diagram in Fig. A.3. *Note:* The data show the percentage dissatisfied in each category.

Analyze the cause and determine corrective action

This was separated into three steps.

1. *Prepare cause-and-effect diagram.* Here we determined the causes of customer dissatisfaction. A cause-and-effect analysis was done for the first three Pareto bars of Fig. A.3. This is shown in Fig. A.4. Our purpose was to understand the most likely causes and to eliminate them.

2. *Prepare hypothesis and verify the most likely causes.* Our team reviewed the various possible causes and selected the 10 most probable causes by voting, based on our experience. These are listed in Fig. A.5. From this list, the most likely causes were verified with separate independent data. This is also shown in Fig. A.5. It can be seen that several probable causes were discarded.

3. *Determine corrective action.* The corrective action, shown in Fig. A.6, was agreed upon. Where there was no direct corrective action, we proposed an alternative.

Do Stage

Implement corrective action

The proposed corrective action was implemented over a period of 5 months.

Check Stage

Check the results

Another survey was held 16 months later. The following effects were observed:

- The overall survey rating was 8.22, an improvement over the previous survey.

- The rating for sales interactions was 7.03, also an improvement over the previous survey.

- The percentage of dissatisfied customers in sales interactions reduced dramatically, as shown in Fig. A.7.

Figure A.7 shows that we met our goal of a 50 percent reduction in dissatisfied customers in the sales interaction category. A marked improvement was also evident in the subcategories of sales interaction. This is shown in Fig. A.8. Note the improvement in the first three Pareto bars, where the impact was greatest. This is where we had focused our efforts. There were no unwanted side effects—in fact, improvements were observed in all categories.

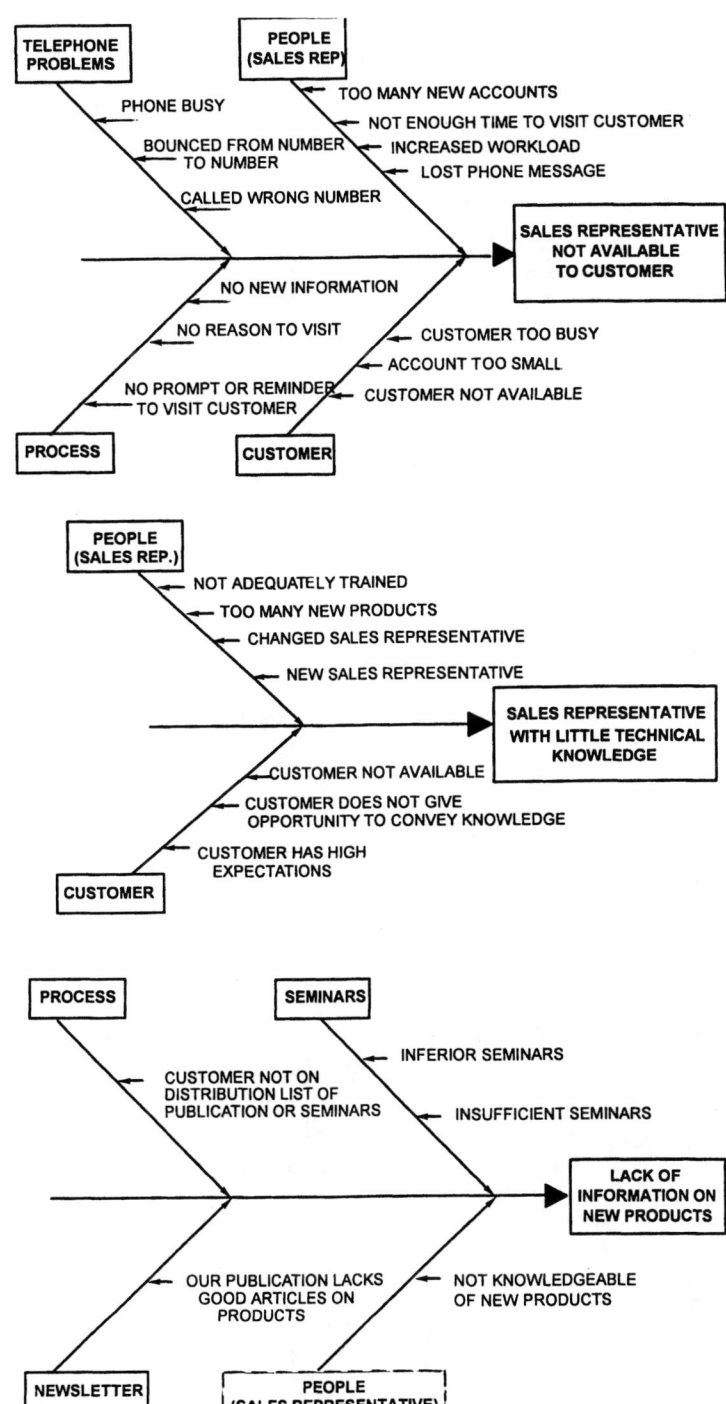

TELEPHONE PROBLEMS
PEOPLE (SALES REP)
PHONE BUSY
TOO MANY NEW ACCOUNTS
NOT ENOUGH TIME TO VISIT CUSTOMER
BOUNCED FROM NUMBER TO NUMBER
INCREASED WORKLOAD
LOST PHONE MESSAGE
CALLED WRONG NUMBER

SALES REPRESENTATIVE NOT AVAILABLE TO CUSTOMER

NO NEW INFORMATION
NO REASON TO VISIT
CUSTOMER TOO BUSY
ACCOUNT TOO SMALL
NO PROMPT OR REMINDER TO VISIT CUSTOMER
CUSTOMER NOT AVAILABLE

PROCESS
CUSTOMER

PEOPLE (SALES REP.)
NOT ADEQUATELY TRAINED
TOO MANY NEW PRODUCTS
CHANGED SALES REPRESENTATIVE
NEW SALES REPRESENTATIVE

SALES REPRESENTATIVE WITH LITTLE TECHNICAL KNOWLEDGE

CUSTOMER NOT AVAILABLE
CUSTOMER DOES NOT GIVE OPPORTUNITY TO CONVEY KNOWLEDGE
CUSTOMER HAS HIGH EXPECTATIONS

CUSTOMER

PROCESS
SEMINARS
INFERIOR SEMINARS
CUSTOMER NOT ON DISTRIBUTION LIST OF PUBLICATION OR SEMINARS
INSUFFICIENT SEMINARS

LACK OF INFORMATION ON NEW PRODUCTS

OUR PUBLICATION LACKS GOOD ARTICLES ON PRODUCTS
NOT KNOWLEDGEABLE OF NEW PRODUCTS

NEWSLETTER
PEOPLE (SALES REPRESENTATIVE)

Figure A.4   Cause-and-effect diagrams for the first three Pareto items.

	PROBABLE CAUSE (Hypothesis)	VERIFICATION OF MOST LIKELY CAUSE	REMARKS	
1a	Increased workload	Compared to last year SR has more workload due to: (a) Collection of sales process data (b) Implementation of Product support plan for every proposal This has caused about 4 hours or 10% more work per week	This is an issue of concern	✓
1b	Too many new accounts	Overall new accounts have increased by 8% but 5% of old accounts have dropped and overall change is 3%	This is not a major cause	X
1c	No prompt or reminder to visit a customer	This is correct. Currently there is no system to remind SR who to visit and when	This is an issue	✓
1d	No reason or new information, hence why visit customer	Data does not support this statement. Last year we introduced over a 100 new products. Also we held 7 seminars	This is not an issue. There are good reasons to visit customers	X
1e	Telephone problems o Phone busy o Never returned call o Called wrong number o Bounced from number to number	We were not sure. We checked with 30 dissatisfied customers via a phone survey: Every customer had experienced one or more of these phone problems	A major revamp of the phone system or process is needed. These are perennial problems.	✓
1f	Account too small	40% of sales are of a value less than $20K per year	SR feels that they do not want to visit these accounts	✓
2a	Too many new products	This contradicts item 1d. There were over 100 new products introduced and 7 new product seminars in the last year.	This is an issue. SR's feel they are not informed and do not understand some of our new offerings.	✓
2b & 3a	Not adequately trained. SR not knowledgeable of new products	Agrees with item 2a. We polled 15 SR's and 80% felt they were inadequately trained	Same as for Item 2a.	✓
2c	New SR not adequately trained	The proliferation of new products and new SR's is causing a training gap.	This is a concern and issue.	✓
3b	Customer not on distribution list of newsletter or seminar	This was difficult to check. So we looked at the dissatisfied customers to see if their company was on the distribution list, but not all customers gave their company name. Of those that did 45% were not on the distribution list.	This is an issue. We lack a good process to keep the list updated, complete and accurate	✓

(Note SR = Sales Representative)

Figure A.5   Verification of most likely cause.

	ITEM	CORRECTIVE ACTION
1a	Increased Workload	o Secretaries to manage paperwork o Automation of data collection and report preparation o Support Administration will provide resources for plan preparation
1c	No prompt or reminder	o SR will prepare a monthly call schedule

1f	Account too small	o District Manager will fill in gaps in SR schedule
		
		o Not all accounts can be visited, hence other small accounts
		  will be contacted by a Telephone Sales force which
		  will be set up during the next 6 months
		
		o Long term plan: Customer data base will be updated based
		  on inputs from telephone sales force and SR visit schedule.
		  Missed visits or calls will be red-flagged every 6 months via
		  exception report to managers.

1e	Telephone Problems	o Revamp phone system as follows:
		
		- Provide a single Customer Information center number
		  for all customer needs.
		- All unanswered calls to SR will get automatically
		  transferred to this number
		- This proposal will take minimum of 6 months to install

2a
 	
2b
 	
3a
	Too many products
	
	Not adequately trained
	
	SR not knowledgeable of new products
		We need to increase the knowledge and
		marketplace competence of Sales Reps via
		the following:
		
		o Conduct monthly, day and night classes on
		  continuous product and sales technique
		  education
		
		o Provide more product information in customer
		  newletters which will also be distributed to
		  sales representatives.

2c	New SR not adequately trained	Revamp new SR training
3b	Customer not on data base list for newsletter and seminar.	Revamp data base by doing the following: o Update current lists o Ensure monthly updates o No deletions allowed of "inactive" accounts o Add field to indicate visit or call in current year and provide 6 monthly printout if no visit occurs.

(Note SR = Sales Representative)

Figure A.6 Determine implementation plan.

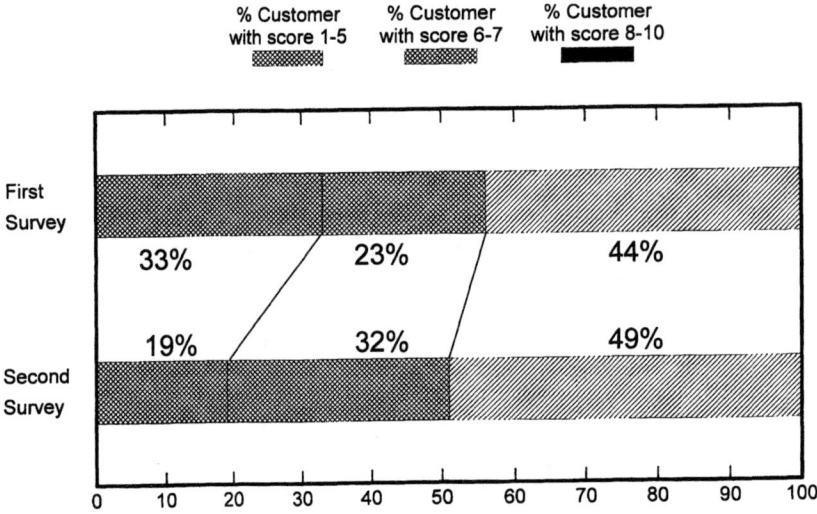

Figure A.7 Improvement in customer satisfaction for sales interaction category.

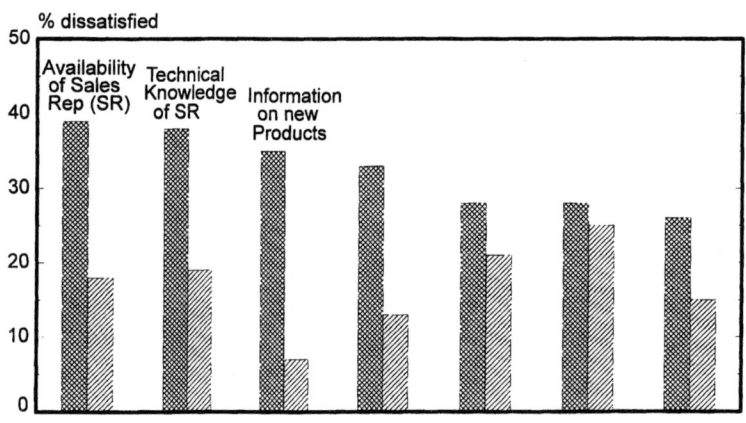

Figure A.8 Paired Pareto of reduction in dissatisfied customers.

Act Stage

Standardization of activity

We will continue all the current activity that was launched after the first survey. In particular, the telephone-sales force and the customer information center will be enhanced; all implemented activities will be documented. Currently more funds are being sought to improve the customer database. We note that training and customer database management are recurring issues; we have proposed a sales process manager job function to manage these on a long-term basis. Our proposal has been accepted.

Future plans

We will now review the second survey data and determine which area to improve.

B

Seven Quality Control Tools and the New Seven Tools

We offer here a very short discussion of the seven tools. This is meant to be an overview. For more information, refer to the recommended readings listed later in the Bibliography. The discussion is followed by a list of definitions of the new seven quality control tools, which was contributed by Khushroo Shaikh, Hewlett-Packard, Direct Marketing Division. For more details, refer to the Bibliography.

In terms of importance, the seven tools are the most useful. Kaoru Ishikawa has stated that the seven tools can be used to solve 95 percent of all problems. Independent data that we have reviewed support his statement.

Check Sheet

The check sheet is used to facilitate the collection and analysis of data. "Garbage in, garbage out" is an old cliché, but true. Therefore, the purpose for which the data are being collected must be clear. Data reflect facts, but only if the data are properly collected. An example of a check sheet used in the soldering process is shown in Fig. B.1. The number of defects and the location where they are found can be recorded and analyzed for causes.

Pareto Diagram

A Pareto diagram is constructed to show the relative importance of different categories in a process, for example, defects, cost, and failure modes. The vital few can be separated from the trivial many. This will serve as a basis for us to select the most crucial items for

Wave Soldering Defects Record

Description	No. of Defects				
	5	10	15	20	25
1. Blow Hole					
2. Pin Hole					
3. Icicle					
4. Etc					

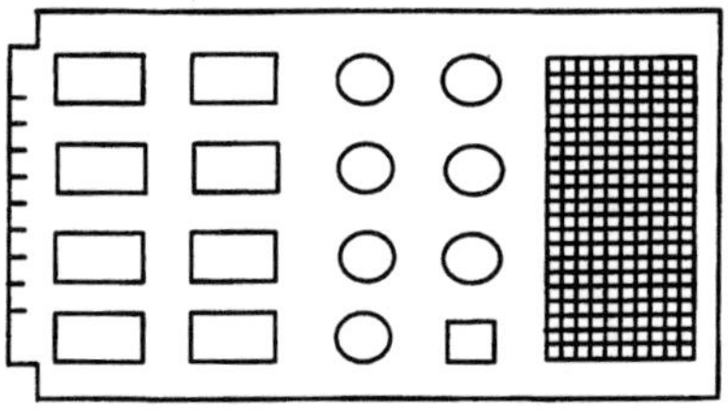

Total no. of Defects =

$$PPM = \frac{Total\ no.}{No.\ of\ Joints} \times 10^6 =$$

Remarks:

Figure B.1 A checksheet for soldering process.

improvement—typically those at the left-hand side of the Pareto diagram. This will ensure that what we try to improve will have the greatest impact. An example of a Pareto diagram for defective printers is shown in Fig. B.2.

Cause-and-Effect Diagram

Causes of problems are numerous. The cause-and-effect diagram helps us to find out all possible causes, to sort them out, and to organize their interrelationship. Typically, the causes are brainstormed in a free-flowing session; the various causes are then sorted into a few main categories. In the manufacturing environment, the main categories are usually people, machine, materials, and method. The cause-and-effect diagram is often called a *fishbone* diagram or an *Ishikawa* diagram, after its creator. Figure B.3 shows a cause-and-effect diagram for tasty pizza.

Stratification

Stratification is the technique of analyzing data by separating them into several groups with similar characteristics. It is central to the effective use of the seven tools. Imagine an engineer trying to sort out operator-related problems in a production line. The data must be stratified by operator to better understand their individual situations.

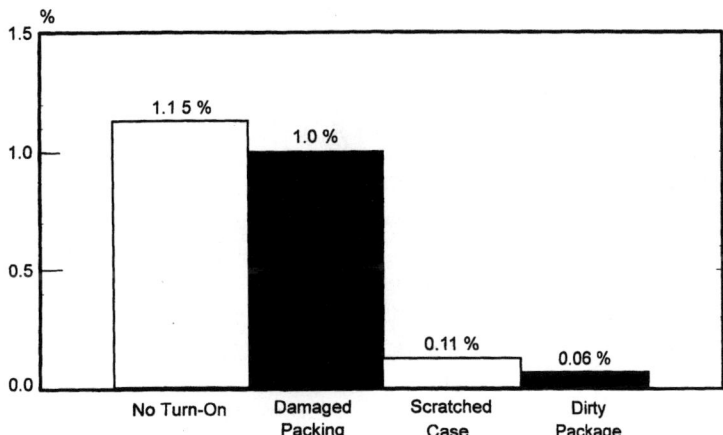

Figure B.2 A Pareto diagram to show major defects.

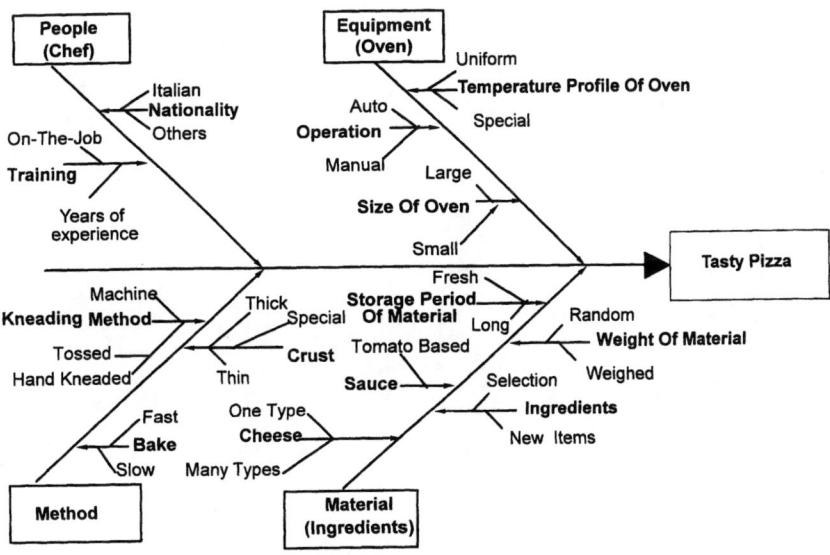

Figure B.3 A cause-and-effect diagram for tasty pizza.

Since individuals behave differently, specific corrective actions tailored to individuals may be required to improve their performance.

Graph and Histogram

A graph is a pictorial way of summarizing data. It gives us a visual display to reveal the message hidden in a maze of data. Some common types of graph we use are the line chart, pie chart, bar chart, scatter diagram, and histogram.

A *histogram* is used to summarize the frequency of occurrence of something, from a sample of data. The shape of the distribution is displayed, and descriptive statistics can be readily calculated. In the histogram shown in Fig. B.4, the proportion of items not meeting specification is easily grasped. A *scatter diagram* is a special type of graph which shows the relationship between two variables. If there is an empirical relationship between the two variables, it will be seen in a scatter diagram. Furthermore, if the relationship is linear, then the correlation coefficient is used to measure the degree of association. In a cause-and-effect relationship, a regression equation is fitted to describe the behavior. A scatter diagram showing the relationship between labor cost and production volume is shown in Fig. B.5.

Control Chart

The basic requirement in a manufacturing process is to establish a state of control and to sustain this state through time. Standardization of working methods is necessary to maintain this state. A control chart enables us to observe if this standardization is correct and whether it is being maintained.

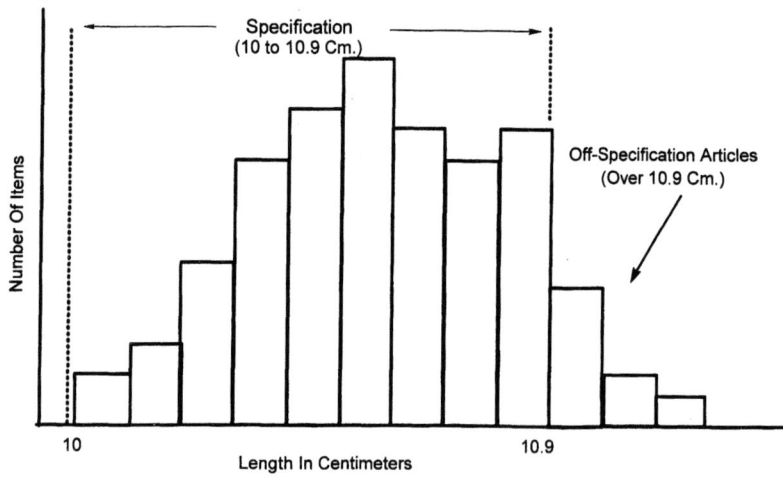

Figure B.4 Histogram showing off-specification articles.

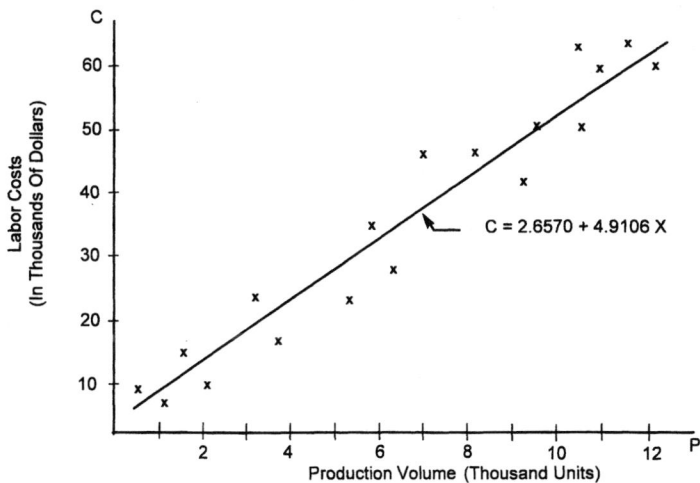

Figure B.5 A scatter diagram to show the relationship between direct labor costs and production volume.

A control chart, whether for measurements, attributes, or defects, has a centerline corresponding to the average quality at which the process should perform. If statistical control of the process exists, there will be two control limits: the upper and lower control limits. The average and the two limits are computed from the data—objectively. An out-of-control situation, trend, cycles, and other unnatural patterns can be detected from the chart easily.

The now-renowned example quoted from Edwards Deming's writing can best demonstrate the essence of a control chart. Figures B.6 and B.7 show the average scores x of a beginning golfer and an experienced golfer.

In Fig. B.6, the average scores before lessons were not in a state of control, as evident from the points outside the two control limits, UCL and LCL. There could be many causes: the way the golfer stood, the clubs used, and so on. The causes were identified by a golf professional, and corrective training was given. The purpose of the training was to remove all the identified or assignable causes. After the lessons, the average scores dropped. Greater consistency was shown as the two control limits were narrower or closer to each other. The lessons improved the score or system.

In Fig. B.7, the experienced golfer tried to improve his game by taking lessons. The control charts for before lessons and after lessons showed similar characteristics. The lessons accomplished nothing. Only a new way of training can improve the score or system. For example, a golf professional could do a detailed swing analysis of the experienced golfer's swing and then propose corrective action.

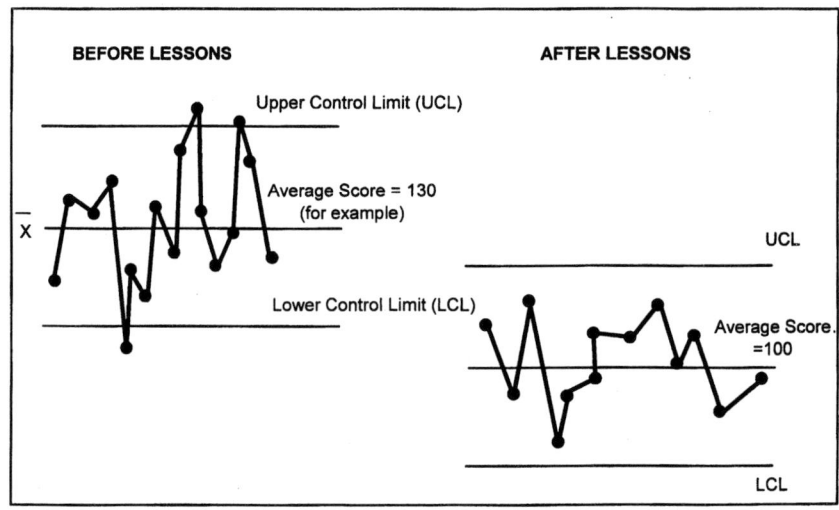

Figure B.6 X-chart for a beginner in golf.

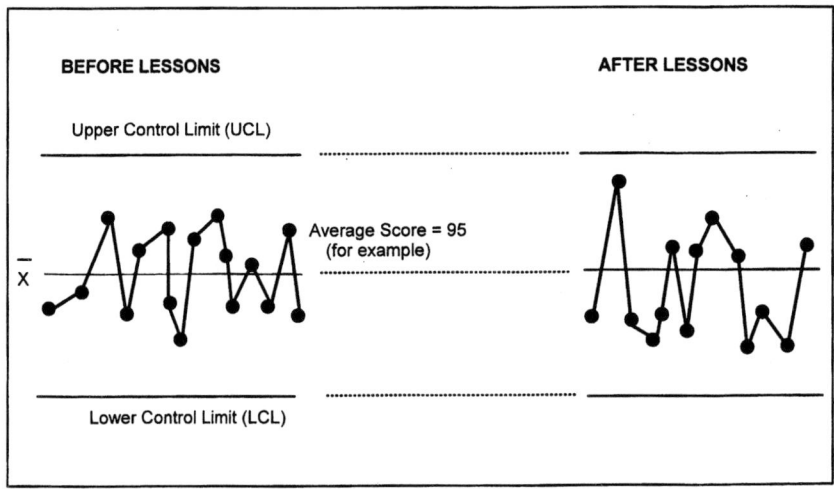

Figure B.7 X-chart for an experienced golfer.

An analogy is easily drawn between the average scores of the golfers and performance of a production process. The inexperienced golfer can be compared to our process which is not under control. After a proper education and work standardization, we will be able to remove all assignable causes and thus bring the process to stability. When the process is in control, as in the case of the experienced golfer, management and engineers must intervene to improve the process—after a thorough analysis.

The New Seven Quality Control Tools, or Seven New Management Tools

The KJ method, or affinity diagram

This technique extracts the essence of the message in a large volume of verbal data by using the mutual affinity method. An example is as follows: There could be an enormous amount of verbal data from a customer satisfaction survey or an organizational problem. The numerous verbal comments are put on paper, cards, or Post-It stickers; they are then sorted into categories of issues. The data can be arranged in a diagram showing the affinity of the issues; insight into solutions can be determined next. This method capitalizes on intuitive reasoning. Kawakito Jiro conceived and developed this tool, hence the name.

A subtle variation of this technique could be to group a large number of verbal data, for example, complaints and comments, by categories from a customer satisfaction survey; this allows the preparation of a Pareto diagram of categories. Hence, under appropriate circumstances, this methodology allows conversion of verbal data to quantitative data.

The relations diagram or interrelationship diagram

This tool is helpful in clarifying intertwined relationships in a complex, or multidimensional, situation in order to find an appropriate solution. This tool is useful for understanding complex relationships in a manufacturing or planning process, or for introduction of new ideas and programs.

The matrix diagram

The matrix diagram is helpful for understanding multilevel, multidimensional relationships and expressing them in a graphical manner. Applications include deployment of quality in a product (for example, quality function deployment, or QFD), tying goals to methods, and understanding complex relationships between processes and failures (for example, via a T-type matrix).

When we have two variables, we get a two-dimensional matrix which consists of rows and columns. A two-dimensional matrix is an L-type matrix; such a matrix displays relationship information between two variables, for example, goals and methods.

A T-type matrix is obtained when two L-type matrices are combined. In a T-type matrix, variable A has a relationship with variable B. Variable A also has a relationship with variable C, but there is no direct relationship between variables B and C. An example of a T-matrix is given in App. C.

A Y-type matrix is a three-dimensional display of information. There is a simultaneous pairwise relationship between the three variables, in other words, a relationship between A and B, between B and C, and between C and A.

An X-type matrix portrays four variables; the relationships, however, are confined to the adjacent pairs but not to random selection of a pair.

The systematic diagram, tree diagram, or dendogram

This is a very systematic approach to identifying means (methods, ways, or approaches) that will be needed to achieve an overall objective or goal. The completed systematic diagram resembles a tree (for example, one main trunk with primary, secondary, and tertiary branches); hence, it is also known as a *tree diagram*. An organizational chart is a good example of a systematic diagram. Another name for this tool is a *dendogram*. It may be familiar to you if you have traced your family roots.

The systematic diagram helps to identify the starting point for problem solving. It is also useful for developing the necessary implementation plan for a given objective. In certain applications it is called a systematic cause-and-effect diagram; in this case we start with the desired effect (or objective), and then we prepare the causes (or implementation steps).

The arrow diagram

This tool uses elements of the program evaluation and review technique (PERT) chart to organize daily operations in a sequence and to monitor their progress.

Process decision program chart (PDPC)

This tool is helpful for developing processes that do not fail. The tool is used to study all links in a given process, searching for weak ones. A good understanding of the failure modes at the weak links facilitates development of countermeasures to make the process strong and foolproof.

Matrix data analysis or multivariate statistical analysis

The pictorial relationships of the matrix diagram can be quantified by using appropriate multivariate statistical techniques. The multivariate techniques facilitate the understanding of complex relationships between variables, resulting in well-informed business decisions.

C

Some Useful Processes for Creating Product and Process Quality

In this appendix we list three useful and documented processes:

- *Quality function deployment* (QFD), which is helpful in the product development process.
- *Failure mode effects analysis* (FMEA), which is useful for predicting and preventing failures—for hardware, software, and services.
- The *T-type matrix*, a very esoteric tool for analyzing the efficiency of the new product development process. This tool is useful in detecting the root cause of failures in a product after it is released. The example given proves that the bulk of failures occur in the design stage and gives suggestions on catching problems early in the design stage.

The Product Development Process and QFD

The product definition process (which is the beginning of the product development process), if done well, will ensure that the right products are manufactured. The succeeding process (design) will ensure that the products are designed correctly—with appropriate features and high reliability.

One of the best ways of ensuring a good product definition process is by using the quality function deployment methodology. QFD is a tool that helps to tightly link the entire product development process. It enhances the product definition process and ensures that the succeeding processes, such as design and manufacturing, are linked back to the product definition. We will discuss QFD in the context of a good product development process. The discussion of QFD is courtesy of Khushroo Shaikh of Hewlett-Packard.

Introduction to QFD

Why QFD?   A lack of a well-defined product development process in research and development can be the root cause for the ruination of a company. QFD offers many features which are helpful in defining this process, something that for many reasons is often very difficult.

Although many companies do not have a well-defined new product development process, they manage to market blockbuster products or services which fill the needs of customers or users. At other times their new products or services may fail because customers do not purchase. In other words, success comes when customer's needs are well understood and met by the products or services. This is another powerful feature of QFD. Hence, learning more about QFD is worthwhile. We mentioned earlier, in Chap. 2, the reasons why products or services succeed or fail the marketplace. A common thread runs through all the studies that we quoted: *To have a successful product, we must understand the user's needs.* QFD provides a formal process to understand user needs and to weave these needs across the entire product development cycle, including manufacturing.

Definition of QFD.   QFD is a very systematic and organized approach of taking customer needs and demands into consideration when one is designing new products and services or improving existing products and services. Another name for this approach is *customer-driven* engineering, because the voice of the customer is diffused throughout the product (or service) development life cycle.

QFD is a planning tool that defines a process for developing products or services. The aptitude to plan is rare in the human race. Managers are evaluated on short-term results, which further inhibits this aptitude for planning. It is difficult to use Deming's PDCA cycle to improve the product development process if the P of PDCA cycle is weak. QFD is applying total quality philosophy to product development by focusing on P. Using QFD counteracts the inherent weakness embedded in human nature—that of avoiding planning.

Where did QFD come from?   Dr. Yogi Akao of Tamagawa University is the key contributor of QFD development in Japan. There are 30 matrices in Akao's approach. The QFD team can pick and choose the matrix which would be of most use for a particular phase of product development. There are many other experts, and for more information on the history of QFD you should read a chapter written by Dr. Akao.[1]

Another expert in the area of QFD is Dr. Fukahara. He is associated with the Central Japan Quality Control Association. He mainly focuses on the house of quality, namely, the product definition aspect. There is also the four-phase (or four-matrix) approach promoted by the American Supplier Institute (ASI). Later we will discuss both the house of quality and the four-phase approach. There are, of course, many other approaches to QFD.

Each approach has its pros and cons. What is the best approach? The answer is not very straightforward. Each company has a unique set of employees, products, customers, systems, and so on, giving birth to unique circumstances and culture. The best solution is to develop a custom-fit QFD model to meet the particular needs of your company. If a cookbook approach to apply a tool like QFD is taken, your company will not get the optimum benefit of QFD, but instead may get bogged down.

The collective brain power of the employees is the most valuable resource of any company. One of the major goals of using total quality philosophy is to enable employees to make meaningful contributions as a result of using their collective brain power. QFD allows this freedom.

QFD in the United States. The principles and ideas used in QFD were suggested by A.V. Feigenbaum when he discussed total quality systems and a cross-functional management way of thinking. Joseph Juran describes quality assurance systems in *Quality Handbook* and touches on the early stages of a QFD system. Dr. Naddler, a systems dean at the University of Southern California, teaches "The Planning and Design Approach" with ideas that could have culminated into QFD as it is today.

The first formal introduction of QFD as whole system came to the United States in 1983, via the Cambridge Corporation in Chicago. The following year Professor Don Clausing from the Massachusetts Institute of Technology introduced QFD to Ford Motor Company. Don Clausing and John Hauser wrote an article in *Harvard Business Review* in May 1988. In the same year, Ford told its vendors that in order to remain on Ford's preferred-vendors list, it would be advisable for them to use QFD. Symposiums on QFD are regularly held these days. QFD is spreading like wildfire in the United States.

What are the benefits of using QFD?

The ultimate benefit of QFD for any company is its contribution to meeting and exceeding customers' needs, thus obtaining greater market share and higher profits. There are, however, many other tangible benefits that participants can experience. Some are listed here:

1. Product definition is firmer and takes place earlier in the new product development life cycle. This minimizes engineering changes and results in better quality.

2. QFD addresses major issues and complaints expressed by customers during the early stages of product definition. Hence, the number of complaints and dissatisfaction with new products decreases with time. This benefit is seen after several product cycles.

3. Cross-functional walls break down with QFD as the team must address issues that affect all departments. Suboptimization of resources in a company is minimized, and communication between departments improves.

4. Team members develop a deeper understanding of customer needs and have the customer's voice as a basis for making trade-offs, resulting in superior decisions for the organization.

5. The analytic vigor of QFD causes streamlining or elimination of many internal processes that do not add value to the new product development process.

6. Customers' needs are evaluated with respect to the product and services of the competition. This allows identification of the internal processes that need improvement to stay competitive.

7. Documentation is an essential ingredient of QFD. Hence, one of its greatest benefits is that it builds product intelligence. This documentation provides the following advantages:
   - It helps new engineers come aboard faster.
   - Easily accessible documentation reduces chances of repeating mistakes of the past.
   - The accumulation of knowledge decreases the need for having a leader with seniority to head the project. A senior leader contributes significantly to the success of the project, as we learned from the project Sappho study in Chap. 2. A well-established QFD process will deliver this critical benefit—even from a younger person who possesses leadership qualities. Also, under a total quality environment the team will feel empowered and will possess the authority generally associated with a senior leader.

8. QFD is beneficial in understanding and identifying a market niche where customers' needs are not being met. This provides opportunities for introducing niche products.

9. QFD provides an excellent framework for cross-functional deployment of quality, cost, and delivery. The importance of this was discussed in Chap. 3.

10. QFD allows for quick changes, which is very important for the new product development process. It is possible to revise previous decisions when new information becomes available during product development, for example, if the competition introduces a new product or a state-of-the-art technology becomes available. Detailed documentation keeps all information visible to the QFD team at all times.

All the above benefits result in a robust new product development cycle and minimize difficulties and problems. QFD is one of the best ways to introduce total quality to the marketing, design, and manufacturing environments.

Implementing QFD is a very strong strategy for propagating total quality in a company. QFD provides the much-needed horizontal weave across the company. Typically this horizontal weave is missing in the current traditional management hierarchy because organizations are managed vertically. The benefits of QFD include a shorter product or service development cycle and higher productivity.

Nuts and bolts: How to begin? The ASI, the Kaizen Institute, and the Greater Opportunity Alliance of Lawrence (GOAL) teach the nuts and bolts of QFD in the United States. The details of how to implement QFD can be learned from attending their classes. Some books have been published on the subject,[2] but we recommend that you attend a QFD class to acquire hands-on knowledge of this subject. We will share a brief overview of the process, using the example of the pencil. We will also share insights developed from working with QFD, and in some cases we bring greater depth to a particular topic. This will enhance your chances of developing a dynamite product or service.

The four phases of the product development cycle are shown in Fig. C.1:

I.   Product planning

II. Part deployment

III. Process planning

IV. Production planning

Figure C.1 illustrates the closely linked phases of QFD as the product develops. The tightly linked characteristic of phases accomplishes delivery of the product as desired by the customer. The design requirements of the first matrix feed into the second matrix so that part requirements are done correctly. Part requirements then drive the process requirements of the third matrix. Process requirements for manufacturing will determine process assembly and the quality checkpoints chart. Details of quality checkpoints are given later in this chapter, and will ensure consistent product quality.

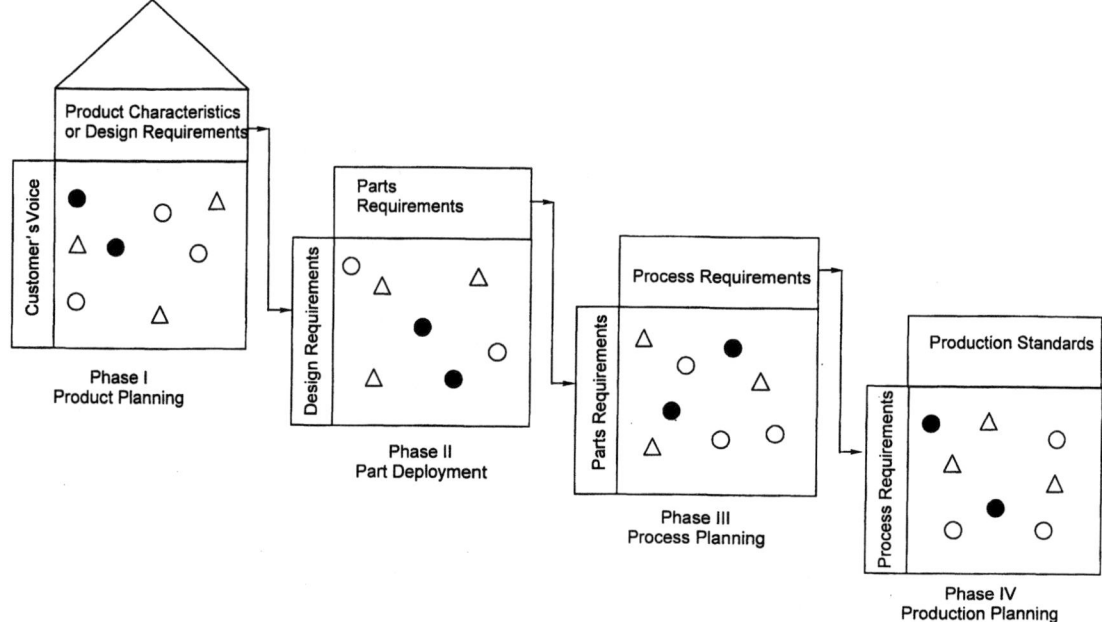

Figure C.1 Phases of customer-driven product development.

The critical planning for the product takes place in the first phase. This phase is also known as the *house of quality matrix* and *matrix of matrices.* The components of this matrix are

1. Customers' voice

2. Importance rating

3. Product characteristics or design requirements

4. Relationship matrix

5. Quality planning using competitive evaluation and strategic positioning

6. Correlation matrix (the triangular roof portion)

7. Technical evaluation

8. Target value determinations

The house of quality with these components is shown in Fig. C.2. The house of quality, which focuses on product definition, is a critical engine that guides all steps of product development. The overall success of the product therefore depends on good product definition.

HOUSE OF QUALITY

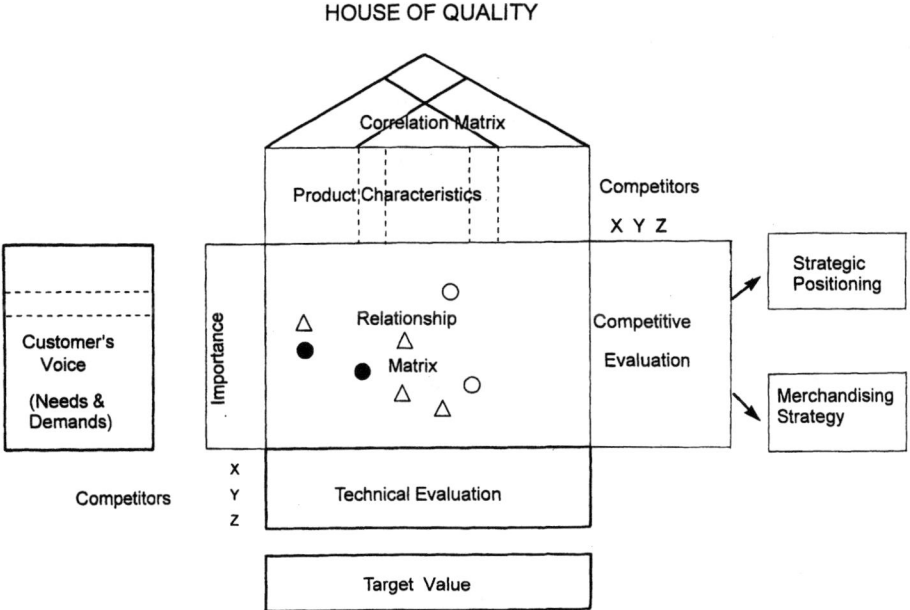

Figure C.2   Details of phase 1: product planning or product definition.

Why is product definition important?   Research indicates that schedule slippage in product development occurs due to either lack of firm definition of the product or changing definition of the product. So product definition is a critical first step in making a good product. QFD models of Fukahara, Makabe, and Akao start with a matrix that facilitates product definition. Freezing the definition helps to develop the product in a timely manner. Additions and deletions not only cause schedule slippage but also have a deteriorating effect on quality. One possible outcome of product definition is a decision not to make the product. This decision reduces the cost of abandoning projects later. Abandoned projects can be a tremendous drain on business, as noted in the Union Carbide study discussed in Chap. 2.

Review of QFD basics via the pencil example

We mentioned earlier that a well-developed house of quality is an essential ingredient for successful product development. Let us develop a house of quality for a pencil. The pencil example does not cover all details of QFD, but will give you a good understanding of QFD basics. The ideas developed in this example were germinated at a class on QFD at Hewlett-Packard, and the ideas have been validated by experts from the pencil industry.[3]

TABLE C.1   Customer's Voice or Demands for a Pencil

Writes	Has nonsmearing eraser
Sharpens	Smells good
Does not roll	Has lead that does not break too easily
Is easy to hold	Tells me type of lead in pencil
Does not smear	Creates visible writing
Can erase	Can photocopy
Has choice of lead	Does not tear paper when I write
Has choice of color	Creates no slivers
Is inexpensive	Is not toxic
Has eraser that does not break	Is soft for biting but does not break the point
Feels good	Sharpened point will last up to a half page of text before it needs resharpening
Spells correctly	

Listing of customer needs.   The students acting as customers listed their needs. These needs or demands, known as the customer's voice, are tabulated in Table C.1. Developing an excellent customer's voice is of the utmost importance, and we will share ideas on this subject later in this chapter.

Next, the house of quality, described in Fig. C.2, was developed for the pencil. The contents of Fig. C.2 were slightly rearranged; all the relevant factors, however, were included. The fully developed house of quality for the pencil is displayed in Fig. C.3. The steps to develop the figure are described here.

Grouping of customer needs.   The needs, gathered in Table C.1, were grouped according to the KJ method. KJ stands for Kawakito Jiro, who developed this tool. Refer to the discussion of the seven new quality control tools in App. B. This tool uses the right, or creative, side of the brain to group verbal data. Moreover, the KJ tool retains the customer's own words in the house of quality as tertiary labels, such as *nonsmearing*.

These groupings were given labels—for example, the secondary labels of *selection* or *quality*—to succinctly describe the ideas captured in those particular groupings. Similarly, secondary labels are grouped to develop primary labels; for example, the secondary labels of *selection* and *quality* can be grouped under the primary label of *lead* (refer to Fig. C.3). By deploying the tree diagram (one of the seven new quality controls) on these primary and secondary labels, the unspoken needs of customers can be ferreted out.

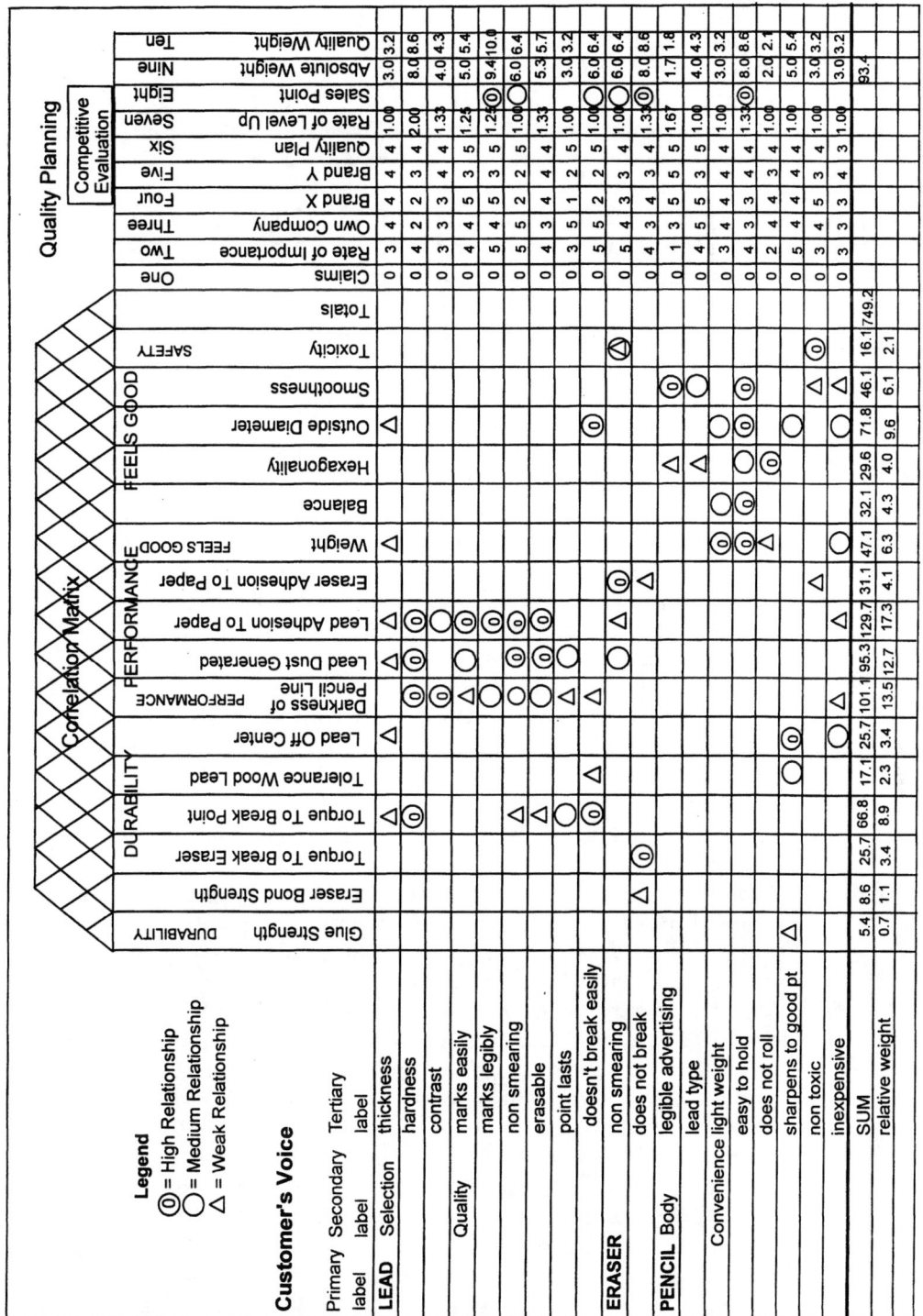

Figure C.3 Product planning (pencil example).

Quality planning. The 10 columns in Fig. C.3 under quality planning are developed. This incorporates competitive analysis, strategic positioning, and merchandising strategy. Let us focus on one demand—nonsmearing (sixth line under customers' voice). The claims column (column 1 under quality planning) contains zero for our example. Claims refer to the ability of our company to provide proper product warranty or service (claims).

The next column, rate of importance, is 5 on the scale of 1 to 5, with 1 being the least important and 5 the most. This rating is provided by the customers during the customer interview phase.

The competitive evaluation columns (columns 3, 4, and 5) indicate that our company is rated 5 whereas both competitors X and Y are rated 2.

Since customers rated nonsmearing to be an important need, the quality plan column (column 6) is targeted at 5, even though there is no threat from the competition. This means we plan for the best possible quality for this item of nonsmearing.

Rate of level up (column 7) measures whether we are improving, maintaining, or decreasing quality, compared with the competition. The rate of level up is defined as Quality plan value ÷ Competitive rating value = 5/5 = 1.00. That is, it is column 6 divided by column 3. In this case, we are rated high and plan to stay high.

The planned sales points are illustrated in column 8. These are decided by management after studying customers' needs and the competition. These will represent competitive strengths that can be conveyed in a sales and marketing campaign. The primary (or important) sales points are indicated by a double circle and carry a weight of 1.5, while the secondary sales points are indicated by circles and carry a numerical value of 1.2. These weights have been arbitrarily set, based on experience.

The absolute weight (column 9) is calculated by multiplying values of rate of importance, rate of level up, and sales point. Thus Absolute weight = Rate of importance×Rate of level up×Sales point. For nonsmearing, Absolute weight = 5×1×1.2 = 6.0.

All the absolute weights are added vertically (in column 9); here the sum is 93.4. Each of the absolute weights is normalized to a base of 100, to obtain quality weight, tabulated in column 10. The quality weight is calculated to be 6.4 for nonsmearing.

Purpose of numerical evaluation. Numerical evaluation helps give the team quantitative information to make correct decisions. Glancing at column 10 we can easily determine which are the important needs. We can show the importance in another way: Figure C.4 shows the competitive data pictorially. The data here are collected from columns 3, 4, and 5 of Fig. C.3. The picture makes it easy to understand our

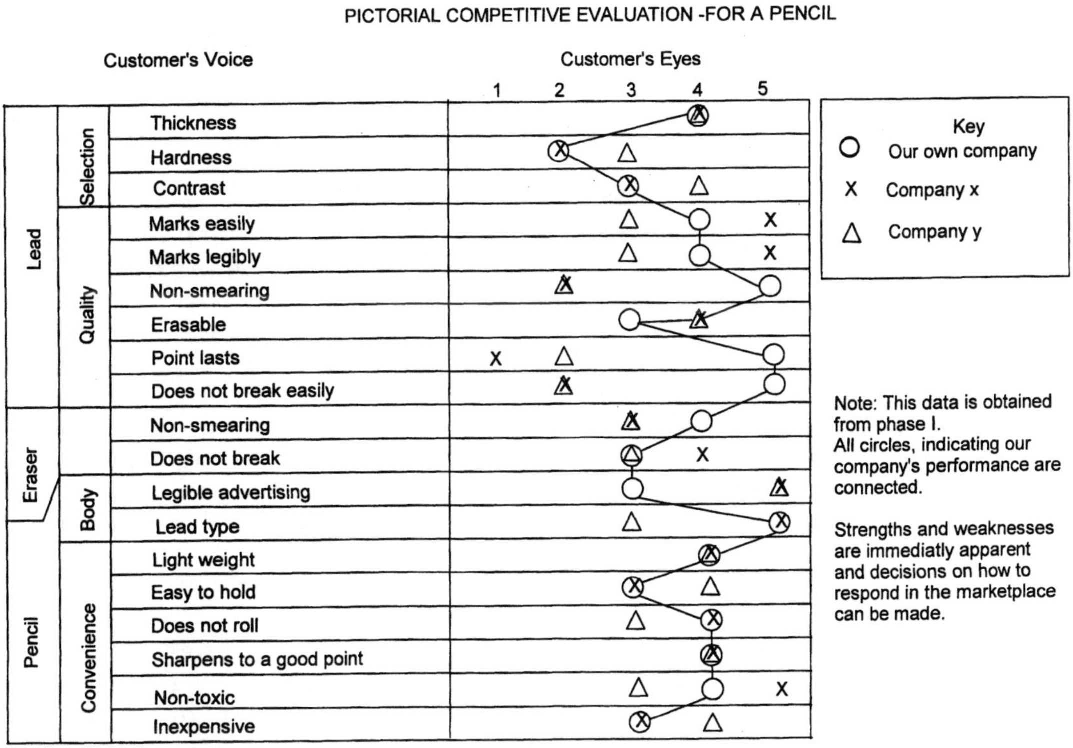

PICTORIAL COMPETITIVE EVALUATION -FOR A PENCIL

Figure C.4 Pictorial competitive evaluation, developed from phase 1.

own strengths with respect to those of the competition; it is very easy to identify challenges and make a conscious decision to either address them or leave them alone.

Defining product characteristics. One method for developing product characteristics or design requirements of a product is by *analytical brainstorming*. These characteristics are then organized, using KJ and tree diagrams, as was done with customers' needs. The results are shown at the top of Fig. C.3 with tertiary (for example, lead dust generated) and secondary labels (for example, performance). Primary labels were not developed for this example.

Analytical brainstorming is a term coined by the author to distinguish this specialized brainstorming from the generally used free-form brainstorming, where knowledge of subject matter is not a prerequisite. Engineering or design knowledge is needed to generate ideas to meet the customers' needs. The brainstorming is carried out sequentially with respect to the needs of customers, which is another attribute of analytical brainstorming.

Another method to generate design requirements is through use of a cause-and-effect diagram. Note that the correlation matrix (the triangular roof area) between design requirements was not carried out for the pencil example. In real life this an invaluable step and must be incorporated.

Developing relationships. Next, the relationship between customer's voice and design requirements is evaluated. This relationship is indicated by symbols, such as a triangle, circle, and double circle.

Once again, we will develop this section by focusing on the nonsmearing quality of lead. When the QFD team sees no relationship, the cell is left blank. The team felt that the torque to break point and nonsmearing had a weak relationship, and so a triangle is placed in that cell. Darkness of pencil line and nonsmearing had a medium relationship according to the team members, so the cell contains a circle. The strong relationship between design requirements, such as lead dust generated and lead adhesion to paper, and nonsmearing is shown by a double circle in the cell.

For our example weights of 1, 2, and 3 are assigned to a triangle, circle, and double circle, respectively. The numerical values in the relationship matrix are then derived by multiplying the quality weight and the numerical value of the symbols in each cell; these are added vertically, and the sum is entered in the sum cell, at the bottom of Fig. C.3. All the columns are added in this way. The sum for adhesion to paper is 129.7, and for torque to break point it is 66.8. These calculations help prioritize design requirements that will generate increased customer satisfaction.

Developing the next phase. In the case of the pencil example, lead adhesion to paper received the maximum weight. Therefore, lead adhesion to paper is carried to the part deployment phase (phase II), as shown in Fig. C.5. Other important characteristics, such as darkness of pencil line, could also be deployed to phase II. In phase II we apply a similar rigorous analysis to that in phase I, and we prioritize parts that are critical for increasing customer satisfaction. Graphite, clay, and gum are identified as these parts. The parts now guide the process planning phase—phase III.

The critical processes are identified in phase III, for example, grinder pressure, to guide production planning in phase IV. A clear understanding of production knob settings (see Fig. C.5) for the critical processes is thus established. Appropriate control charts are designed before manufacturing begins—specifically, a quality checkpoints document is prepared. Refer to the section "Quality Checkpoints Chart System" in Chap. 5 for more details.

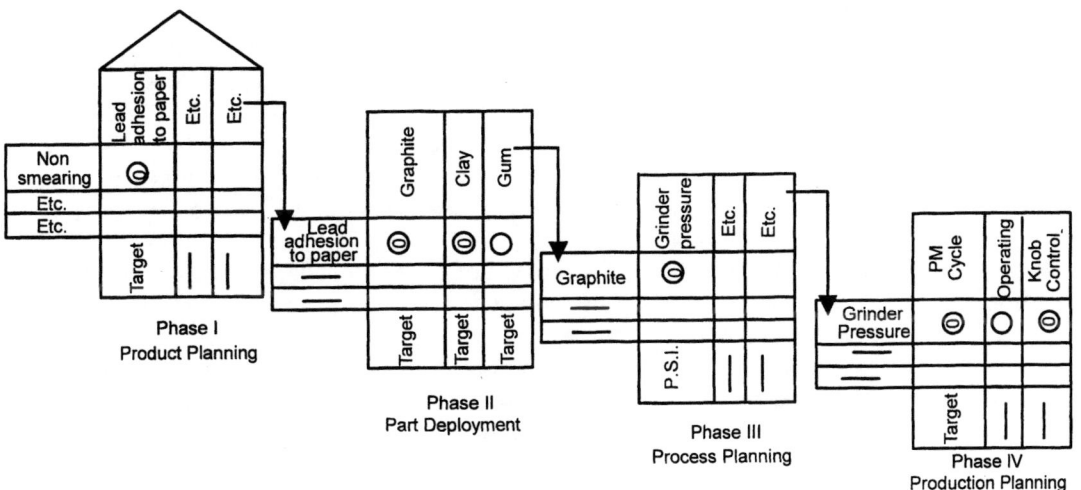

Figure C.5 Phases of customer-driven product development.

We have given you a glimpse of the steps taken to address the customers' need for nonsmearing throughout the product life cycle, in order to deliver that need to the customer. Finally, when the product reaches the market, the customers' voice should be updated and validated. This would complete the entire PDCA cycle for developing the next generation of pencil.

How many phases or matrices should you have? In Fig. C.5, we have shown four phases. Other proponents of QFD suggest up to 30 (for example, Yoji Akao). We suggest you begin with one initially: Phase I is shown in Fig. C.5 and also is known as the house of quality. The house of quality is illustrated in detail in Figs. C.2 and C.3. Once you have mastered the house of quality, proceed to the other phases.

With this fundamental understanding of QFD, we now study two critical success factors in depth:

1. Development of the customer's voice

2. A process for increasing QFD success in your company

Developing the customer's voice for QFD

Who is the customer? It is critical to properly select the customer. This can be done via market analysis, market segmentation, and market potential.

Developing the customer's voice. After the customer is identified, the next step is to develop the customer's voice. Developing the voice of the customer is the most important and most difficult task. Since the voice of the customer drives all other activities of the development process, its importance is easy to comprehend. The voice may be spoken or unspoken, direct or indirect. One of the challenges is to combine many voices to glean the essence. It is critical to tap all sources of the customer's voice.

Some of the direct sources are input (words) from a customer (this could be accomplished by using surveys or focus groups); repair records; replacement records of worn-out or defective parts; identification of defects; enhancements or requests for enhancements; market research data; list of things gone right in previous products; list of things gone wrong in previous products; manufacturing problems due to poor design; requests for engineering changes; suppliers' difficulties in the past; and even a test engineer's voice, which could include the parameters that were not tested in previous designs. More indirect sources come from observing customers' difficulties when using your product; observing the difficulties customers experience when they are solving their own work problems; capturing murmurs at trade shows; ideas from technical media; the voice of some customers experiencing the product or service from their point of view; anticipated future needs of customers by capturing comments such as "I only wish."

We want to make one important point. Every one of us holds a set of paradigms by which we view the world around us. Our ability to hear what is being communicated is heavily affected by the paradigms we hold. If a customer's need does not fit our paradigm, we may fail to hear the need. In order to minimize missed customers' needs, use as many tools as possible, including tape and video recorders. We recommend that you view a videotape by futurist Joel Barker called "Discovering the Future Business of Paradigms"[4] before contacting your customers. This will make you sensitive to the fact that we all have our own paradigms. This awareness will facilitate a conscious effort to capture the customer's voice.

Steps for Developing a High-Quality Customer's Voice

1. Collect information from every conceivable customer source. Many sources are listed in the previous section.

2. Express the voice in terms of need. Do not include what is not required by the customer.

3. Use the exact words of the customer whenever possible. The "unfiltered expression" of the customer's voice is critical.

4. Express customer's need in brief, positive phrases with at least a subject and a verb. Use no double negatives.

5. List only one need per line.

6. Focus on the target market only.

7. Stratify types of customers when applicable. Software products, for example, have customers who are programmers, systems integrators, systems maintainers, managers who are decision makers, and end users. Needs will vary with the type of customer strata; the degree of importance will vary even if the needs are identical.

After all the needs are collected, a tool called the *affinity diagram* or KJ method is very useful in grouping the customer's needs. This is one of the seven new quality control tools. The use of tools can help increase management participation in total quality activities.

A company invests in market research to be able to market products and services that would thrill current and future customers and would result in increased market share. The objective of research is to develop a deeper and more imaginative understanding of customers' needs, but as explained in Noriaki Kano's *two-dimensional quality* model (Chap. 1), this is not a trivial effort. The KJ method allows a team to be more imaginative in capturing customers' needs.

The KJ method should be used to organize the customer's voice. Being imaginative also facilitates capturing of customers' emotions. Since the buying decision of customers includes emotions, it makes sense to include emotions in product development. In the traditional management environment, predetermined categories are used to organize the customers' voice. But categories do not help capture emotions. Moreover, we are filtering customers' voice through our preexisting bias, for example, by adding categories which are conceived by the designer rather than the customer, hence biasing the product definition. The unfiltered customer's voice has a higher probability of being included when the KJ method is used correctly.

Another very useful tool in the seven new quality control tools is the tree diagram. Unlike the KJ method, the tree diagram is a very logical and systematic tool. Using the tree diagram along with the KJ method helps the QFD team to decipher the needs which customers may have neglected to mention. The topics developed when the KJ method is combined with the thoroughness of the tree diagram enable us to get at the unspoken words of the customer. Understanding and capitalizing on unexpressed needs can be a key competitive advantage. The QFD process steps will help you move toward this goal.

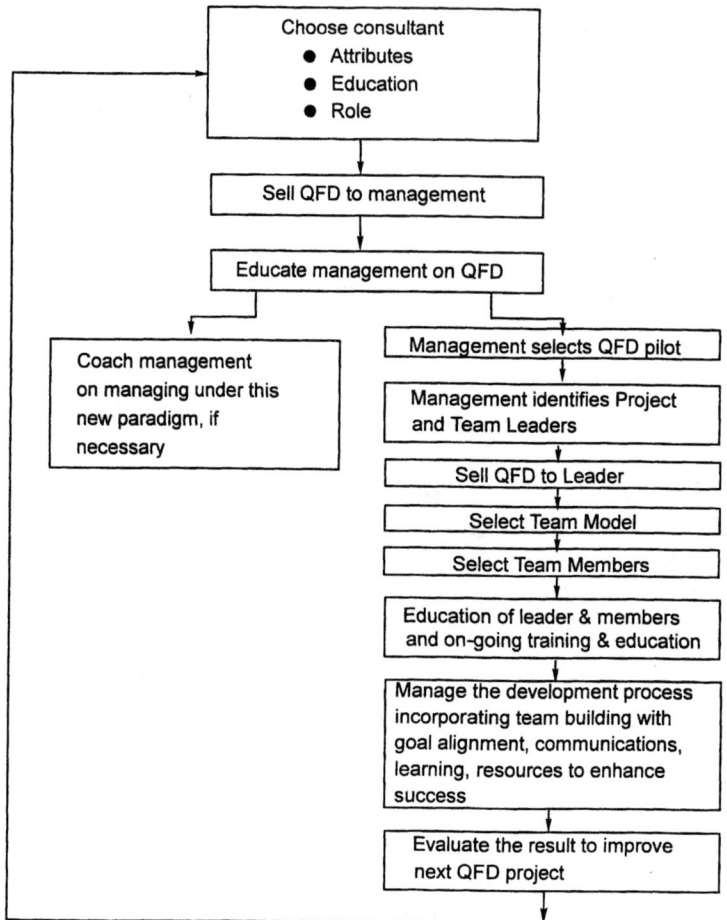

Figure C.6   Process flow for QFD success.

Recommendations for increasing QFD success in your company

We give some guidelines for ensuring success in implementing QFD in your organization. During this discussion, refer to Fig. C.6.

Choose a consultant or facilitator

Attributes. In Fig. C.6, we show a flowchart for ensuring the success of QFD in your company. The consultant can be from an internal or external source. A good consultant is critical to the success of this effort. The consultant should have the respect of all team members and should be perceived as an expert by the team. The consultant should be analytical and should understand QFD, believe in the quality-oriented management philosophy, and work well with people. The consultant must establish and meet time lines, work on a win-win philosophy, proceed only when team consensus is established, know how to facilitate successful meetings, and build well-functioning teams.

Education. There are some fundamental educational requirements for the consultant. Total quality philosophy and the PDCA cycle should be an integral aspect of his or her education. The consultant must recognize times when experts from other disciplines are needed by the QFD team. A thorough understanding of QFD is assumed.

Role of the consultant or facilitator. A consultant must be a fast thinker and two steps ahead of the team. The person must be a process expert and help the team stay focused on the task. A refresher, just-in-time training for the QFD team by the consultant will alleviate the stress of not remembering details learned in the past. Training should directly address the task at hand.

It is well known that venture capitalists do not invest in an idea, but invest in a person who has the idea. The team's confidence in a consultant will increase the prospects for success. The consultant also functions as a change agent when QFD is initiated in the traditional business environment. After the transformation has taken place and the total quality philosophy is pervasive in an organization, the significance of the consultant will diminish.

Sell QFD to management. A package to sell QFD to management is presented in the paper "Does Success Have a Secret?"[5] You may want to consult the paper for more ideas on selling. Managers tune in to economic benefits, and so these benefits should be explained with data. Alternatively, you can use the data on why products succeed and fail from Chap. 2 and the benefits of QFD listed earlier in this section.

Educate management on QFD. This session should contain fundamentals of QFD like the pencil example described earlier. In addition, the session should answer any questions on time line, resources, and learning curve. All the benefits articulated earlier in the chapter may not be achieved for the first pilot project.

This session should develop a realistic expectation of resources needed for success. This education is a combination of selling the concept, developing realistic expectations, and obtaining firm commitment from management. An honest communication that return on investment (ROI) will not be instantaneous should be given. Part of the problem is that many businesses or companies do not know how to connect the cost of losing customers or market share with generally acceptable internal accounting figures. The QFD process, however, will allow management a clearer understanding of what went wrong. More importantly, it will help identify how to improve the next time by eliminating root causes of process malfunction. Today an individual designer or manager is blamed for failure, but no steps are generally taken to eliminate recurrence of the failure in the future.

Coach management.   Correct management behavior will enhance the chance of success tremendously. Some managers will be able to adopt this behavior easily, but others will need coaching. Here are some examples of desirable behavior: Listen to the team's progress by attending reviews and offer support by facilitating or reallocating resources needed for the success of the QFD team. Do not discourage team members by asking, "Why are you doing such and such?" or "Do this and that." Instead the comments should be, "What did you learn from doing this?" or "How will you use this information for your project?" Management should encourage documenting of things gone right as well as those gone wrong, to create a more open environment which is free of blame and focuses on learning from the process.

Management selects QFD project.   A high-level management team should be involved with the project selection process for QFD. Management should reach consensus on the selection. The project should be a product (or service) critical to the success of the business. Correct project selection will increase the probability that good market research is available or will be funded for the project. Good market research is a critical starting point for a QFD project. Management should also identify the team leader.

Sell QFD to the leader.   The same process of selling the concept and developing interest should continue to the level of project leader and team members. A successful QFD project will require hard work by the leader and team. QFD should not be demanded by the management; rather, the team should be sold on the idea.

Team models.   There are many open questions in implementing QFD. One question is, What is the best model for team structures? Here are some possibilities:

- An assigned leadership with the core team having members from all functions for the duration of the product development cycle.

- Dynamic membership and fluid leadership in tune with the product developmental phase of the cycle. In other words, in the beginning when the customer's voice is being developed, the team is led by marketing with members of the next phases, R&D, and manufacturing. The leadership then switches to R&D with membership from the previous and preceding links of the process, that is, marketing and functions involved with manufacturing.

An organization should study this question with appropriate data to adopt the best model for its culture and product type (for example, a simple product or a complex product).

Criteria for Selecting Members

Here are some guidelines on selecting members for your first QFD project:

1. Members should come from approximately equal rank to minimize position power influence and to facilitate good communication, exchange of ideas, and team dynamics.
2. Choose those with a sense of adventure.
3. Take calculated risk takers.
4. Choose those who are open to new ideas.
5. Select those who are learning and growing.
6. Select those who are not too vested in the established procedures or in how things are done.

Six to eight persons constitute a team that functions well in terms of size. Should every expert be on the team, or just those charged with the responsibility of developing the product? The second alternative has merits provided we bring in experts to address the special needs of the team as they arise. Having limited representatives from each critical functional area will ensure that they assume the responsibility to obtain pertinent information.

Who should be on the QFD team? The membership comes from the areas listed below. This is a long list, and priorities will have to be set. During the life of the team, however, many of the personnel listed will have to be contacted.

1. Functions that are in touch with customers
   - Marketing: product or technical
   - Sales: tele-sales, direct sales, catalogers
   - Service: repairs, personnel connected with warranty management
   - Hot (problem) site personnel
   - Personnel connected with customers' visits at customers' sites and company sites
   - Market researchers

2. Functions involved with design
   - Design engineers
   - Technical researchers
   - Academic experts in a specific technical field
   - Test engineers
   - Process design engineers
   - Cost engineers
   - Design of experiment experts for scientific study of break-through engineering or tradeoff effects

3. Manufacturing functions
   - Manufacturing line personnel
   - Maintenance personnel of manufacturing equipment
   - Manufacturing process engineers
   - Packaging engineers
   - Material engineers
   - People in charge of keeping the manufacturing process under statistical control

Training and education.   Training the team is very important. Everybody should learn the correct vocabulary; communications improve as a result of common language, which in turn helps with team dynamics. Goal alignment is achieved as all team members focus on customers' need for carrying out their own tasks and activities in order to develop a dynamite product which will thrill the customer. This goal alignment also helps with team building. The 2-day training version provided by organizers such as ASI will be adequate.

What Else Is Needed for Success?

1. A realistic time line
2. Regular project reviews
3. Milestone celebrations to keep interest high and to develop a sense of closure
4. Recognition from management
5. Sharing with other teams to facilitate deeper learning

Summary of QFD

We have discussed the benefits of QFD. A pencil example was used to describe the key steps for developing QFD tables. Ideas on developing high-quality customers' voice and a flowchart to increase QFD success have been explored. Three success factors are an accurate customer's voice, strong management commitment, and a good consultant. In closing, we quote Chris Fosse of Omark Industries: "QFD does not give you the answers. It is a tool to organize thinking and activity to help you ask the right questions."

Failure Mode Effects Analysis

In Chap. 4 we discussed the problem-solving hierarchy. Ultimately the stage you should be in is the third stage—problem prediction and prevention. One of the tools that can help is FMEA, which is a very effective method of design reliability analysis. This method originated with National Aeronautic and Space Agency (NASA) engineers and is specified in U.S. Military Standard, MIL-STD-16291 (Procedures for Performing a Failure Mode, Effects and Criticality Analysis).

This is not a commonly used method, so we will give you a taste of what it is, but further details are best obtained from the recommended readings listed in the Bibliography. The following discussion on FMEA comes from Rajan Vrittamani of Hewlett-Packard, to whom we are deeply indebted.

Definition and benefits of FMEA.
FMEA is a systematic process to evaluate failure modes and causes associated with the design and manufacturing processes of a new product. It is somewhat similar to the potential problem analysis phase of the Kepner-Tregoe program.

The FMEA process.
A list of potential failure modes of each component or subassembly is made, and then each mode is given a numeric rating for frequency of occurrence, criticality, and probability of detection. Finally, these three numbers are multiplied together to obtain the *risk priority number* (RPN), which is used to guide the design effort to the most critical problems first. Typically only the high-RPN items (say, 50 percent of total items) are eliminated. Refer to Fig. C.7, a simple flowchart of FMEA activity.

Two aspects of FMEA are particularly important: the team approach and timeliness. The team approach is vital because the broader the expertise that is brought to bear on making and assigning values to the failure mode list, the more effective the FMEA will be. Timeliness is important because FMEA is primarily a preventive tool, which can help steer design decisions between alternatives. FMEA is equally applicable to hardware or software, to components or systems.

The RPN calculated by FMEA allows us to prioritize the failure mode list, guiding design effort to the most critical areas first. It also provides a documentary record of the failure prevention efforts of the

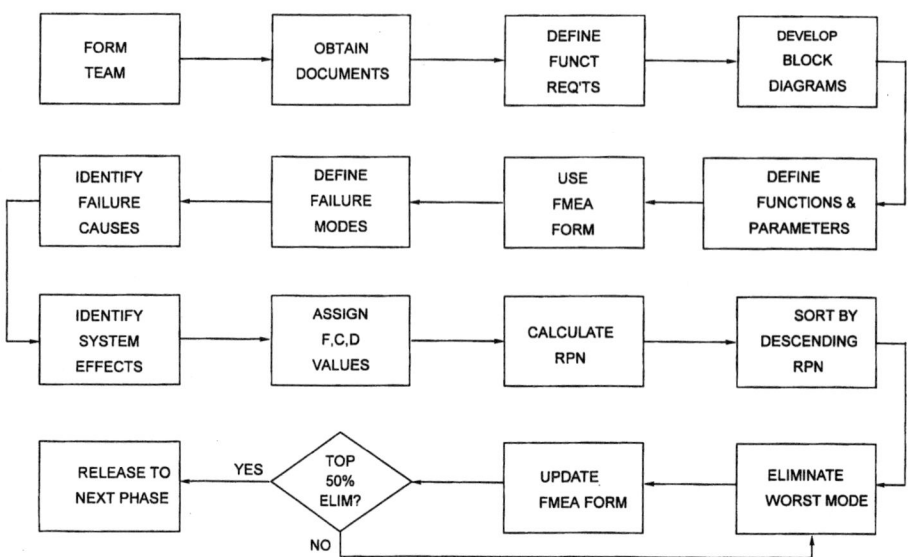

Figure C.7 Flowchart of FMEA activity.

design team. This will be helpful to management in gauging the quality and extent of the effort, to production in solving problems which occur despite these efforts, and to future projects which can benefit from all the work and thinking which went into the failure mode and cause lists.

Benefits. Eliminating potential failure modes has both short-term and long-term benefits. The short-term benefit is most often recognized because it represents savings of the costs of repair, retest, and downtime. The long-term benefit is much more difficult to measure, since it relates to the customer's satisfaction with the product and perception of its quality. That perception affects future purchases of products and is crucial in creating a good image for your products.

FMEA supports the design process. FMEA also strengthens and supports the design process by

- Aiding in the selection of alternatives during design
- Increasing the probability that potential failure modes and their effects on system operation have been considered during design
- Providing additional information to aid in the planning of thorough and efficient test programs

- Developing a list of potential failure modes ranked according to their probable effect on the customer (these failure modes can then be minimized or eliminated by design effort)
- Providing an open, documented format for recommending and tracking risk-reducing actions
- Identifying known and potential failure modes which might otherwise be overlooked
- Detecting primary but often minor failures which may cause serious secondary failures
- Detecting areas where failsafe or "fail-soft" features are needed
- Providing a fresh viewpoint to understand a system's functions

Requirements to Perform FMEA

1. A team of people with a commitment to improve the ability of the design to meet the customer's needs.
2. Schematics and block diagrams of each level of the system, from subassemblies to the complete system.
3. Component specifications, parts lists, and design data.
4. Functional specifications of modules, subassemblies, etc.
5. Manufacturing requirements and details of the processes to be used.
6. FMEA forms (paper or electronic) and a list of any special considerations, such as safety or regulatory, applicable to this product. Refer to Fig. C.8 for a sample FMEA form. The form has been completed for a portion of a medical product, for patient monitoring. If during design—after setting priorities—we decided to prevent only one failure mode, we would select the mode with the highest RPN.

Summary of FMEA

We recommend the use of the FMEA technique for all product design, preventive maintenance schedules, and test equipment. Initially, to get started, you can use this in products that have potential liability problems, for example, power supplies, fuel equipment, products that cannot be easily tested, totally new designs, or test equipment. With time, this should be a standard design for reliability and quality technique.

| | | | | **FAILURE MODES AND EFFECTS ANALYSIS** | | | | | |

Product: MAGIC _____   Assembly: XMTR RF Module _____

Engineer: Rajan _____   Date: 2/29/89 _____

Assy ID	Functional Block	Failure Mode	Failure Cause	Effect of Failure on System	F	C	D	RPN	Corrective Action or Remark
1	Oscillator tank circuit	No output frequency	Drift trim cap Defect induc cap, or Q1. Shorted res.	No RF sig from XMTR. Error msg on the screen	2	4	1	8	Error message to be defined. Trim caps should be sealed with epoxy.
		Wrong freq. Wrong dev. level	Drifted trim cap						
2	RF tank circuit	Output pwr level low	Def transistor	Weak patient sig. Possible bit errors in that channel	2	3	3	18	Invalid data message on the screen
		Output pwr level high	Shorted CR3 Shorted res						
			Drifted trimpot Open CR3	High current drain from XMTR battery	2	3	5	30	Trim pots should be sealed with epoxy

KEY:

1 = Remote   2=Low   3=Moderate   4=High   5=Very High

F: Frequency)
C: Criticality }
D: Detection)

RPN: Risk Priority Number = FxCxD

Figure C.8   A completed FMEA form.

The T-Type Matrix—A Tool for Analyzing the New Product Development Process

The T-type matrix is developed from one of the matrices proposed in the seven new management tools. Much of the discussion here originates from the paper by Dr. Noriaki Kano.[6] The example, however, comes from Hewlett-Packard Singapore.

Objectives of using the T-type matrix. The T-type matrix can be used to identify the root cause of problems or design changes that are generated in new product development. Proper use of the matrix will help in the following areas:

- Understanding why the problem was generated, resulting in a design change

- Understanding why the problem was not detected when it was generated

- Proposing where and how to improve our design process, so as to eliminate or diminish such problems or design changes in the future

Mechanics and preparation of the T-type matrix. After a new product is designed and released to manufacturing, data on design changes (anything requiring rework) are collected and displayed in a T-type matrix. For each design change or problem, the data are displayed in phases:

Phase A: the problem generation phase, that is, when the design error was created.

Phase B: When the problem could have been discovered. This can be during phase A or later.

Phase C: the actual problem detection phase. This can be in phase B or later.

Glance again at the three phases listed above. Obviously, we never want problems or errors to occur. And if they do, we want to detect them immediately, in the problem generation phase. Alas, this cannot happen every time, hence we need to identify the three phases of a problem or design change. The three phases (A, B, and C) are displayed clockwise in the T-type matrix in Fig. C.9.

Completing the Matrix

1. For a new product that is being developed, collect all problem and error data as and when they occur. Such errors can include a design change, a rewritten specification, problems detected in the prototype stage, all the way to problems detected after purchase by the customer. These data should be recorded in a central record book or computer file, which is accessible to the entire marketing, design, and manufacturing team. This activity calls for a cooperative effort by the entire team.

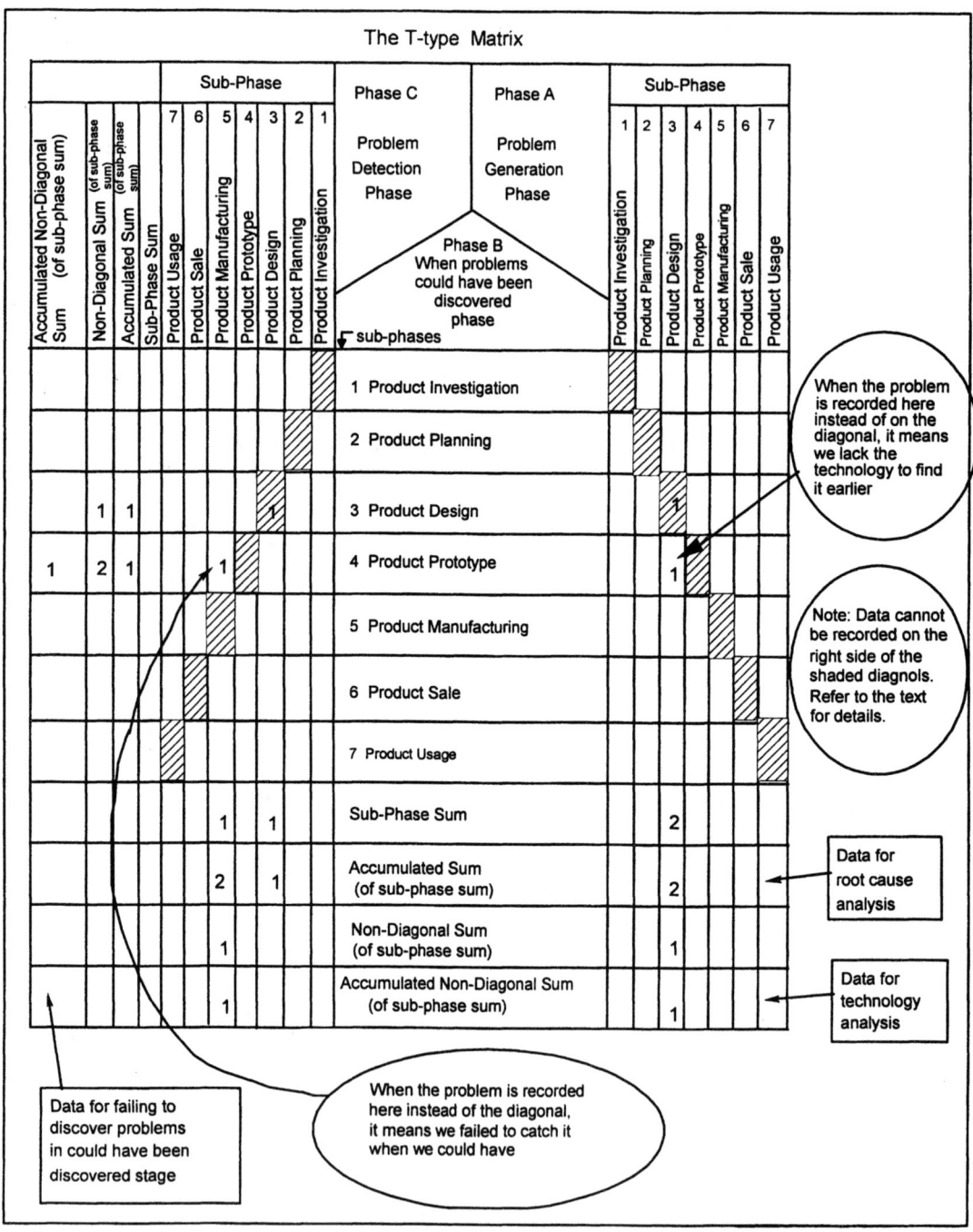

Figure C.9 The T-type matrix.

2. We use the following nomenclature:

n = the maximum number of subphases in new product development. In Fig. C.9, we show seven.

Phase Ai = actual subphase (1, 2, 3, etc.) in phase A (problem generation phase). In our example there are seven subphases in each of phases A, B, and C.

Phase Bj = actual subphase (1, 2, 3, etc.) in phase B (could-have-been-discovered phase)

Phase Ck = actual subphase (1, 2, 3, etc.) in phase C (problem detection phase)

When we fill in the data, we will display all the problems generated in phase A. We will also determine when the problems were detected—immediately or later.

3. First, complete the left-hand side of the matrix in Fig. C.9. We call this area phase BC, and it denotes the relationship between phases B and C. Here we enter the following data: subphase BC, which means problems discovered in subphase C but which could have been discovered in subphase B. There are three possibilities: a problem discovered after generation, a problem discovered during generation, and a problem discovered before generation. Clearly, the last phase is impossible.

  a. Phase BC, where $j<k$. This means a problem was discovered in subphase C, later than when it could have been discovered in subphase B. For example, in Fig. C.9, a problem was discovered in manufacturing, subphase 5, but could have been discovered in product prototype, subphase 4. In such a case, put a 1 (for one problem or error) where subphase 5 (in phase C) intersects subphase 4 (in phase B) on the matrix. We have done so in Fig. C.9.

  b. Phase BC, where $j = k$. This means a problem was discovered in subphase C at the same time that it could have been discovered in subphase B. For example, in Fig. C.9, the problem was discovered in the design, subphase 3, and could have been detected in design, subphase 3. In such a case, put a 1 (for one problem or error) at the intersection of subphase 3 (in phase C) and subphase 3 (in phase B). We have done so in Fig. C.9. In this condition of $j = k$, you will be filling in the diagonal (the shaded boxes) in Fig. C.9. This is the desired condition, where problems are detected when they are created.

  c. Phase BC, where $j>k$. This means a problem was discovered in subphase C before it was made in subphase B. Clearly this cannot happen. An example is discovering a problem during product design, but one which only the customer could discover after use. Hence, there will be no data entered on the right of the shaded diagonal in Fig. C.9.

4. Next complete the right-hand side of the matrix in Fig. C.9. We call this area phase AB, and it denotes the relationship between phases A and B. Here we enter the following data: subphase AB, which means problems that were generated in subphase A but were discovered in subphase B. There are three possibilities: a problem discovered after generation, a problem discovered during generation, and a problem discovered before generation. Clearly, the last phase is impossible.

   a. Phase AB, where $i<j$. This means the problem could have been discovered in subphase B later than it was generated in subphase A. For example, in Fig. C.9, a problem could have been discovered in the prototype, subphase 4, but was generated in the design, subphase 3. In such a case, put a 1 (for one problem or error) where subphase 3 (in phase A) intersects subphase 4 (in phase B). We have done so in Fig. C.9.

   b. Phase AB, where $i = j$. This means a problem could have been discovered in subphase B at the same time it was generated in subphase A, for example, a problem generated in design, subphase 3, which could have been discovered in design, subphase 3. We put a 1 (for one problem or error) at the intersection of both design subphases in Fig. C.9. In this condition $i = j$, you will be filling in the diagonal (the shaded boxes) in Fig. C.9. This is the desired condition in the matrix, where problems are discovered as they are created.

   c. Phase AB, where $i>k$. This means a problem could have been discovered in subphase B before it was generated in subphase A. Clearly this cannot happen. Hence there will be no data entered to the right of the shaded diagonal.

5. Sum the rows and columns in Fig. C.9. There are four types of summing; these are listed in the last four lines at the bottom and extreme left of the figure. The following are the four types of sums:

   a. Subphase sum: This is the sum of the problems in each subphase. Each phase is summed vertically (for phases A and C) or horizontally (for phase B).

   b. Accumulated sum (of subphase sum): This is the accumulated sum of the number of problems in the subphases, indicated in (a). As we move from left to right in the subphase sum boxes, we add the numbers and put each addition in the accumulated sum box just below.

  c. Nondiagonal sum (of subphase sum): This is the sum of the
problems in each subphase, excluding the problems in the diagonal boxes. These sums are the same as the number in (*a*) above
but minus any numbers in diagonal shaded boxes.

  d. Nondiagonal accumulated sum (of subphase sum): This is the
accumulated sum of the nondiagonal sum [item (*c*) above]. As
we move from left to right in the nondiagonal sum boxes, we
add the numbers and put each addition in the box just below.

If necessary, we can also convert the summed data to percentages and
show them in the same box by dividing each box into two. This is
optional and shown in Fig. C.10. The reason for conversion to percentages is to facilitate communication to management.

Analysis and interpretation of data. Analysis and interpretation will be
segmented into three categories: why the problem was not discovered
when it should have been, understanding what test technology is
needed to discover problems, and getting to the root cause of problem
generation.

1. *Reason for not discovering a problem in the could-have-been-discovered phase.* Here we want to understand why the problem was not
discovered when it could have been discovered. That is, why did we
fail to catch it when it occurred? These data are collected from the
left-hand side of the T-type matrix, specifically from the accumulated
nondiagonal sum data. Refer to Fig. C.9. The data can be displayed in
the categories shown in Table C.2. A Pareto diagram analysis should
be prepared, followed by a plan to reduce the major causes of the
problems. (We give an example that will make this clear.)

2. *Lack of evaluation or test technology.* Here we want to understand
why the problem could-have-been-discovered phase (phase B) occurs
later than expected, that is, later than the problem generation phase
(phase A). In other words, why was the problem not discovered when
it was created? The reasons are typically lack of evaluation or test
technology.

  The data are collected from the right-hand bottom of the T-type
matrix, specifically from the accumulated nondiagonal sum data.
Refer to Fig. C.9. The data can be displayed in the categories shown
in Table C.3. A Pareto diagram analysis should be prepared, followed
by a plan to reduce the major causes of the problems.

3. *Root cause of problem generation.* Here we want to understand
why the problems are generated in the first place. The data from
phase A, the problem generation phase, are collected from the right-
hand bottom of the T-type matrix, specifically the accumulated sum
data. Refer to Fig. C.9. They are then analyzed for root cause. The
data can be displayed in the categories shown in Table C.4. A Pareto

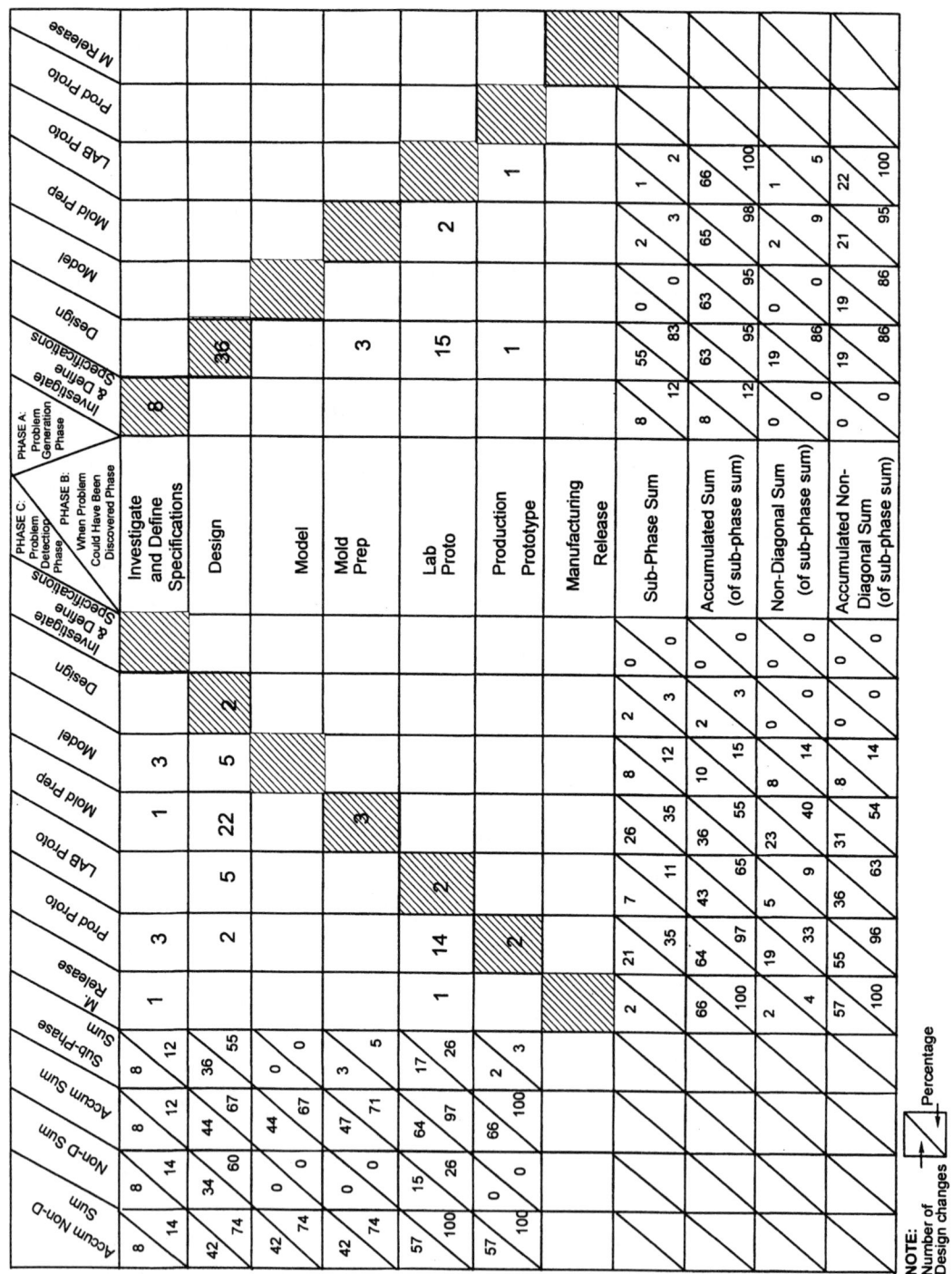

Figure C.10 Application of a T-type matrix to track design changes.

TABLE C.2   Common Causes for Not Discovering Problems

A.   Checked but missed the problem
     A-1 Checker's mistake (careless, inexperienced)
     A-2 Inappropriate criteria
     A-3 Inappropriate checking methods
     A-4 Inappropriate item to check
     Note: Item A-1 is useful in separating human error from incorrect criteria or
     methods.

B.   Did not check for problem.
     B-1 Checking judged unnecessary
     B-2 New problem, never planned to check
     B-3 Impossible to check for

TABLE C.3   A List of Common Causes for Lack of Test Technology

C.   Technology exists inside the company or division.
     C-1 Insufficient time to use technology
     C-2 No plan to use the technology
     C-3 No understanding of the technology by engineers

D.   Technology exists outside the company or division.
     D-1 Insufficient time to use technology
     D-2 No plan to use the technology
     D-3 No understanding of technology by engineers

E.   Technology does not exist.
     E-1 Planned to develop technology
     E-2 Did not plan to develop technology

TABLE C.4   A List of Common Causes for Problem Generation

F.   Inappropriate new product targets
     F-1 Inappropriate survey of user's needs
     F-2 Inappropriate survey of competitor's capability
     F-3 Inappropriate decision on user needs or competitive capability

G.   Inappropriate design despite having technical design manual or design guidelines
     G-1 Error in spite of using manual or guidelines
     G-2 Designed by experience, that is, ignored design manual or guidelines

H.   Technical manual or design guidelines did not exist.
     H-1 Manual or guideline planned
     H-2 Manual or guidelines not planned due to lack of technology

diagram analysis should be prepared, followed by a plan to reduce the major causes of the problems.

An application of the T-type matrix.   In Fig. C.10, we show an actual application of a T-type matrix for a computer keyboard. This example comes to us from Hewlett-Packard Singapore. It was the first application by a young but enthusiastic design team, and they benefited tremendously.

The data in Fig. C.10 were collected for seven subphases, from "investigate and define specifications" to "manufacturing release." In theory, we recommend that you collect data up to the point where customers use the product. In our example, this was not done mainly because the enthusiastic designers wanted to quickly analyze the data.

Note that in Fig. C.10, the percentage of errors is also given in the lower portion of the box, in addition to the various sums. Look at the right-hand bottom of the matrix, at the accumulated sum box. Here we are reviewing the problem generation phase. *You will notice that 95 percent of the errors occurred by the second subphase (the design phase). These data support our statement that 50 percent to 80 percent of errors discovered in manufacturing and at the customer's site are generated in the early stages of design.* Therefore we reiterate: The project postmortem, including a T-type matrix, is an excellent tool to analyze and uncover the source of errors in new product design, in order to improve the product development process.

Analysis of T-matrix data.   In Fig. C.11, the data from Fig. C.10 are analyzed after being segmented into the categories listed in Tables C.2, C.3, and C.4. Note that the specific line for obtaining the data is indicated in Fig. C.9. In Fig. C.11, the data are divided into the following broad categories:

1. Reason for not discovering in the could-have-been-discovered phase

2. Lack of test technology

3. Root cause of problem generation

The errors in each of these categories from Fig. C.10 were analyzed as to why they were not caught and then entered in Fig. C.11. Let us review an example in the "Reason for not discovering in the could-have-been-discovered phase." Look at Fig. C.11 and read the product design components column for PCB (printed circuit board). We had a total of four problems that were not discovered. Of these, three occurred because of inappropriate checking criteria, and one occurred because checking was judged unnecessary. Now, go all the way down in the PCB column, until you reach the "root cause of problem generation" category. Three of these errors occurred despite having a technical manual; apparently the manual was not helpful. One of the errors occurred because the designer decided to use his experience instead of the manual. The corrective action will be to improve the manual and to ask designers to use it.

From Fig. C.11, the data have been converted to Pareto diagrams for each of the three categories. This is shown in Fig. C.12. The

T-MATRIX APPLICATION: KEYBOARD PROJECT ANALYSIS

AREA / PROBLEMS	CAUSE CATEGORY	DETAILED CATEGORY	IC	PCB	MEM-BRANE	KEY-CAP	OUT-SERT	DOME-PAD	CASE PART	PAPER PART	DETAIL TOTAL	CAUSE TOTAL	TROUBLE TOTAL
REASON FOR NOT DISCOVERING IN COULD HAVE BEEN DISCOVERED PHASE	A - CHECKED	A-1 Checkers Mistake			1	1					2		57
		A-2 Inappropriate Criteria		3		1		1	9		14	16	
		A-3 Inappropriate Methods											
		A-4 Inappropriate Items											
	B - NOT CHECKED	B-1 Checking Judged as unnecessary					2		1		4	41	
		B-2 No Checking Item Existed	2	1		13	1	1	22		37		
		Component Total	2	4	1	15	3	1	32		57	57	57
EVALUATION or TEST TECHNOLOGY ISSUES	C - EXISTED INSIDE	C-1 No Time to Apply				7					8	11	22
		C-2 No Plan to Apply					3				3		
	D - EXISTED OUTSIDE	D-1 No Time to Apply											
		D-2 No Plan to Apply											
	E - DIDN'T EXIST	E-1 Development Was Planned							2		2	11	
		E-2 Development Not Planned		2		7		1	2		9		
		Component Total		2		14	3	1	2		22	22	22
ROOT CAUSE OF ERRORS OR DESIGN CHANGES	F - R&D TARGETS	F-1 Inappropriate Survey of Users Needs											66
		F-2 Inappropriate Survey of Competitors Capability											
		F-3 Inappropriate Decision on User Needs of Competitive Capability											
	G - TECHNICAL MANUAL EXISTED	G-1 Error in spite of using manual	3	3					15		21	24	
		G-2 Designed by experience instead of Manual		1		1	1				3		
	H - TECHNICAL MANUAL DIDN'T EXIST	H-1 Manual was Planned											
		H-2 Manual was not Planned				22	2	1	17		42	42	
		Component Total	3	4		23	3	1	32		66	66	66

Figure C.11 Analysis of T-matrix data.

351

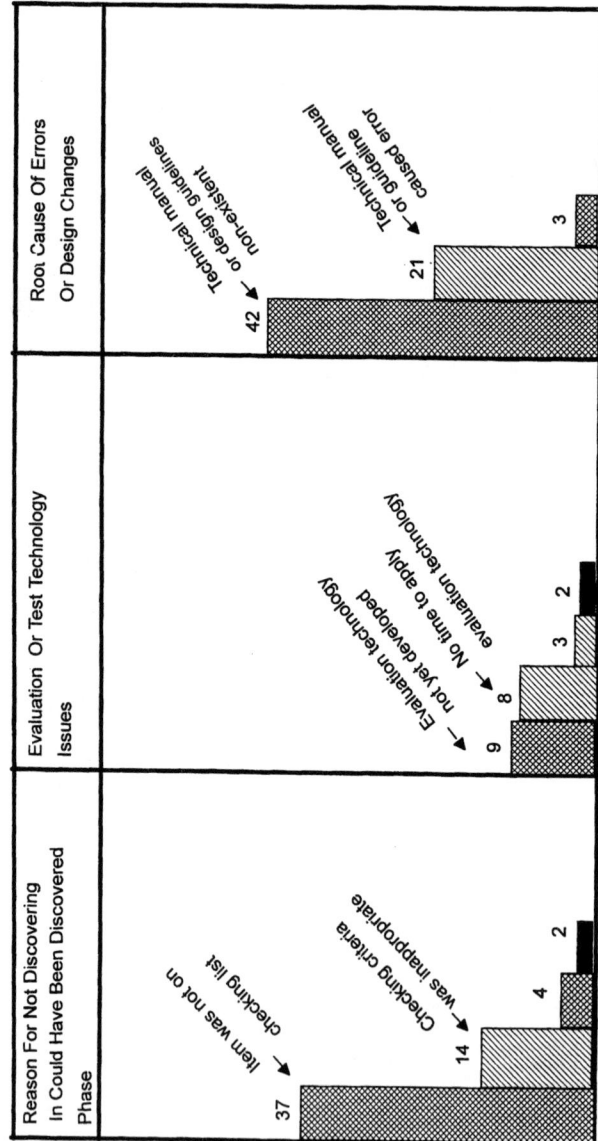

Figure C.12 A Pareto diagram analysis of weaknesses in the design process (based on data from Fig. C.11).

Pareto diagrams display the weaknesses uncovered in the analysis. Based on the weaknesses, corrective action was taken in three areas:

- Improved inspection and checklist
- Introduced new test and evaluation technology and enforced the use of existing test and evaluation technology
- Improved technical manual with design guidelines for all designers

The information was conveyed to the entire design team, and training was provided where necessary. Subsequent analysis of the next keyboard design showed a marked reduction in design changes.

When and how to use the T-type matrix.   Collecting, displaying, and analyzing the T-type matrix data takes some effort. The benefits, however, are tremendous. Here are some hints that make data collection easier for this important analysis.

1. Provide a central record book for each project, to record every problem as soon as it is discovered. Where possible, direct entry into a computer database is preferable.

2. Convey the reasons and benefits of doing this analysis to one and all. Nobody should be reprimanded; instead, R&D, marketing, and manufacturing staff should understand that this is a method for ensuring continuous improvement in the new product development environment.

3. The matrix can be entered and computed in a computer spreadsheet such as Lotus 1-2-3.

4. It is difficult to commit to doing this analysis for all your new products, so try it on a few each year. The lessons learned in most cases are generic and will steer you in the right direction; for example, raising the awareness for better design guidelines, ensuring automatic applications of design guidelines via computer-aided design software, and improving testing and evaluation technology.

5. Use the T-type matrix within a project postmortem to get the maximum benefit.

6. The T-type matrix will be most useful when applied to the first-generation product in a series, for example, your first keyboard, workstation, optical disk drive, or digital recorder.

References

1. Yogi Akao, "History of Quality Function Deployment in Japan," from course notes of the Kaizen Institute of America, 1989, pp. V-A-1–13.
2. Yoji Akao, *Quality Function Deployment* (Methuen, MA: GOAL, 1990).
3. Khushroo Banu, the contributor of the QFD section, visited California Cedar Company in Stockton, California, in April 1988. This company conducts research on pencils to help their customers. They verified the house of quality on pencils was accurate.
4. Joel Barker, "Discovering the Future, the Business of Paradigms," Chart House, 1989.
5. "Does Success Have a Secret? Can LDC (Lesser Developed Countries) Compete in Free Market?" Stanford University, *Transactions PEASCON-87,* Second Biennial Conference, Santa Clara, CA, 1987, chap. 8, pp. 1–20.
6. Dr. Noriaki Kano, "Problem Solving in New Product Development: Application of T-Type Matrix," *World Quality Congress Proceedings,* 1984, vol. 3, pp. 45–55.

Bibliography

We provide here a list of recommended readings. Where appropriate, comments have been provided.

1. Robert C. Camp, *Benchmarking: The Search for Best Practices that Lead to Superior Performance,* ASQC Quality Press, Milwaukee, Wisconsin, 1989. This is a good book, the first we know of that provides a detailed discussion of the benchmarking process.
2. Kaoru Ishikawa, *Guide to Quality Control,* Asia Productivity Organization, Tokyo, 1986. Available from the ASQC Quality Press in the United States. It provides an excellent and easy-to-understand discussion of the seven quality control tools.
3. Dr. Hitoshi Kume, *Statistical Methods,* Association for Overseas Technical Scholarship, Tokyo, 1985. This provides an excellent and easy-to-understand discussion of the seven quality control tools.
4. *Statistical Quality Control Handbook,* Western Electric, Mack Printing Company, Pennsylvania, 1982. This is an excellent, easy-to-read book on statistical quality control. The discussion focuses on control charts, design of experiments, and process capability.
5. E. L. Grant and R. S. Leavenworth, *Statistical Quality Control,* McGraw-Hill, New York, 1982. This is a good book that gives a detailed and theoretical discussion of statistical quality control. For a simpler and friendlier discussion, references 2, 3, and 4 are preferred. Many technical people, however, will prefer the technical discussion provided here.
6. D. J. Klinger, Y. Nakada, and M. A. Menentez (eds.), *AT&T Reliability Manual,* Van Nostrand Reinhold, New York, 1990. It gives details on component derating and thermal design.
7. D. Brownell and E. Guyer, *Handbook of Applied Thermal Design,* McGraw-Hill, New York, 1989. Details on component derating and thermal design are given.
8. P D. T. O'Conner, *Practical Reliability Engineering,* 2d ed., John Wiley & Sons, New York, 1988. It provides details on component derating, thermal design, and FMEA applications.
9. *How to Operate QC Circle Activities,* Japanese Union of Scientists and Engineers. This can be obtained from ASQC Quality Press, Milwaukee, Wisconsin, in the United States.
10. *The Quality Circle Process: Elements for Success,* SQC technical committee, 1986. This can be obtained from the ASQC Quality Press in the United States.
11. Edwards Deming, *Out of the Crisis,* MIT Press, Cambridge, MA, 1986. This book by the redoubtable Dr. Deming explains his philosophy, via his 14 points. It is not an easy read, but nevertheless a must for all managers and professionals.

Index

ABOUT THE AUTHOR

Sarv Singh Soin is currently general manager for Hewlett-Packard's Asia Pacific Distribution Operation based in Singapore. He was formerly Quality Manager for Hewlett-Packard's Intercontinental operations based in Palo Alto, California. Altogether he has been a quality manager for more than a dozen years in manufacturing as well as sales in Hewlett-Packard.